Troubleshooting and Repairing Computer Printers

2nd edition

Stephen J. Bigelow

TAB Books
Imprint of McGraw-Hill

New York   San Francisco   Washington, D.C.   Auckland   Bogotá   Caracas   Lisbon   London
Madrid   Mexico City   Milan   Montreal   New Delhi   San Juan   Singapore   Sydney   Tokyo   Toronto

McGraw-Hill

A Division of The McGraw·Hill Companies

pbk   9 0 DOC/DOC 9 0 9 8 7 6 5 4 3 2
hc   2 3 4 5 6 7 8 9 DOC/DOC 9 0 0 9 8

Product or brand names used in this book may be trade names or trademarks. Where we believe that there may be proprietary claims to such trade names or trademarks, the name has been used with an initial capital or it has been capitalized in the style used by the name claimant. Regardless of the capitalization used, all such names have been used in an editorial manner without any intent to convey endorsement of or other affiliation with the name claimant. Neither the author nor the publisher intends to express any judgment as to the validity or legal status of any such proprietary claims.

Library of Congress Cataloging-in-Publication Data
Bigelow, Stephen J.
     Troubleshooting and repairing computer printers / by Stephen J.
 Bigelow.—2nd ed.
        p.   cm.
     Includes bibliographical references and index.
     ISBN 0-07-005731-1.   ISBN 0-07-005732-X (pbk.)
     1. Printers (Data processing systems)—Maintenance and repair.
 I. Title
 TK7887.7.B55   1996
 681'.62—dc20   96-16467
                                                              CIP

McGraw-Hill books are available at special quantity discounts to use as premiums and sales promotions, or for use in corporate training programs. For more information, please write to the Director of Special Sales, McGraw-Hill, Professional Publishing, Two Penn Plaza, New York, NY 10121-2298. Or contact your local bookstore.

Acquisitions editor: Roland S. Phelps
Editorial team: Melanie Brewer, Book Editor
               Andrew Yoder, Managing Editor
               Lori Flaherty, Executive Editor
               Joann Woy, Indexer
Production team: Katherine G. Brown, Director
               Ollie Harmon, Coding
               Wanda S. Ditch, Desktop Operator
               Lorie L. White, Proofreading
               Jeffrey Miles Hall, Computer Artist
Design team: Jaclyn J. Boone, Designer
               Katherine Lukaszewicz, Associate Designer   EL3
                                                              005732X

Troubleshooting and Repairing Computer Printers

Other Books by Stephen J. Bigelow

Troubleshooting, Maintaining, & Repairing Personal Computers
Troubleshooting and Repairing Computer Monitors
Troubleshooting and Repairing Notebook, Palmtop, and Pen Computers
Maintain and Repair Your Computer Printer and Save a Bundle

This book is dedicated to my wonderful wife, Kathleen.
Without her loving encouragement and support,
this book would still have been possible,
but not nearly worth the trouble.

Contents

6 Troubleshooting guidelines   *127*

Acknowledgments

I WOULD LIKE TO ACKNOWLEDGE THE COOPERATION AND support of the following organizations in the preparation of this book. I appreciate their generous reprint permission:

☐ Hewlett-Packard Company, Vancouver Division

☐ Hewlett-Packard Company, Boise Division

☐ B+K Precision

☐ ROHM Corporation

☐ NEC Technologies, Inc.

☐ Okidata

☐ Singer Data Products, Inc.

I also would like to extend my particular thanks to Ron Trumbla at Tandy Corporation. Their generous permission to reprint so many schematics and diagrams has contributed significantly to the success of this second edition.

Finally, I want to thank my Acquisitions Editor, Roland Phelps, my Executive Editor, Lori Flaherty, and the entire crew at McGraw-Hill for their continued enthusiasm and support.

xvii

Preface

MY FATHER INTRODUCED ME TO ELECTRONICS WHEN I WAS in my early teens. Through the GI bill and a lot of hard work, he made his way through an array of home-study courses. It didn't take long before he could follow just about any circuit. The cellar was his shop. A small arsenal of home-built test equipment took up a better part of the workbench. Any sick or dying television, radio, or tape deck was considered fair game for his soldering iron. I can remember long evenings assisting Dad in "emergency surgery," trying to save ailing home electronics units from certain destruction. Some were rescued, others simply went on to "hardware heaven." Although I was too young to understand the math and science of electronics at the time, I caught the bug—troubleshooting and repair was fun.

In the years that followed, I've learned a lot about electronics and troubleshooting. Perhaps the most important thing of all is the need for information. Without a clear understanding of how and why circuits and systems work, tracking down trouble becomes much more difficult. That is why I wrote this book. Because computers and printers have become so prevalent in our everyday lives, it seemed only natural to provide a thorough, comprehensive text on printer technology and repair. It is written for electronics enthusiasts and technicians who want to tackle their own printer problems as quickly and painlessly as possible.

After all, troubleshooting and repair should be fun.

Thanks, Dad!

Introduction

PRINTERS HAVE BECOME AN INVALUABLE TOOL IN THE high-tech revolution. It seems that just about every computer owner has a printer on hand or has access to one at home or work. They can be found in offices, laboratories, stores, classrooms, factories, and homes throughout the United States (and much of the world). Printers have evolved into flexible, reliable devices using any one of several mature printing technologies. This book discusses those technologies in detail and shows you how many of today's printers work.

No matter how reliable the printer might be, it will eventually need some sort of maintenance and repair. This book is intended to provide the essential background information on printer mechanics and electronics, along with the techniques that will guide you through the troubleshooting and repair of just about any commercial printer. It is written for the hobbyist or repair technician with an intermediate knowledge of electronics and some working knowledge of mechanics. A knowledge of elementary troubleshooting would be helpful, but it is not a prerequisite. This book describes the basic operation and use of several common test instruments. Troubleshooting procedures are presented in a discussion format that will aid your overall understanding of the printer and how it works.

Changes in the second edition

This second edition incorporates some important changes and improvements over the original book. Probably the most notable change is the use of technical illustrations. Rather than the many sketches and interpretations that appeared in the first book, this edition uses an array of schematics, schematic fragments, block diagrams, and exploded views from actual printers. Such "real life" examples provide unprecedented coverage of a printer's assembly, operation, and repair concerns. In addition to the improved illustrations, this edition also expands the coverage of symptoms and solutions of the original. This edition provides 150 complete symptoms and solutions, whereas the original book offered 60.

A chapter has been added to deal with Windows and Windows 95 printing problems. While purists might criticize this decision as "polluting" a good hardware book, it is virtually impossible today to separate the hardware from the operating system—even the slightest kink in a configuration can result in printing problems.

Another chapter deals with PRINTERS—the new companion disk for this book. PRINTERS provides an inexpensive, handy, multipurpose printer troubleshooting tool capable of testing impact, ink jet, and EP printers with equal ease. For the first time, you have a source of "standard" test patterns, which have been specifically designed to help you resolve the major problem areas of most printers. You also can enter escape codes manually to test even the most obscure feature. You can find the order form for PRINTERS at the back of this book, along with information on our premier troubleshooting newsletter *The PC Toolbox*.

How to contact us

Every possible measure was taken to ensure a thorough and comprehensive book. Your comments, questions, and suggestions about this book are welcome at any time, as well as any personal troubleshooting experiences that you might like to share. Feel free to write to me directly:

Dynamic Learning Systems
Attn: Stephen J. Bigelow
P.O. Box 805, Marlboro, MA 01752 USA
Tel: 508-366-9487
Fax: 508-898-9995
TechNet BBS: 508-366-7683 (up to 28.8KB 8 data bits, 1 stop bit, no parity bit)
CompuServe: 73652,3205
Internet e-mail: sbigelow@cerfnet.com

Or visit our home page on the World Wide Web:
http://www.dlspubs.com/

A modern printer

COMPUTER PRINTERS ARE AS INDISPENSABLE TO OUR modern world as automobiles, telephones, radios, or television (figure 1-1). Printers can type business or newsletters, address mail, plot drawings and illustrations, generate long listings of information, and many other boring, redundant jobs that people once had to do by hand or by typewriter. But the printer has evolved beyond the role of a simple workhorse. High-resolution and color printers have revolutionized desktop publishing, commercial art, and all aspects of engineering. Just think of what life would be like if you did not have your printer on hand. This chapter provides you with an introduction to printer concepts, specifications, and major assemblies.

Hewlett-Packard Co.

■ **1-1** *Hewlett-Packard DeskJet printers.*

The sheer variety of printer sizes, shapes, technologies, and features simply staggers the imagination. Yet in spite of all this diversity, every printer ever made performs the same task—a computer printer is a device that transcribes the output of a computer into some permanent form. It sounds simple enough, right? In reality, however, it takes a complex interaction of electrical and mechanical parts all working together to make a practical printer. Stop for a moment and consider some of the things that your own printer must be capable of.

First, the printer is a peripheral device (like a monitor or a modem). It can do nothing at all without a *host computer* to provide data and control signals, so a communication link must be established. To operate with any computer system, the printer must be compatible with one or more standard communication interfaces that have been developed. A printer must be able to use a wide variety of paper types and thicknesses that can include such things as multipart forms, envelopes, and specially-finished papers. It must be capable of printing a vast selection of type styles and sizes, as well as graphics images, then mix those images together onto the same page. The printer must be fast. It must communicate, process, and print information as quickly as possible. Printers must be easy to use—many features and options are accessible with a few careful strokes of the control panel. Paper input and output must be convenient. Expendable supplies (such as toner, ribbons, and ink) should be quick and easy to change. Finally, printers must be reliable. They must produce even and consistent print over a long working life, which can easily exceed 50 million characters (expendable items have to be replaced more frequently). The first place to start any study of printers is to understand their features and specifications.

Features & specifications

Make it a point to know your printer's specifications and features before you begin any repair operation. This gives you a good idea as to what the printer is capable of, and might help you to test it more thoroughly during and after your repair. A listing of specifications is usually contained in an introductory portion of the owner's manual, or at the end in an appendix.

Remember that there is no standard format for listing a printer's specifications. It is up to the individual preferences of each manufacturer. Some specifications will be listed depending on which printing technology is in use. For example, an electrophotographic (EP) printer will be specified a bit differently than a dot-matrix impact printer. Regardless of the particular technology, however, a list of specifications will almost always contain the following points: power requirements, interface compatibility, print capacity, print characteristics, reliability/life information, environmental information, and physical information. Each of these areas has some importance to a technician, so you should be familiar with them in some detail.

Power requirements

As with any electrical device, a printer requires power in order to function. Voltage, frequency, and power consumption are the three typical specifications that you will find here. Domestic U.S. voltages can vary from 105 to 130 volts ac (Vac) at a frequency of 60 Hz. European voltages can range from 210 to 240 Vac at 50 Hz. Because virtually all commercial printers are designed for global sales, designers accommodate this variation in voltage by providing a voltage selection switch (usually located near the printer's ac cord). The switch sets the printer's power supply for proper operation at 120 Vac or 220 Vac. If this switch is set improperly for the local ac voltage, the printer will function erratically (if it runs at all), and can even result in damage to the printer's power supply. The wise technician will always start a repair by checking for this voltage selection switch, and verifying that it is set to the proper position.

Power consumption is rated in terms of watts (W). Power demands vary widely in printer designs. Contemporary impact dot-matrix printers use up to 100 W during printing. Ink printers use much less power (25 W), even less for portable ink jet models. On the other hand, EP printers require substantially more power, with peak demands exceeding 800 W. Keep in mind that power demands are much less while the printer is idle (not actually printing). Chapter 8 discusses the operation and repair of printer power supplies.

Interface compatibility

Remember that a printer is a peripheral device, that is, it serves no purpose at all unless it can communicate (or *interface*) with a computer. A communication link can be established in many different ways, but three predominate interface techniques have become standard: RS 232, Centronics, and IEEE 488. Only a properly wired and terminated cable is needed to connect the printer and computer. Printer communication is discussed in Chapter 9.

RS 232 is a serial interface technique used to pass binary digits (*bits*) one at a time between the computer and printer. Serial links of this type are very common, not just for printers, but for other serial communication applications such as modems and simple digital networks. RS 232 is popular due to its moderate speed, the physical simplicity of its wiring, and its potential to handle data over relatively long distances. Unfortunately, serial communication links require a surprising number of variables to be set. This makes the serial approach more difficult to configure (and adjust later).

Centronics is the standard for parallel printer communication. It is a "de-facto" standard, so it is not officially endorsed by standards organizations such as the IEEE, EIA, or ITU. Instead of passing one bit at a time, entire characters are passed from the computer to the printer as complete sets of bits. Centronics is popular due to its functional simplicity. Even though it requires more interconnecting signal wires than an RS 232 cable, the hardware required to handle parallel information is simpler. There are also no variables to consider, just connect the printer to the PC and you're on your way. While parallel ports were once the exclusive realm of printers, contemporary peripherals (i.e., external tape drive, CD-ROM drive, and so on) are competing for space on the port.

IEEE 488, also known as GPIB (General Purpose Interface Bus), is an official standard for parallel communication that serves printers and plotters, along with automated instrumentation. It is not as widely used as Centronics, but GPIB supports network and bidirectional communication (where Centronics is traditionally a one-way, single-peripheral technique). The GPIB technique was originally developed by Hewlett-Packard Company, where it is still widely used in their line of printers and plotters.

4 Print capacity

Print capacity is actually a set of individual specifications that outline the mechanics of a printer's operation. Some of the characteristics that you should be familiar with are: CPL, CPS, print direction, dot configuration, resolution, pages per minute, warm-up time, and paper feed.

The term *CPL* stands for Characters per Line, and is typically related to moving-carriage printers, such as impact or ink jet dot-matrix printers. This is the number of characters that can fit onto a single horizontal line of text. You will likely find several entries under this heading—one for each type size (or style) that the printer can produce. A larger type size will result in fewer characters per line, and vice versa. For example, a type size that yields 10 CPI will fit 80 characters on an 8-inch horizontal line. At a setting of 15 CPI, 120 characters could fit on that same 8-inch line. CPL is also widely known as *character pitch*.

CPS means characters per second. It is also related to moving-carriage printer designs and indicates just how fast your printer will generate complete characters. This specification is closely related to character pitch, so you will probably find one entry for each pitch setting. Smaller characters that are composed of fewer dots can be produced faster, so they will have a higher CPS rate. Larger characters (or letter-quality characters) are composed of more dots, so their CPS rate will be lower.

Print direction is related to moving-carriage printers, and defines the way in which a printer's carriage can move during operation. Virtually all moving-carriage printers that produce simple text operate in a bidirectional mode as shown in figure 1-2. The carriage moves in one direction across a page, paper advances, then the carriage moves across the page in the opposite direction. Paper advances, then the carriage moves in its original direction again. Unidirectional printing is often used for more calculation-intensive printing applications such as bit-map graphics. The carriage moves in one direction across the page (usually left to right), the page advances, and the carriage moves back to its starting point before printing a subsequent line. Figure 1-3 shows a basic unidirectional pattern.

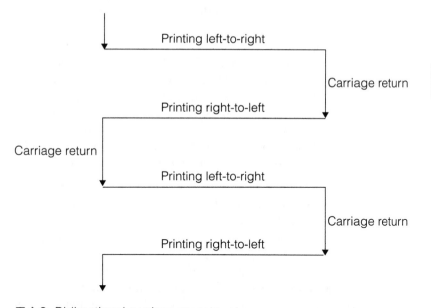

■ **1-2** *Bidirectional carriage movement.*

Individual dots placed on a page are defined by several different specifications depending on whether the printer is working in a text or graphics mode. Moving-carriage printers typically define a dot configuration, which indicates the physical size of each dot being placed on the page. Impact dot-matrix print heads frequently provide a dot configuration of $\frac{1}{127}$" (0.2 mm) in diameter. Ink jet print

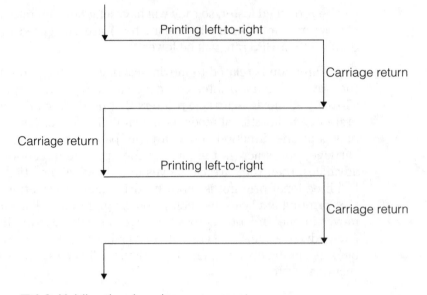

■ **1-3** *Unidirectional carriage movement.*

heads often sport much smaller dot sizes. The dot alignment details how many dots are placed (vertically and horizontally) to form a character. For example, an impact printer dot alignment of 24 × 9 shows that a draft mode character is composed of up to 24 vertical dots, and 9 horizontal dots. In letter quality mode (LQ), this increases to 24 × 30. On the other hand, an ink jet printer in draft mode runs at 50 × 15, where LQ mode offers 50 × 30. This is why many ink jet printers provide better-looking text even in draft mode.

Finally, dot pitch indicates how precisely dots can be placed in the horizontal and vertical direction. A typical impact printer can achieve a dot pitch of ¹⁄₁₂₀" (0.21 mm) in the horizontal direction, and ¹⁄₁₈₀" (0.14 mm) in the vertical direction. In other words, the impact printer can place 120 horizontal dots per inch (dpi) and 180 vertical dots per inch. This is the printer's graphic resolution. While some impact printers are capable of up to 360 × 360 dpi, this produces a tremendous amount of print head heat and mechanical work, which results very slow printing. From a practical perspective, impact printers work best at low resolution (60 × 60 dpi or so). Ink jet printers are also capable of very fine resolutions (up to 360 × 360 dpi), and they provide faster performance since the print head does not get hot. Still, the slow rate at which ink dots can be ejected limits printing speed at high resolutions. As you will see in later chapters, EP printers are not limited by such constraints, and resolutions are available from 75 × 75 dpi to 1,200 ×

1,200 dpi (and higher). The "standard" operating resolution for current, low-end, commercial EP printers is 300 × 300 dpi. In the printer world, the reference to resolution always indicates that impact or ink jet printer is working in the graphics mode (EP printers work only in a graphics mode).

Where moving-carriage printers focus on specifications, such as characters per second and lines per inch to establish a printing rate, EP printers simply define their printing speed as pages per minute (ppm). This rate is largely determined by the printer's mechanical design. Low-cost, low-end EP printers will work at 4 ppm, while high-cost, high-end EP models can operate at 12 ppm or higher. Keep in mind that ppm is an optimum printing rate—large graphic images might require additional time to transfer the corresponding data, which easily can slow printing.

Warm-up time is another specification applied to EP printers. Impact and ink jet dot-matrix printers do not require warm-up time; they're ready to go as soon as you turn the printer on. However, EP printers must melt toner to the paper. This requires the EP printer to reach a stable working temperature. When the printer is operating correctly, a working temperature can be reached in under 90 seconds.

Paper handling

Paper handling specifies the method(s) used to transport paper through the printer. There are two classical techniques for transporting paper through a printer; tractor-feed and friction-feed. The tractor-feed is the most widely accepted type of mechanism for use in general-purpose or high-volume printing, especially for impact dot-matrix printers, but friction-feed is quickly growing in popularity for single-sheet "letter-quality" printers, such as ink jet. Electrophotographic printers use friction-feed techniques exclusively, but the EP paper handling system is far more involved than those used in moving-carriage printers (as you will see in Chapter 4).

The tractor-feed technique uses a continuous length of folded paper with holes evenly perforated along both sides. This paper is threaded into the paper transport, and inserted into sprocket wheels located just over the platen. Teeth on each sprocket wheel mesh exactly with holes in the paper. The platen and sprocket wheels are linked mechanically, so as the platen is advanced, both sprockets turn to pull paper through evenly. Some tractor-feed configurations can be threaded somewhat differently so that paper is pushed through the printer instead of pulled through.

The friction-feed technique is effective for single sheets of paper. In most cases, any fresh, dry sheet of standard 20-lb bond xerography-grade paper will do, though ink jet printers might require specialized ink jet paper for best results. A sheet is started into the mechanism, then locked down against the platen using one or more pressure (or "pinch") rollers. Paper is now fixed in place. As the platen advances, friction between the platen and pinch rollers literally push the paper along. Chapter 10 discusses paper transports (and other mechanical systems) in more detail.

Print characteristics

Print characteristics specify the way in which printer images will appear, how they will be produced, or how characters sent from the computer will be interpreted. Font, software emulation, and character sets are the three specifications that you should be most familiar with.

A *font* is essentially a style of type with certain specific visual characteristics (sometimes very subtle characteristics) that set it apart from other type styles. Such characteristics might include differences in basic character formation, accents, and decorative additions (e.g., Courier versus Helvetica type). Figure 1-4 is an overview of several basic fonts found in impact dot-matrix or ink jet printers. Most commercial moving-carriage printers are capable of producing at least two fonts—often a simple "draft" style and a more detailed "letter-quality" style. Contemporary ink jet and laser printers are typically fitted with dozens of fonts. Printers also can produce font enhancements, such as underlining, bold, italic, superscript, or subscript. Keep in mind that not all enhancements are available in all fonts.

There are two methods of extending a printer's font selection. The first method is to install one or more font cartridges in the printer itself. The font cartridge is basically just a ROM that contains the data needed to form different type styles. The desired font (sometimes referred to as a hard-font) can then be selected from the printer's control panel. When subsequent ASCII characters are received, they will be printed using the selected font. While contemporary font cartridges are available containing dozens of fonts, the font-cartridge technique is still limited by flexibility; if you need to print in a font that's not in the printer, you need to change the font cartridge or select an available font.

Such hardware limits are overcome using soft fonts. The soft-font technique allows the document processor in the PC (i.e., your word

This is standard Courier Font

This is standard Sans Serif Font

This is standard Times Font

This is Bold PS Font

This is 10 characters per inch (cpi)

This is 12 characters per inch (cpi)

This is 15 characters per inch (cpi)

This is 17 characters per inch (cpi)

This is Emphasized print

This is Double Height Print

This is Double Width Print

This is Underlined Print

This is Italic Print

■ **1-4** *Comparison of basic printer fonts.*

processor) to select the fonts used in the document, then download those fonts from the PC hard drive to the printer. The document processor can then specify a font in the document and send along the ASCII characters without concern for whether the font will be available or not. Of course, soft fonts have their disadvantages. The major drawback to soft fonts is their use of printer memory; each font can demand several kilobytes (KB) of printer memory. When several fonts are downloaded, you can imagine how a moving-carriage printer with only a few KB of memory could quickly be overloaded by soft fonts. It is possible to add a bit of memory to most moving-carriage printers, but the expense is rarely worth the lower print quality. As a consequence, soft fonts are typically reserved for large-memory printers, such as EP printers.

Hard fonts and soft fonts still play a vital role in DOS-based printing. However, it is important to realize that printer fonts are only important when the printer is working in its text mode (i.e., receiving ASCII characters). Windows and Win95 print exclusively in the graphics mode, so the fonts and images that appear in a printed page rely on the fonts contained in the operating system itself, not the printer. The information you see in a WYSIWYG display (what you see is what you get) for Windows applications such as Microsoft Publisher is translated directly to graphic data in the PC, then sent on to the printer as a graphic. This approach offers almost limitless potential for creating very sophisticated documents, while removing the need for any printer fonts.

Of course, just sending an ASCII character to the printer is not enough; the printer has to know how to respond to that character. Printers use an internal printer language to establish and maintain standard operations. The language resides in the printer's permanent memory, and specifies such things as how to recognize and respond to control codes embedded in the string of ASCII characters. Most of the popular printer languages were originally developed by leading printer manufacturers, such as Hewlett-Packard, IBM, and Epson. While many printer manufacturers offer nonstandard languages to take advantage of the particular printer's unique capabilities, printers can almost always switch languages to achieve compatibility.

Other manufacturers that wish to make their printers functionally compatible must write their own internal software that emulates one or more of the existing language standards. As you become familiar with printers you will see that the Epson LQ-2500 and FX-80, IBM Proprinter X24 and Proprinter III, Postscript, and various levels of Hewlett-Packard's printer control language (referred to as PCL and often found in EP printers) have been embraced as "standard" printer languages. For example, a Panasonic impact dot-matrix printer might offer software emulation for an Epson LQ-2500. This means even though the Panasonic printer is physically and electrically different from an Epson, it will respond as if it were an Epson LQ-2500 when connected to a host computer. An Okidata EP printer will typically emulate an HP LaserJet III.

When a character code is sent to a printer, it is processed and delivered as a fully formed alphanumeric character or other special symbol. However, because a character code is not large enough to carry every possible type of text or special symbol (e.g., foreign language characters or block graphics), characters are grouped into character sets that the printer can switch between. Switching a character set can often be accomplished through a series of computer codes defined by the printer language, or control panel commands. By switching character sets, the same ASCII code can be made to represent completely different symbols. A standard character set consists of 96 ASCII (American Standard Code for Information Interchange) characters (26 uppercase letters, 26 lowercase letters, 10 digits, punctuation, symbols, and some control codes). Other character sets can include 96 italic ASCII characters, international characters (German, French, Spanish, etc.), and unique block graphics.

Reliability/life information

Reliability and life expectancy information expresses the expected working life of the printer or its print head in terms of pages, time, or characters. For example, a typical impact dot-matrix print head can have a life exceeding 100 million characters. A good-quality electrophotographic printer could have a rated service life better than 300,000 pages. You might see this same information expressed as MTBF (Mean Time Between Failures) or MPBF (Mean Pages Between Failures). Expendable material lifetimes are also included under this heading. Ribbon, ink cartridge, and electrophotographic cartridge life is usually quoted in terms of pages or characters.

You should realize that reliability figures are hardly absolute. A printer's working life is greatly influenced by factors such as regular maintenance, the working environment, and media quality. In other words, a dot-matrix printer processing three-part forms all day long in the middle of a dirty factory floor isn't going to last nearly as long as the same printer delivering a few dozen text pages each day in a clean, air-conditioned office.

Environmental information

Environmental specifications indicate the physical operating ranges of your printer. Storage temperature and operating temperature are the two most common environmental conditions. A typical printer can be stored in temperatures between 0°F and 100°F, but it can only be used from 50°F to 90°F (on the average). As a general rule, it is a good idea to let your printer stabilize at the current temperature and humidity for several hours before operating it. Relative humidity can often be allowed to range from 10% to 90% during storage, but it must be limited from 30% to 80% during operation. Keep in mind that humidity limits are given as "noncondensing" values; water vapor cannot be allowed to condense into liquid form. Liquid water in the printer would certainly damage its motors and electronics. Moist paper will also result in poor print quality. Your printer might also specify physical shock or vibration limits to indicate the amount of abuse the printer might sustain before damage can occur. This is usually rated in terms of *g-force*. As a rule, the printer should be free of shock and vibration.

Physical information

Finally, physical information includes such routine data as the printer's height, width, and depth, as well as its weight. In some cases, an operating noise level specification is included to indicate just how loud the printer will be while in operation and standby modes. Acoustic noise level specifications are usually given in dBA (audio decibels).

Understanding the typical assemblies

No matter how diverse or unique printers might appear from one model to the next, their differences are primarily cosmetic. It is true that each printer might use different individual components, but every printer must perform a very similar set of actions. As a result, most moving-carriage printers can be broken down into five typical subsections, or functional areas: the paper transport assembly, the print head assembly, the print head transport (or carriage), the power supply, and the electronic control unit. These key elements are illustrated in the detailed block diagram of figure 1-5. This part of the chapter introduces you to each of those subassemblies.

Paper transport assembly

Every printer must handle paper. Paper must be picked up, carried in front of the print head, then delivered out of the printer intact. There are two classical methods of achieving paper transport: tractor-feed and friction-feed. Figure 1-6 illustrates a typical tractor-feed paper transport assembly. All paths lead paper around a rubber platen (in front of the print head) and into a set of plastic sprocket wheels. Any additional rollers, such as the "bail" rollers, are used to apply light pressure, which keeps paper flat and even

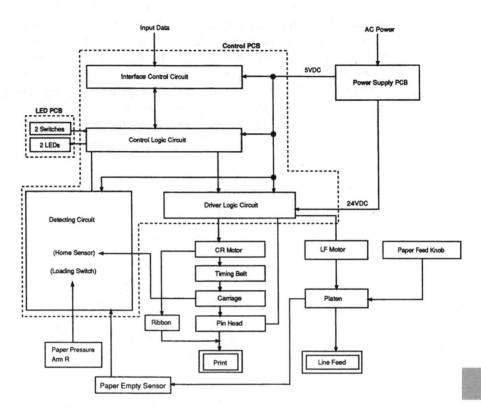

■ **1-5** *Block diagram of a moving-carriage printer.* Tandy Corporation

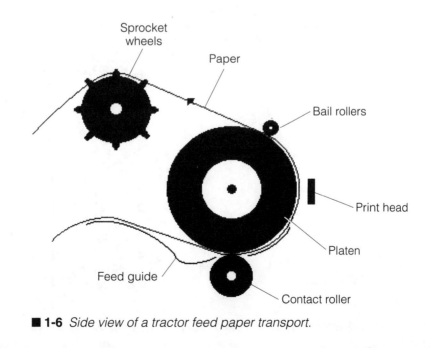

■ **1-6** *Side view of a tractor feed paper transport.*

Understanding the typical assemblies

against the platen. Each sprocket wheel is fitted with teeth that mesh exactly with holes perforated along both sides of the paper. Both wheels are linked together, then linked mechanically to the platen by a gear train or pulley system. When the platen advances, it will turn both sprockets an equal amount to pull paper up and out of the printer. Tractor-feed systems require some type of continuous form paper, but because paper is pulled evenly at all times, it should always feed evenly. Figure 1-7 shows you the exploded view of a typical tractor-feed paper transport assembly.

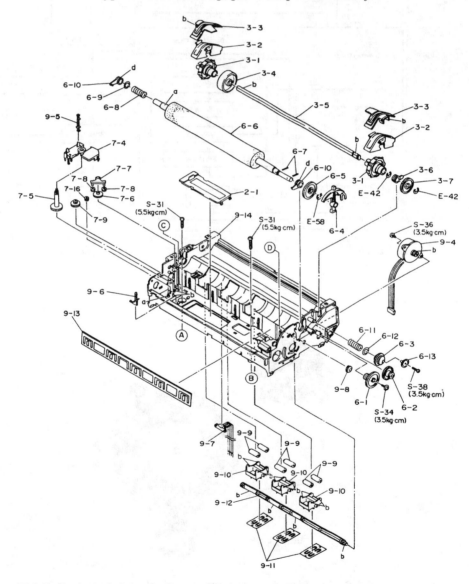

■ 1-7 *Exploded view of a tractor feed paper transport.* Tandy Corporation

A modern printer

A friction-feed paper transport system is shown in figure 1-8. A mechanical lever is often used to pry pressure rollers away from the platen while paper is being threaded and aligned. When the lever is released, pressure rollers are brought into tight contact with the paper and platen. As the platen advances, contact forces between the platen and rollers move the paper through. Bail rollers only serve to keep paper flat and even against the platen as paper leaves the printer.

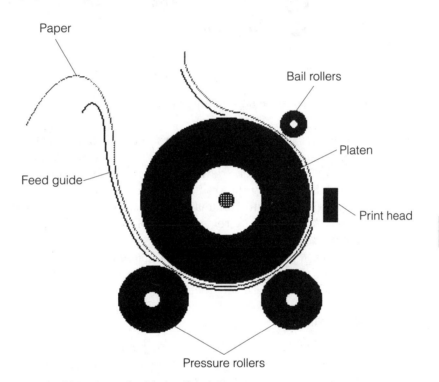

Paper

Bail rollers

Platen

Feed guide

Print head

Pressure rollers

■ **1-8** *Side view of a friction feed paper transport.*

Friction-feed systems can use just about any single-sheet, medium-weight paper or multipart form. This can make friction feed very flexible, but not without its problems. Because paper is clamped into place under pressure, it will invariably travel the direction in which it is clamped. If paper is not perfectly straight, it will "walk" to the left or right as the platen advances. Even if the sheet was inserted straight, old or misaligned rollers can cause walking.

A variation of the friction-feed system can be found in most electrophotographic printers as shown in figure 1-9. Conventional friction-feed printers require you to insert and align single sheets of

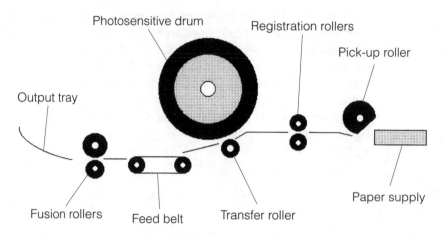

■ **1-9** *View of an EP paper transport system.*

paper, one at a time. An EP paper feed system uses a notched rubber pick-up roller to grab individual sheets as needed from a central paper tray. Paper is always started reasonably straight because of its alignment by the tray. A set of registration rollers holds the sheet stationary until an image is prepared on a photosensitive EP drum. At that point, registration rollers push the paper to the drum where its image is transferred to paper. A feed belt gently carries the page to a set of heated fusing rollers, which fix the image permanently. Chapter 4 presents a much more complete examination of EP printing technology.

Print head assembly

A print head is a discrete device responsible for actually applying permanent print to the page surface. Four major technologies have evolved to accomplish this process: impact, thermal, ink jet, and electrophotographic. Printers are typically categorized by the particular technology that they use (e.g., thermal dot-matrix printer, ink jet dot-matrix printer, and so on). As discussed later, however, EP printers form images through a process rather than a print head.

Impact technology is just what the name implies; characters, symbols, and (sometimes) graphics are literally struck onto a page surface through an inked ribbon of fabric or plastic. The force of impact leaves an ink impression of whatever was to be printed. You might encounter two types of impact print heads: the character print head and the dot-matrix print head.

Character print heads (also called daisy wheels) are the simplest and most straightforward type of print head design, as shown in figure 1-10. Modeled after a conventional typewriter, a daisy wheel is little more than a print wheel containing a fixed selection of preformed letters, numbers, punctuation, or other symbols. Each is reverse-molded onto a single plastic support structure. When an ASCII character is received, the print wheel rotates so that the desired character is positioned in front of the platen. The character is then rammed against the page by a solenoid. Impact takes place through an inked ribbon, so the character's image is transferred to the page.

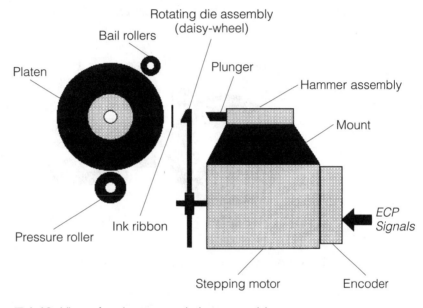

■ **1-10** *View of a character printing assembly.*

Although the daisy-wheel printer produced some excellent print, the drawbacks were easy enough to recognize. First, the continuous spinning and ramming of the daisy wheel created a serious clatter. Also, mechanical limitations of the daisy wheel limited the printer's maximum printing speed. Constant flexing of daisy wheel "petals" frequently resulted in broken print wheels; so regular print-wheel replacement was a must. Next, the print size and font was static, so it was impossible to select different fonts or type sizes without replacing the daisy wheel (graphics were out of the question). Character printers have long since been obsoleted by dot-matrix designs, and it is highly unlikely that you will ever find a character printer still in service. This book will not cover character printers further.

The impact dot-matrix print heads use a series of individual metal wires arranged in a vertical pattern as shown in figure 1-11. Each wire is operated by its own small solenoid, which can be driven independently. By firing groups of print wires in predetermined combinations, any letter, number, symbol, or graphic image can be produced while a dot-matrix head moves across the page. Impact occurs through an inked ribbon, so the image is transferred to paper. Figure 1-12 shows some typical dot formations for 9- and 24-pin print heads. The face of an impact dot-matrix print head can be flat or slightly curved to match the radius of a platen. Print wires are ground flat against the print head's face once they are installed to ensure an even and consistent dot formation for every print wire.

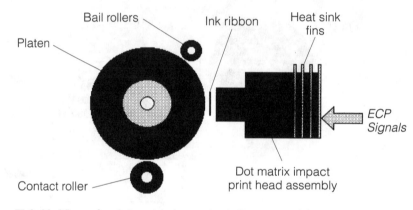

■ **1-11** *View of a dot-matrix impact printing assembly.*

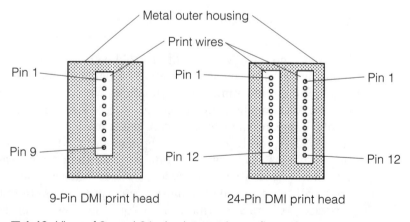

■ **1-12** *View of 9- and 24-pin dot-matrix configurations.*

While impact dot-matrix print heads offer the versatility and reliability that daisy wheels do not, the technology also suffers from
some drawbacks. First, impact print heads need a substantial amount
of energy to operate. This demands a relatively large power supply,
and the print head becomes extremely hot during operation. Dot-
matrix printing is not terribly fast. Per-character printing rates are
faster than those of daisy wheels, but to achieve a comparable-
quality print, the dot-matrix head might have to make several
passes to complete a single line of text. Impact dot-matrix printers
also make a great deal of noise. The high-frequency chatter of continuous wire impacts can become very annoying.

Thermal dot-matrix print heads overcome some of the limitations
of impact print heads. A serial thermal print head is shown in figure 1-13. Instead of physically moving print wires in and out, a
thermal print head uses an array of microscopic heater elements.
An electrical pulse from driver circuits will cause a dot heater to
warm very quickly. This leaves a corresponding dot on the page
surface. As with impact dot-matrix heads, an array of dot heaters
can be fired in any desired sequence to produce virtually any letters, numbers, or graphics. Because there are no moving parts in
the print head, its operation is totally silent.

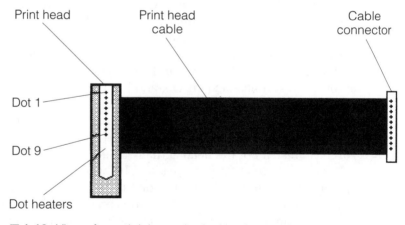

Print head   Print head   Cable
cable   connector

Dot 1

Dot 9

Dot heaters

■ **1-13** *View of a serial thermal print head assembly.*

An alternative to a serial (or moving) thermal print head is the
line-head technique shown in figure 1-14. A line-head is essentially
a thermal print bar. It contains a horizontal row of dot heaters—
one heater for every possible dot along a horizontal line. Instead of
a vertical column of dots forming full characters as the head moves
across a page, an entire line of text or graphics can be formed one

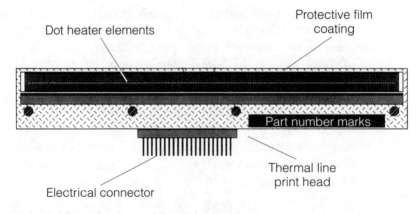

Dot heater elements

Protective film coating

Part number marks

Electrical connector

Thermal line print head

■ **1-14** *View of a thermal line print head.*

row at a time. Line-head printers offer the advantage of simplicity; no carriage transport assembly is needed to carry a print head back and forth. Facsimile machines typically use thermal line-head printing.

Thermal print head technologies have proven to be a handy, reliable, and quiet alternative to impact print heads, but there are some serious trade-offs. Because a finite amount of time is required to heat and cool a heater element, thermal printing is a relatively slow process. Thermal printing is also limited to a single pass (such as fax printing), so the print is generally quite legible, but its overall appearance is poor. Another strike against thermal print heads is their need for heat-sensitive paper, which has proven to be fragile, expensive, and hard to handle or store. Normal paper could not be used because the temperatures required to mark normal paper would burn it. This restriction can be overcome by using a "transfer" version of a thermal print head (serial or line). Instead of the head actually contacting a page, a transfer ribbon is inserted between the two. The head heats corresponding points of wax on the transfer ribbon that melt onto the page. While this made thermal printing a bit more practical for "plain paper" printers, the technology never really gained broad acceptance.

Ink jet dot-matrix print head technology evolved to offer an alternative method of quick, quiet, inexpensive printing that would provide print quality superior to impact print heads and still work on almost any type of paper. This method of printing draws liquid ink from a central reservoir, then "spray paints" the desired characters or graphics onto a page surface through a series of independent nozzles. Ink jet printing speeds can easily match that of an impact dot-matrix printer. There are two primary methods of commercial ink jet printing: piezoelectric jet and bubble jet (also called thermal jet).

A piezoelectric jet print head uses liquid ink from a small, local reservoir to fill a series of ink channels illustrated in figure 1-15. While older ink jet heads typically provided 9 or 24 ink channels, contemporary ink jet heads can provide 50 ink channels or more. Each channel is covered by a nozzle, which is nothing more than a microscopic hole drilled into a metal plate. A small piezoelectric crystal in each channel acts as an "ink pump." A short, high-energy pulse from the printer's driver circuits vibrates the crystal, which ejects a single droplet of ink from the corresponding nozzle.

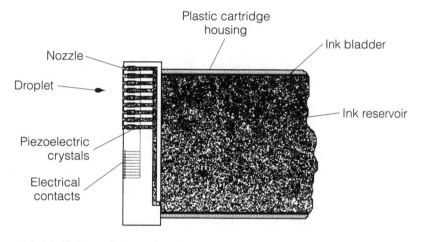

■ **1-15** *Enlarged view of a piezoelectric ink jet print head.*

Bubble jet print heads are very similar to piezoelectric jet devices, but the resonating piezoelectric crystals are replaced by small heater elements. An electrical driver pulse heats ink in a channel. A bubble forms and expands until it bursts. The force of a bursting bubble ejects a droplet of ink from the nozzle.

Modern ink jet print heads offer an astounding mix of advantages. They require relatively little power (an ideal attribute for mobile printers), yet they can generate high-quality text and graphics at resolutions that rival EP printers. Ink jet operation can be slow when printing high-resolution graphics, but the low cost and disposable print head design often makes the trade-off worthwhile. Ink jet technology has also been adapted to inexpensive color printing. A remaining disadvantage of ink jet technology is the need for specialized paper; most standard papers will work, but clay-coated papers provide the best ink drying characteristics.

The electrophotographic printing technique is radically different from any of the previous print technologies. As you will see in Chapter 4, an electrophotographic (EP) "print head" is actually a combination of devices that incorporate light, static electricity, chemistry, optics, heat, and pressure to place a permanent image on a page. Images are formed as an array of dots written to a light-sensitive drum. A laser beam has been used to write dots, but they are rapidly being replaced by smaller, lighter, and more reliable writing devices. EP printing offers many advantages over other technologies; it is fast and quiet, it works at resolutions up to 1,200 × 1,200 dpi and higher, and it works with single sheets of paper.

Print head transport (carriage)

Most serial print heads only provide a single vertical column of dots. As a result, the print head must be carried back and forth across the page to form characters, words, and sentences. This movement is accomplished by a print head transport mechanism (also known as a carriage). A basic head transport system is shown in figure 1-16. The key assembly here is a slide that is free to move along a set of rails. The print head itself mounts securely to the carriage with either a series of screws or tight hold-down clips. A simple pulley system driven by a stepping motor is used to supply the left/right motion through a belt or threaded rod. Figure 1-17 presents an exploded view of an actual carriage transport.

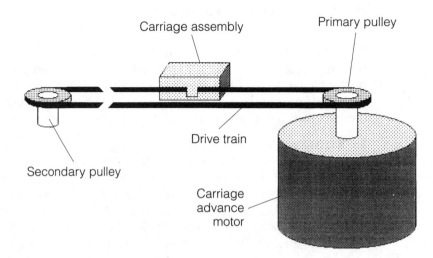

■ **1-16** *Simplified view of a carriage transport system.*

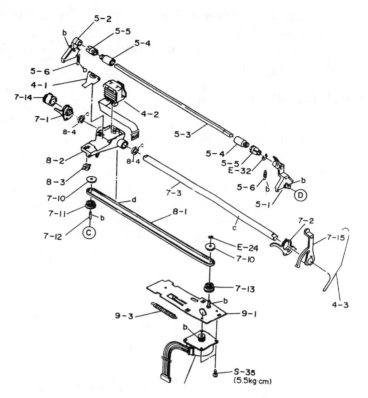

■ **1-17** *Exploded diagram of a carriage transport assembly.*
Tandy Corporation

Two sensors are usually incorporated into the head transport system (not shown in figures 1-16 or 1-17). A print head home sensor is located at the left end stop. This tells the printer's electronics when a carriage has reached its home position. An optical encoder is often added to the motor assembly. It sends distance and direction information back to the control electronics that allows constant monitoring of the print head's current location.

Power supply

A power supply is responsible for converting ac line voltage into one or more of the dc voltage levels that are needed by the printer's electronic and electrical components. Typical voltage outputs are 5, 12, and 24 Vdc. High-voltage supplies used in electrophotographic printers can provide ionizing voltage levels of −6,000 V or more. There are both linear and switching power supplies. Each is distinguished by the method used to control output voltages. Linear supplies are relatively straightforward devices as discussed in Chapter 8. They are simple to understand and troubleshoot, but they are inefficient. Switching supplies can be much more complex to follow and repair, but they are smaller, and can be very efficient.

Electronic control unit

The electronic control unit (ECU) represents a combination of electronic components and circuits that direct every detail of a printer's operation. A typical ECU is made up of five major sections: a communications interface, driver circuits, a control panel, memory, and main logic. In spite of the diversity between printers and print head technologies, every printer model must include circuitry to handle these areas. Figure 1-18 illustrates each of these key areas in the ECU for an HP DeskJet 500 printer. The exploded diagram of figure 1-19 shows you the ECU (marked 19), the sepa-

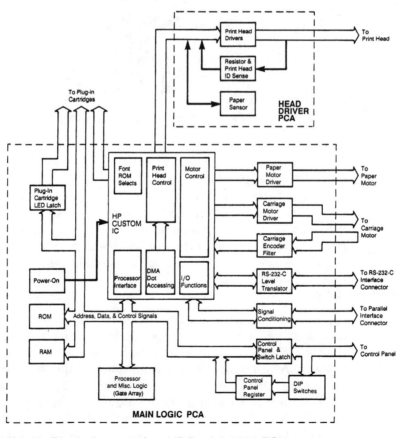

■ **1-18** *Block diagram of an HP DeskJet 500 ECU.* Hewlett-Packard Co.

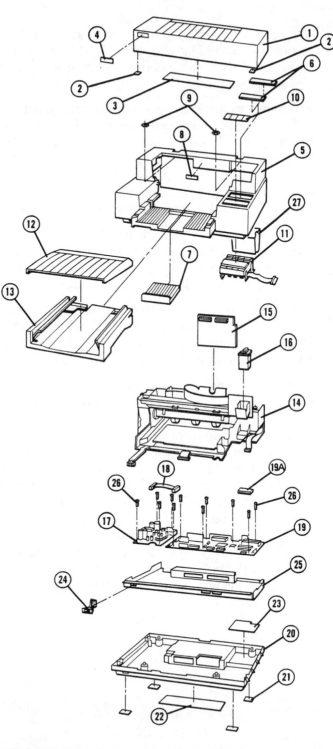

■ 1-19 *Exploded view of an HP DeskJet 500 printer assembly.* Hewlett-Packard Co.

Understanding the typical assemblies

rate head driver circuit (marked 15), and the system power supply module (marked 17) in the overall DeskJet assembly. The detailed operation and troubleshooting techniques for an ECU are presented in Chapter 9.

A communications interface handles the transfer of all data between the printer and computer. It also manages and coordinates signals ("handshaking") that synchronize data transfer. Data can be transferred over either a "parallel" or "serial" interface. Both interface methods work equally well for most general-purpose printers. Your printer's particular compatibilities are listed in its user's manual. Digital electronics alone is not enough to run a printer.

Drivers convert low-power digital signals into voltage and current levels that are needed to operate real-world devices, such as motors, solenoids, and print heads. Drivers also might be used to condition sensor signals for use by digital logic.

A control panel allows you to work with the printer directly by operating manual functions (e.g., line feed, form feed, etc.). It also lets you alter various operating modes like font style or character pitch. Indicators are included to display various printer conditions.

Your printer's overall operation is organized and managed by the main logic circuits. Decision making is handled by programmable processing components such as microprocessors or microcontrollers. There might be more than one processing component depending on the printer's sophistication. New printers tend to replace a second microprocessor with one or more application-specific integrated circuits (ASICs), which can optimize printer communication and driver functions. Memory devices are needed to retain permanent and temporary information used within the printer. The printer's command language and microprocessor instructions are stored in permanent memory (ROM). Temporary memory (RAM, also called the buffer) is used to hold character and graphic data received by the printer's communication interface.

Typical components

THIS CHAPTER DETAILS A CROSS-SECTION OF ELECTRICAL and mechanical components that can be found in just about any moving-carriage or EP printer. Your troubleshooting efforts will be simplified greatly if you are able to identify important components on sight, understand their purpose, and spot any obvious defects. This is by no means a complete review of every possible type of component, but it will give you a good idea of what to expect.

Mechanical components

Mechanical parts basically serve a single purpose: to transfer force from one point to another. For example, a paper transport system must transfer the physical force of a paper advance motor to the paper itself to move it through the printer. This is accomplished through a series of gears, pulleys, rollers, and belts. Whenever mechanical parts are in contact with one another, they produce friction, which causes wear. Lubricants, bushings, and bearings work to minimize the damaging effects of friction on mechanical parts.

Gears

A gear performs several important tasks. Their most common application is to transfer mechanical force from one rotating shaft to another. The simplest arrangement uses two gears in tandem as shown in figure 2-1. When two gears are used, the direction of secondary rotation is opposite that of the primary shaft. If secondary direction must be the same, a third gear might be added as shown in figure 2-2. It is possible to change the direction of applied force by using angled gears. By varying the angles of both gears, force can be directed just about anywhere. Several secondary gears can be run from a single drive gear to distribute force to multiple locations simultaneously, a common tactic in complex mechanical systems like the bubble jet printer in figure 2-3.

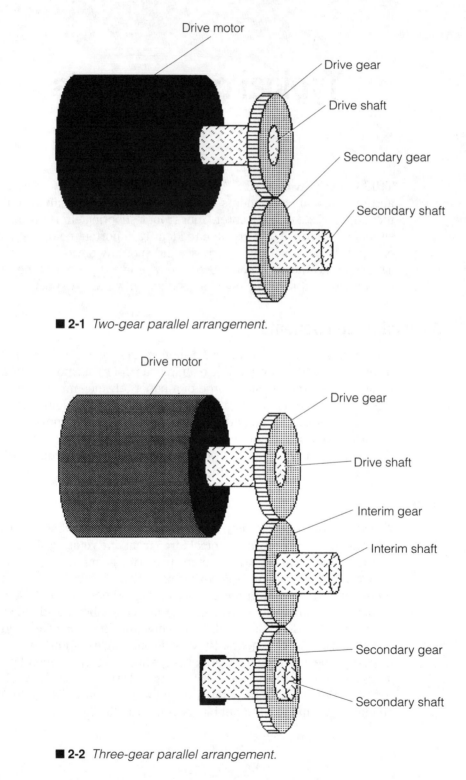

■ **2-1** *Two-gear parallel arrangement.*

■ **2-2** *Three-gear parallel arrangement.*

Typical components

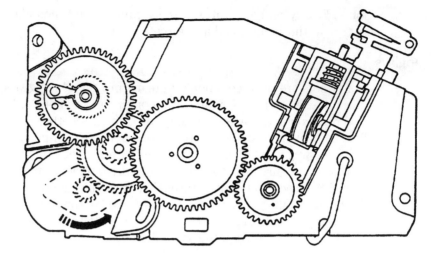

■ **2-3** *An example of multiple gear operation.* Tandy Corporation

Not only can gears transfer force, they can also modify the speed and force that is applied at the secondary shaft. Figure 2-4 illustrates the effects of simple gear ratios. A gear ratio is usually expressed as the size ratio of the primary gear versus the secondary gear. For a high ratio, the primary gear is larger than the secondary gear. As a result, the secondary gear will turn faster, but with less force. The effect is just the opposite for a low ratio. A small primary will turn a larger secondary slower, but with more

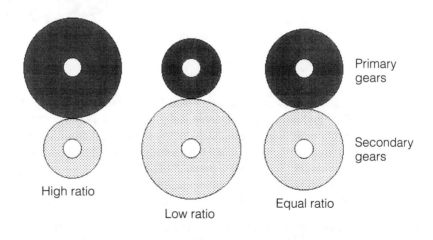

■ **2-4** *The concept of gear ratios.*

force. Finally, an equal ratio causes a primary and secondary gear to turn at the same speed and force.

Pulleys

Pulley sets are used commonly in many printers. Like gears, pulleys are used to transfer force from one point to another. Instead of direct contact, however, pulleys are joined together by a drive linkage, which is usually composed of a belt, wire, or chain (much the same way as a belt in your automobile). A basic pulley set is illustrated in figure 2-5. A motor turns a drive pulley, which is connected to a secondary pulley through a drive linkage under tension. As the drive pulley turns, force is transferred to the secondary pulley through the linkage, so the secondary pulley also turns. Notice that both pulleys turn in the same direction.

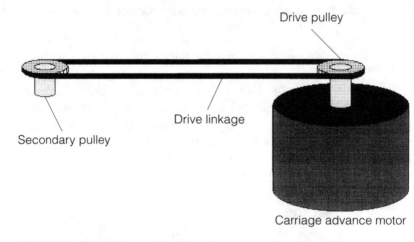

Drive pulley

Drive linkage

Secondary pulley

Carriage advance motor

■ **2-5** *A basic pulley set.*

In addition to transferring force to a secondary shaft, the linear (left/right) motion of a drive linkage can be used to move a load. For example, a carriage transport system must transfer the force of a carriage advance motor to a sliding mount. If a point on the linkage were connected to a carriage, the left/right motion of the drive linkage would carry a print head back and forth. The type of pulley system used depends on the physical load that must be moved.

Pulleys and drive linkages will vary depending on their particular application. Low-force applications can use narrow pulleys (little more than a wheel with a groove in it) connected with a wire linkage. Wire is not terribly rugged, and its contact surface area with both pulleys is relatively small. Therefore, wire can slide when it stretches under tension, or if load becomes excessive. Belts and their pulleys are wider, so there is much more surface contact around each pulley. Belts are usually much stronger than wire, so there is less tendency to stretch under tension. This makes belt-driven pulleys better suited for heavier loads. In practical applications, notched belts and notched pulleys are used in the carriage transport system. The belt is light and strong, while the notches eliminate slipping around the pulleys.

Rollers

For a printer to work at all, paper must be handled gently but firmly. Rubber-coated rollers are ideal for this purpose; they provide the high friction needed to carry paper through the printer reliably and consistently, yet the pressures applied by the rollers will not damage the paper. As you might imagine, damaged, old, or dirty rollers can have an adverse effect on the paper transport system.

There are three types of rollers found in most moving-carriage printers: a platen, pressure (or contact) roller(s), and bail rollers. All three rollers are illustrated in figure 2-6. Your main roller is the platen. It is driven by a paper advance motor, and it provides paper support in front of the print head. The rubber also serves as a firm but pliable surface for impact printing. Pressure rollers hold paper against the platen and ensure positive traction and even paper advance. The actual position and contact force of your pressure roller(s) will depend on the particular paper transport technique

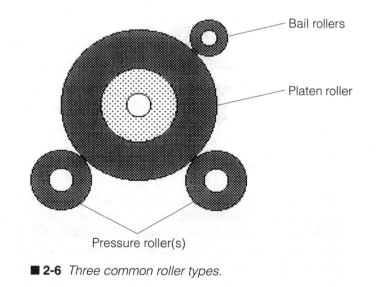

Bail rollers

Platen roller

Pressure roller(s)

■ **2-6** *Three common roller types.*

used. Bail rollers are a set of small, adjustable rollers that help to hold paper flat while it travels around the platen on its way out of the printer.

Reducing friction

As with all mechanical systems, parts that are in contact with one another will wear while the system operates due to unavoidable friction that occurs between parts. It follows, then, that reducing friction will extend your printer's working life. Lubrication, bushings, and bearings are three commonly accepted methods of reducing friction.

Oils or grease have always been one answer (and might prove quite effective in small doses), but this lubrication must be replaced on a regular basis for it to remain effective. Otherwise, it can wear away, dry out, or harden into thick sludge. Lubricants are also notorious for collecting dust and debris from the environment, which eventually defeats any benefits that the lubricant can provide. When applying lubricants, remember to apply them sparingly.

Bushings are essentially "throw-away" wear surfaces as shown in figure 2-7. A bushing is made of softer materials than the parts it is separating, so any friction generated by moving parts will wear out the bushing before allowing the parts to make contact themselves. When a bushing wears out, simply replace it with a new one. Bushings are much less expensive and easier to replace than major mechanical parts such as slides or frames.

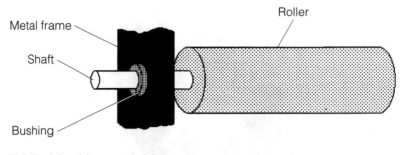

■ 2-7 *A bushing assembly.*

Probably the most effective devices for reducing friction between parts are bearings. Bearings consist of a hard metal case with steel balls or rollers packed inside as shown in figure 2-8. Because each steel ball contacts each load-bearing surface at only one point, friction (and wear) is substantially lower than it is for bushings.

32

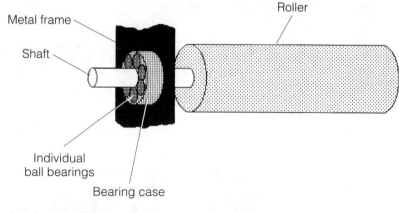

■ **2-8** *A bearing assembly.*

Unfortunately, bearing assemblies are often much more expensive than bushings, so bearings are employed only to handle heavy loads, or in places that would be too difficult to change bushings.

Electromechanical components

Electromechanical components are a particular class of devices that convert electrical energy into mechanical force or rotation. Relays, solenoids, and motors are three common electromechanical components that you should understand and be familiar with. Each of these important devices relies on the principles of *electromagnetism*.

Electromagnetism

Whenever electrical current passes through a conductor, a magnetic field is generated around the circumference of that conductor as shown in figure 2-9. Such a magnetic field is capable of exerting a physical force on permeable materials (any materials that can be magnetized). The strength of a magnetic field around a conductor is proportional to the amount of current flowing through it. Higher amounts of current result in stronger magnetic fields, and vice versa.

Unfortunately, it is virtually impossible to pass enough current through a typical wire to produce a magnetic field that is strong enough to do any useful work. The magnetic field must somehow be concentrated. This is accomplished by coiling the wire as shown in figure 2-10. When arranged in this fashion, the coil takes on magnetic poles, just like a permanent magnet. Notice how the

33

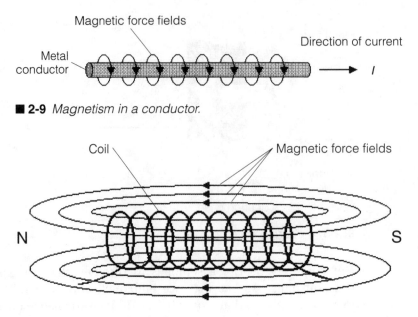

■ 2-9 *Magnetism in a conductor.*

■ 2-10 *Magnetism in a coil.*

direction of magnetic flux always points to the north pole. If the direction of current flow were reversed, the coil's magnetic poles would also be reversed.

To concentrate magnetic forces even further, a permeable core material can be inserted into the coil's center as in figure 2-11. Typically, iron, steel, and cobalt are considered to be the classical core materials, but iron-ceramic composite blends (sometimes called *ferrite*) are used as well. It is coils of wire such as these that form the foundation of all electromechanical devices.

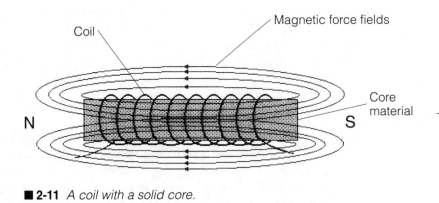

■ 2-11 *A coil with a solid core.*

Relays

A relay is simply a mechanical switch that is actuated with the electromagnetic force generated by an energized coil. A typical relay diagram is shown in figure 2-12. The switch (or contact set) might be normally open (N.O.) or normally closed (N.C.), while the coil is de-energized. When activated, the coil's magnetic field will cause normally open contacts to close, or normally closed contacts to open. Contacts are held in their actuated positions as long as the coil is energized. If the coil is turned off, contacts will return to their normally open or closed states. Keep in mind that a coil might drive more than one set of contacts.

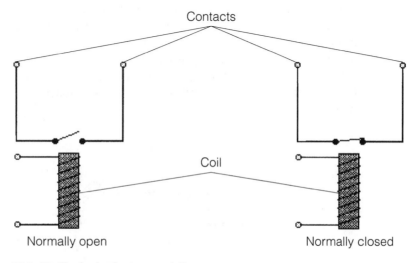

■ **2-12** *Typical relay assemblies.*

Relays are not always easy to recognize on sight. Most relays used in electronic circuits are housed in small, rectangular containers of metal or plastic. Low-power relays can be fabricated into over-sized IC-style packages and soldered right into a PC board just like any other integrated circuit. Unless the relay's internal diagram is printed on its outer case, you will need a printer schematic or manufacturer's data for the relay to determine the proper input and output functions of each relay pin.

A relay rarely fails to actuate; a simple wire coil is notoriously reliable. The problem with relays is in their contact set. Because a relay is often used to switch relatively large amounts of current, the contacts tend to arc a bit each time the connection is made and broken. Over time, this arcing effect damages the contacts. Eventually, the contact set will become intermittent and sometimes fail to make a good electrical connection, even though the actuating coil fires just fine. To combat the arcing effect, some high-quality relays place their contact set(s) in an enclosure filled with inert gas (such as nitrogen), which will inhibit sparks. The only way to correct an intermittent or failed relay is to replace it.

Solenoids

The solenoid converts electromagnetic force directly into motion as illustrated in figure 2-13. Unlike ordinary electromagnets whose cores remain fixed within a coil, a solenoid core is allowed to float back and forth without restriction. When energized, the magnetic field generated by a coil exerts a force on its core (called a plunger), which pushes it out from its rest position. If left unrestrained, a plunger would simply shoot out of its coil and fall away. Plungers are usually tethered by a spring or some other sort of mechanical return assembly. That way, a plunger will only extend to some known distance when the coil is fired, then automatically return to its rest position when the coil is off.

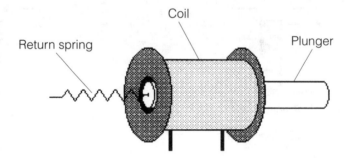

Coil

Return spring

Plunger

■ **2-13** *A typical solenoid.*

A common concern when using relays and solenoids is the potentially damaging effect of "flyback voltage." Remember that some energy is stored in the coil as a magnetic field. While current flows to energize the coil, magnetic strength remains constant. However, when current stops and the coil de-energizes, its magnetic field collapses very quickly. This sudden collapse induces a brief, potentially high-voltage spike. Its polarity is reversed to that of its energizing voltage, thus the term *flyback*.

Such a spike would almost certainly be harmless to the coil itself (ignition coils in automobiles rely on the flyback effect to produce the high-voltage spikes that fire spark plugs), but delicate electronic circuits are extremely sensitive to voltage spikes. Any transient voltage spikes that are conducted into an electronic circuit can easily result in permanent circuit damage. A rectifier-type diode placed across a coil will work to short-out any spikes. Diodes used in this fashion are called *flyback diodes*.

Solenoids have come to serve a variety of uses in printers. Advances in materials and assembly techniques have given rise to smaller, more efficient, and more reliable solenoids. As detailed in Chapter 3, solenoids are the key components of impact printing technology. They drive hammers or print wires that actually strike a page. Solenoids are also used in some printers with multicolor ribbons. By energizing and de-energizing a solenoid under printer control, a ribbon can be positioned at the desired color. In EP printers, solenoids are typically used as electromagnetic clutches that can engage and disengage force from the main motor to various printer mechanisms. You might encounter even more exotic applications.

Like the relay, a solenoid is renowned for its reliability. Chances are that the solenoid's driving circuit will fail before the solenoid does. However, movement of the plunger within the solenoid can easily be affected by accumulations of dust and debris, as well as mechanical wear and failure. For example, the print wires of an impact dot-matrix print head can be jammed by accumulations of dried ink and paper dust. Do not presume that the solenoid is invulnerable. Often, just cleaning away accumulations of gunk can restore a jammed plunger, but a damaged solenoid must be replaced.

Motors

The motor is an absolutely essential part of every printer manufactured today. Motors operate the various transport systems that you will find. All motors convert electrical energy into rotating mechanical force. In turn, that force can be distributed with mechanical parts (e.g., gears and pulleys) to turn a platen or move a carriage. An induction motor performs this task by mounting a series of powerful electromagnets (coils) around a permanent magnet core, as shown in figure 2-14. The core (known as a rotor) is little more than a shaft that is free to rotate as its poles encounter electromagnetic forces. Each coil (also called a phase or phase winding) is built into the motor's stationary frame (or stator).

By powering each phase in its proper order, the rotor can be made to turn with a predictable amount of force. The amount of angular

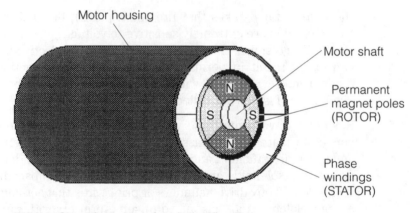

Motor housing

Motor shaft

Permanent
magnet poles
(ROTOR)

Phase
windings
(STATOR)

■ **2-14** *Simplified view of an induction motor.*

force generated by a motor is known as *torque*. Induction motors
generally require two ac signals separated by a 90° phase difference. These sinusoidal driving signals vary the strength of each phase
evenly to cause a smooth rotation. Induction motors are rarely used
in today's commercial printers because they do not lend themselves easily to the precise positioning requirements that most
printers demand. Instead, a close variation of the induction motor
is used. This is called a *stepping motor*.

Physically, a stepping motor is very similar to figure 2-14, but instead of sinusoidal driving signals, a stepper is driven by a series of
square wave pulses separated by a 90° phase difference. Square
waves cause the rotor to jump (or *step*) in predictable angular increments, not a smooth, continuous rotation. Once the rotor has
reached its next step, it will hold its position as long as driver signals maintain their conditions. A typical stepping motor can
achieve 1.8° per step. This means a motor must make 200 individual steps to complete a single rotation.

Stepping motors are ideal for precise positioning. Because the motor moves in known angular steps, it can be rotated to any position
simply by applying the appropriate series of driver pulses. For example, suppose your motor had to rotate 180°. If each step equals
1.8°, you need only send a series of 100 pulses to turn the rotor exactly that amount. Logic circuits in the printer generate each pulse,
then driver circuits amplify those pulses into the high-power signals that actually operate the motor.

Motors are not quite so reliable as relays or solenoids. First, there
are typically several individual coils in the motor. A weakness or
failure in any one of these coils will render the motor useless. Be-

cause the coils generate a relatively large amount of heat under continuous use, heat is frequently the catalyst that causes phase winding problems. The rotor is also mounted on bearings or a high-quality bushing. If the bushing or bearings wear out, the rotor will shift position and move unevenly (if at all). Damaged motors must be replaced. Fortunately, this is often a simple procedure.

Electronic components

You will find a wide variety of electronic components contained within your printer. Most circuits contain both active and passive components working together. Passive components include resistors, capacitors, and inductors. They are called passive because their only purpose is to store or dissipate a circuit's energy. Active components make up a broader group of semiconductor-based parts such as diodes, transistors, and all types of integrated circuits. They are referred to as active because each component uses a circuit's energy to perform a specific set of functions; they all do something. It might be as simple as a rectifier, or as complex as a microprocessor, but active parts are the key elements in modern electronics. This section is intended to familiarize you with each general type of component, how they work, how to read their markings, and how they fail.

Resistors

All resistors ever made serve a single purpose: to dissipate power. Although resistors appear in many circuits, they are used primarily for such things as voltage division, current limiting, volume adjustment, etc. Resistors dissipate power by presenting a resistance to the flow of current. Wasted energy is then shed as heat. Resistance is measured in ohms using the Greek symbol Omega (Ω).

In a typical carbon-composition resistor, two component leads are insulated by a packed carbon filling. It is much harder for electrons to pass through carbon than copper, so the flow of current is limited. The material composition of carbon filling can be altered in manufacturing to provide many different levels of resistance. Carbon-composition resistors are rarely used in modern electronics because they are large, and they lack the precision needed to support contemporary circuits. The carbon-composition resistor has been replaced by the carbon-film type, as shown in figure 2-15. This kind of resistor uses a carbon film deposited onto a ceramic or glass core. Metal caps on both ends provide the electrical connections. The entire assembly is encapsulated in a hard epoxy ma-

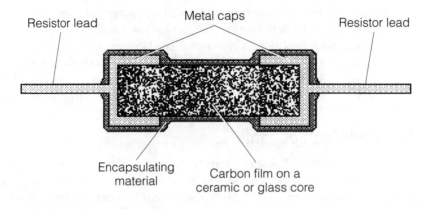

Resistor lead

Metal caps

Resistor lead

Encapsulating material

Carbon film on a ceramic or glass core

■ **2-15** *A carbon-film resistor.*

terial. Carbon film resistors are far more accurate than carbon-composition devices, because films are applied with more precision and control.

Resistors are also rated in terms of the power that they can handle. Common values are ⅛ W, ¼ W, ½ W, 1 W, and 2 W. As long as the power being dissipated by a resistor is less than its rating, the resistor should perform as expected and last indefinitely. However, if a resistor is forced to exceed its power rating, it cannot shed heat fast enough to maintain a stable temperature. Ultimately, the resistor will overheat and burn out (often damaging the printed wiring board and printed copper traces as well). In all cases, a burned out resistor forms an open circuit. A faulty resistor might appear only slightly discolored, or it might look burned and cracked. It really depends on the severity and duration of its overheating. Replace any faulty resistors wherever you might find them. If the printed wiring board is also damaged, the entire circuit assembly should be replaced to maintain peak reliability.

Adjustable resistors (called potentiometers or rheostats), as shown in figure 2-16, employ a movable metal wiper blade resting along a ring of resistive film. While total resistance of the film remains constant, resistance between the wiper and either end can be varied by turning a knob. Failures among potentiometers usually take the form of intermittent connections between the wiper blade and resistive film. Remember that film will wear away as the wiper moves across it. Over time, enough film might wear away at certain points that the wiper might not make good contact there. This can result in all types of erratic or intermittent circuit operation. Replace any intermittent potentiometers or rheostats.

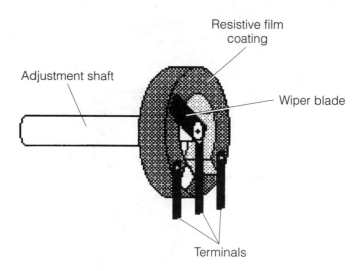

Resistive film coating

Adjustment shaft

Wiper blade

Terminals

■ **2-16** *A variable resistor.*

Capacitors

The capacitor is used to store an electrical charge. This might sound simple enough, but it has important implications when combined with other components in filters, resonant circuits, or timing circuits. Capacitance is measured in farads (F). In reality, a farad is a very large value of capacitance. Practical capacitors are normally found in the microfarad (µF) or picofarad (pF) range. A capacitor is little more than two conductive plates separated by an insulator (called a *dielectric*), as shown in figure 2-17. The amount of capacitance is determined by plate area, the distance between each plate, and the specific dielectric material. Large values of capacitance can be achieved by rolling up a plate-dielectric assembly and housing it in a canister.

When voltage is applied across a capacitor, current flows in and electrons are stored as a static charge. As the capacitor charges, its current flow decreases. This continues until the capacitor is fully charged. At that point, no additional current flows, and the voltage across the capacitor equals the applied voltage. Keep in mind that a capacitor will remain charged even after charging voltage is removed. Large capacitors can store enough energy to present a shock hazard. Ideally, charge should last indefinitely, but internal resistance through the dielectric will eventually bleed off any accumulated charge.

There are two general capacitor types that a technician should be familiar with: fixed and electrolytic. Fixed capacitors are nonpo-

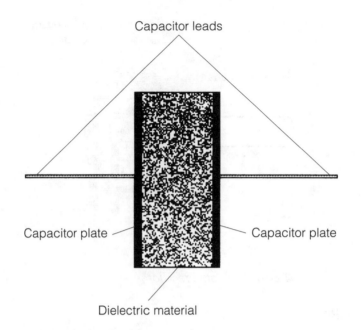

Capacitor leads

Capacitor plate ⟋ ⟍ Capacitor plate

Dielectric material

■ **2-17** *A simple plate capacitor.*

larized devices; they can be inserted into a circuit in any orientation. Many fixed capacitors are assembled as small wafers or disks. Conductive plates are typically aluminum foil. Common dielectrics include paper, mica, and ceramic. The assembly is then coated with a hard plastic, epoxy, or ceramic housing to keep out humidity. Larger devices can be assembled in cans that are sealed hermetically. Electrolytic capacitors are polarized components; they must be inserted in the proper orientation with respect to applied voltage. An aluminum electrolytic capacitor is made using two strips of aluminum alloy (one strip is coated to prevent oxidization) separated by a layer of gauze soaked with an electrolyte paste of boric acid, glycerin, and ammonia. The assembly is rolled up tightly and sealed into a metal canister. Each plate is welded to leads that connect the component. Tantalum electrolytic capacitors are constructed differently. An anode of sintered, porous tantalum powder is housed in a silver-plated container with an electrolyte of sulfuric acid or some other electrolyte. The small devices are then dipped in ceramic or epoxy.

Like resistors, capacitors tend to be rugged and reliable devices. Because they only store energy (not dissipate it), it is virtually impossible to burn them out. Remember that capacitors do have some internal resistance. Ideally, this would be infinite, but it is actually somewhat less. Under heavy loads, a capacitor's internal resistance can dissipate power and cause heating over time. Capacitors are subject to dielectric breakdown due to electrical causes (like severe voltage overloads) or environmental causes (such as plate corrosion due to humidity penetration). These failures generally manifest themselves as open or short-circuit conditions within a capacitor. Most breakdowns are invisible from the part's exterior, but are measurable with test instruments.

Electrolytic capacitors are not only subject to dielectric breakdown and evaporation, but they can also explode if enough energy is applied in the reverse polarity. A failure elsewhere in the circuit (or your incorrect placement) can reverse the voltage across a capacitor. This causes temperature and pressure to rise inside until its enclosure ruptures. The explosion is rather like a firecracker, a bang with smoke and shards of foil and electrolyte. Take care that aluminum shards do not settle back onto the circuit and short out any other components. Electrolytic capacitors also fail from "fatigue" due to frequency stress. For example, it is not uncommon to find filter capacitors that have failed in switching power supplies. Reversed tantalum capacitors can also rupture or shatter, but most often its outer casing will crack. Heat and pressure might force out some internal material. Make sure that there is no tantalum "spatter" bridging across other components.

Inductors

An inductor is used to store a magnetic charge for much the same reasons that a capacitor is used to store an electrical charge. Advances in solid-state electronics have rendered inductors essentially obsolete for traditional applications such as resonant circuits and filters, but they remain invaluable for such high-energy components as transformers, motors, and solenoids. All inductors are measured in henries (H), although some small inductors are measured in millihenries (mH) or microhenries (μH).

A transformer is actually a combination of inductors working together. It is composed of several important parts (as shown in figure 2-18), a primary winding, a secondary winding, and a core structure. An ac voltage is applied across the primary winding. The ac voltage is constantly changing its value and reversing its polarity over time. As a result, the magnetic field generated in a primary coil also fluctuates. When the fluctuating magnetic field intersects a secondary winding, an ac voltage is created (or *induced*) across it. This principle is known as *magnetic coupling*. Notice that the primary and secondary coils are wound around the same core. A

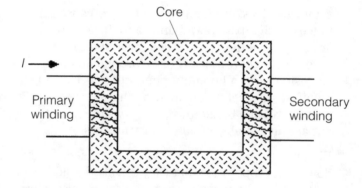

2-18 *Diagram of a transformer.*

common core concentrates magnetic energy and provides efficient coupling to the secondary coil. Although figure 2-18 shows only one secondary coil, there might be several secondary coils in a transformer.

The actual voltage generated in a secondary coil depends on the ratio of primary windings to secondary windings. This is known as the *turns ratio*. If your secondary coil contains more turns than the primary coil, then ac induced on the secondary will be greater than primary voltage by a factor of the turns ratio. For example, if a transformer has 250 turns in its primary and 500 turns in its secondary, its turns ratio is 1:2 (0.5 tr). This means that 10 Vac applied to the primary will yield [10/0.5 tr] 20 Vac. Such an arrangement is known as a *step-up* transformer. If the situation were reversed with 500 turns in the primary and 250 turns in its secondary, your transformer's turns ratio would be 2:1 (2 tr). If 30 Vac were applied to the primary, its secondary voltage would be [30/2 tr] 15 Vac. This is called a *step-down* transformer.

Current is also stepped in a transformer, but opposite to the direction of voltage steps. If voltage is stepped down, current is stepped up by the same ratio, and vice versa. For the step-up transformer above, if your 10-Vac input carries a current of 1 A, the 20-Vac output would only supply a current of [1 × 0.5 tr] 0.5 A. With a step-down transformer, an input of 30 Vac at 1 A would supply an output of 15 Vac at [1 × 2 tr] 2 A. Ideally, output power should equal input power. For example, an input of 30 Vac at 1 A is [30 V × 1 A] 30 W, and the output of 15 Vac at 2 A is [15 V × 2 A] 30 W. In reality, output power will always be slightly less than input power due to losses in the windings and core; no device is 100% efficient at transforming power.

Because inductors are energy storage devices, they should not dissipate any power themselves. However, the resistance of wire in each coil, combined with magnetic losses through the core, does allow some power to be lost as heat. Heat buildup is the leading cause of inductor failure. Wire used in transformer windings is solid and insulated with a thin, tough coating of enamel. Long-term exposure to heat can break down enamel and begin to short-circuit the windings. This lowers a coil's overall resistance and draws more current, creating even more heat. Breakdown accelerates until the coil is completely shorted, or until heat burns a winding apart to open the circuit. You cannot see a failing transformer, but be suspicious of any transformer or coil that seems unusually hot, especially if the circuit has not been on for very long. You can also detect a developing heat failure by a strong, pungent odor of burning enamel.

Diodes

Diodes are two-terminal semiconductor devices that allow current to flow in one direction only, but not in the other. This property is known as *rectification*. As detailed in Chapter 8, rectification is essential to the operation of every power supply.

Because diodes only operate in one direction, they are polarized devices as shown in figure 2-19. General-purpose rectifier-type diodes are available in a variety of case styles depending on the amount of current that must be carried. Glass-cased diodes are often used for low-power (or "small-signal") applications. Plastic or ceramic-cased diodes are moderate power devices, generally used for power supplies, circuit isolation, and inductive flyback protection. Metal-cased, stud-mounted devices are for high-power rectification. The schematic symbol representing rectifier-type diodes is also shown in figure 2-19.

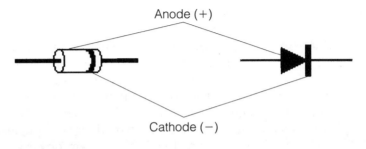

■ **2-19** *A semiconductor diode.*

You should be concerned with two key diode specifications: forward current (IF) and peak inverse voltage (PIV). When a diode is forward-biased, it develops a constant voltage drop of about 0.6 V (for silicon diodes) and current begins to flow. This is called *forward current*, and it is limited by resistance in the circuit. If excessive current flows, the diode will be destroyed by heat. Diodes that are biased in the reverse direction are off. A reversed diode acts like an open switch, so it will not conduct current. Applied voltage appears across the diode. Even if this reversed voltage were increased, the diode will not conduct, up to a point. If the reversed voltage exceeds your diode's PIV, its semiconductor junction can rupture and fail as either an open or short-circuit.

Zener diodes are a unique species designed to operate in a reverse voltage environment, as illustrated in figure 2-20. Notice the new schematic symbol for a zener diode. When applied voltage is below the zener diode's "breakdown voltage" (usually 5, 6, 12, or 15 V), voltage across the zener equals the applied voltage and no zener current will flow. As applied voltage exceeds the zener's breakdown rating, current will begin to flow and zener voltage remains clamped at its rated breakdown voltage. Excess voltage is then dropped across the current-limiting resistor. This makes zener diodes perfect for use as simple voltage regulators or signal level clamps.

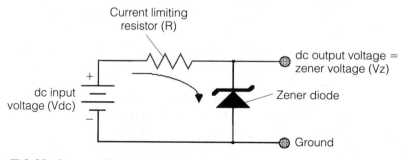

■ **2-20** *A zener diode circuit.*

Electrons must bridge a semiconductor junction in order to operate. If the junction is constructed with special materials and encapsulated inside a diffuse plastic covering, electrons moving across the junction will liberate photons of light. This is the basic principle behind light emitting diodes (LEDs). A typical LED circuit is shown in figure 2-21. Like rectifier-type diodes, LEDs can only be powered in one direction, but their forward voltage drops are higher (0.8 to 2.8 V is common), and they require 20 to 30 mA of current for proper lighting.

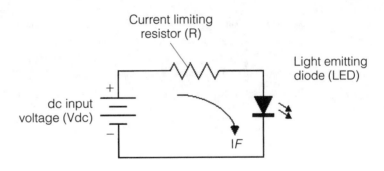

■ 2-21 *An LED circuit.*

Sudden or severe overloads can cause a diode to crack or shatter; these are usually easy to spot with some careful observation. Under most circumstances, however, they show no outward signs of failure, so you will require test instruments and patience to isolate a defective diode. Diodes generally fail as open or short-circuits.

Transistors

Transistors are three-terminal semiconductor devices whose output signal is directly controlled by its input signal. This kind of operation makes transistors particularly well suited for signal amplification and switching tasks. Two major families of transistors have been developed, as shown in the schematic diagrams of figures 2-22 and 2-23. Bipolar transistors are common, inexpensive, general-purpose amplifiers and switches. You will probably encounter bipolar transistors in most commercial printer circuits. Field effect transistors (FETs) and metal oxide semiconductor FETs (MOSFETs) are also used, but not as often. Unfortunately, there is just not enough room in this book to discuss the characteristics and operation of each transistor family, but it is important that you realize they are all transistors, and that you can identify them on sight should they ever appear on a schematic.

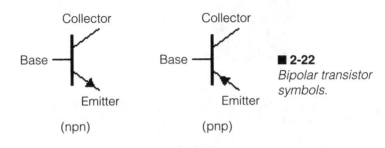

■ 2-22
Bipolar transistor symbols.

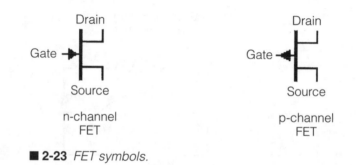

n-channel
FET

p-channel
FET

■ **2-23** *FET symbols.*

Because computer printers are almost exclusively digital devices, discrete transistors are primarily used in switching circuits, such as the one shown in figure 2-24. Unlike an amplifier whose output varies in proportion to its input, a switch is either totally on or totally off. Digital signals from simple logic gates are often used as control signals to operate a transistor switch. A logic "0" (0 V) input leaves the transistor off, while a logic "1" (5 V) input turns the transistor on. A resistor is added to the base lead for current limiting purposes. When a transistor is on, current flows through the load, into the collector lead, then through the emitter lead to ground. Additional current limiting might be needed to restrict current flow in the collector. You will see switching circuits used in a wide range of driver applications.

48

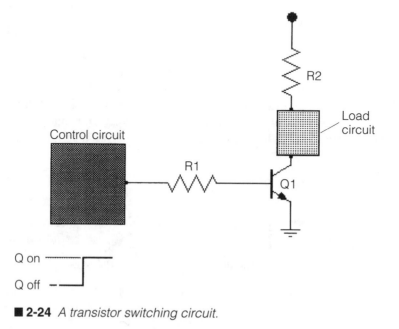

■ **2-24** *A transistor switching circuit.*

There are many electrical specifications that describe particular transistors, but there are four that you should understand when looking for replacement parts. *Base current* (I_b) refers to the maximum current that can flow into a base lead. I_b is usually under 1 mA. *Collector current* (I_c) specifies the limit of current that can flow through the collector. Small-signal transistors might allow up to 100 mA through the collector, while high-power transistors might allow collector currents as high as 10 A or more. *Gain* (beta, hfe, or β) indicates how much base current will be amplified in the transistor. Finally, *collector-emitter voltage* (V_{ce}) describes a voltage drop from collector to emitter when the transistor is turned on fully.

Transistors from any family are available in several case styles depending on the amount of power they can dissipate. Figure 2-25 shows several typical case styles. Low-power devices are often packaged in TO92 cases. Medium-power transistors use larger TO220 cases with a metal mounting frame. A heat sink can be attached to help dissipate excess power. The TO3 case is an all metal casing used for high-power transistors. Two mounting holes allow the case to be attached to a metal heat sink or chassis for best power dissipation. As you can see, the size of a case is relative to the amount of power that a transistor dissipates. The rapid trend toward smaller circuitry has also introduced surface mount transistors.

49

TO-92 case

TO-220 case

TO-3 case

■ **2-25** *Common transistor case styles.*

As with diodes, transistors rarely show any outward signs of failure, but excess heat or electrical stress can cause semiconductor junctions to become open or short-circuited. You must determine the failed transistor by measuring its operation with test equipment in a running circuit, or remove the questionable device and measure its static operation with a transistor checker.

Optoisolators

Optical isolators, also known as optoisolators, are used as position and proximity sensors within the printer. They can also be used to electrically isolate signals between circuits. A basic optoisolator is shown in figure 2-26. The signal to be isolated (usually digital) is fed to the "transmitter" portion of the device. Transmitters can use visible light or infrared (IR) LEDs. Light travels across a physical gap to the "receiver." Instead of current flowing into a transistor's base lead, a phototransistor's base generates current when stimulated by light. Notice the schematic symbol for a standard phototransistor. When the LED turns on, its light activates the phototransistor, which reproduces the original signal in its output circuit.

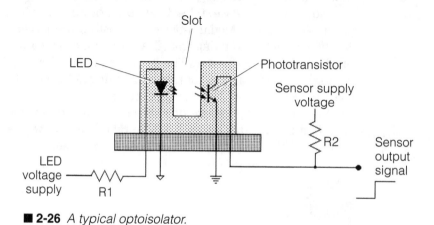

■ **2-26** *A typical optoisolator.*

The advantage of this type of circuit is that input and output circuits are completely isolated. Note that there is absolutely no electrical connection between both circuits, only a light signal. This is ideal for isolating low-power digital circuits from high-power driver circuits, or vice versa.

Suppose a slot was provided between the transmitter and receiver of figure 2-26 to form a position sensor. The transmitter is usually turned on at all times. It is the presence or absence of an object in the slot that turns the receiver on or off. Suppose the slot in figure 2-26 was clear (unobstructed). Transmitted light would simply cross to the receiver and turn it on. With the output circuit shown, an excited transistor pulls its output low. If an opaque object should enter the slot, light will be interrupted and the receiver will turn off. Signal voltage will rise to indicate that something has been detected. This type of sensor is commonly used in printers to sense the presence of paper, or the home position of a print head.

Optoisolators are also key components in optical encoders. Moving-carriage printers must keep track of the print head's position at all times. Optical encoders provide feedback to the printer control circuits that confirm the carriage is in fact where it is intended to be. As a carriage moves, the encoder generates digital pulses that are interpreted as distance and direction by main logic circuits.

Optoisolators do not carry significant amounts of current, so breakdowns due to electrical overloads or heat are virtually impossible. What you should be concerned with is any accumulation of dust or debris that might interfere with the transfer of light from transmitter to receiver. Severe interference can easily cause sensor errors. Electrical problems must be isolated with test equipment while the printer is in operation.

Integrated circuits

Integrated circuits (ICs) are the most powerful and diverse group of electronic components that you will ever deal with. They are the "building blocks" of modern electronics that can take the form of amplifiers, memories, microprocessors, digital logic, oscillators, regulators, or a myriad of other analog or digital IC functions. It is impossible to determine the specific function of an IC just by looking at it. You must refer to the printer's schematic (or manufacturer's data for the particular part) to determine the function of each IC pin.

Dual-in-line package
(DIP)

Dual-in-line package
surface-mount (SMT DIP)

Plastic-leaded chip carrier
(PLCC)

Single-in-line package
(SIP)

■ 2-27 *IC package styles.*

ICs are manufactured in a variety of package styles as illustrated in figure 2-27. Dual in-line packages (DIPs) are the oldest and still most common IC package found in electronic circuits. Single in-line packages (SIPs) are often found on densely packed printed circuits where board space (or "real estate") is limited. Densely packaged ICs such as microprocessors or complex ASICs are sometimes packaged in plastic leaded chip carrier (PLCC) packages. ICs intended for surface mounting are manufactured in specialized surface-mount packages.

Integrated circuits generally do not show any outward signs of failure, so it is necessary to check suspect ICs carefully using the appropriate test equipment while the IC is actually operating in a circuit. As a rule, whenever an IC must be replaced, insert an IC socket in its place on the printed circuit board, then plug in the IC. This prevents you from having to desolder the printed circuit again should the IC need to be replaced in the future. If there is no test equipment available, your best course is to replace the subassembly containing the suspect part.

Conventional printing technologies

THIS CHAPTER DISCUSSES THE THREE CONVENTIONAL printing technologies: impact, thermal, and ink jet, as well as the construction and detailed operation of their print head mechanisms (figure 3-1). These technologies are called "conventional" because their print heads are clearly defined devices, easy to recognize on sight, and easy to replace as an individual assembly. Electrophotographic (EP) printing is accomplished as a "process" of many interacting mechanisms, rather than the actions of any one device. This makes EP technology more complex and difficult to understand, so it is considered to be a "nonconventional" printing method. EP printer technology is discussed in Chapter 4.

■ **3-1** *An NEC PinWriter P3200 dot-matrix impact printer.* NEC Technologies, Inc.

Impact printing

Impact printing is the oldest, simplest, and probably the most dependable form of printing ever developed. Images are literally slammed against a page as a matrix of individual dots; the dot patterns form letters and graphics. Impact printers are known collectively as dot-matrix impact (DMI) printers. Figure 3-2 shows a comparison between character (e.g., typewriter) and dot-matrix print. Notice how the DMI figure appears rougher and less defined than its die counterpart.

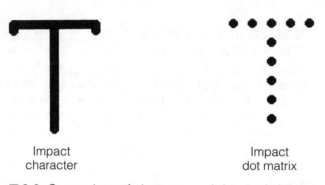

Impact
character

Impact
dot matrix

■ **3-2** *Comparison of character and dot-matrix impact printing.*

The dot mechanics

Each dot is generated by an individual metal print wire driven through a solenoid as illustrated by figure 3-3. When an electrical pulse reaches the solenoid, it energizes the coil and produces a brief, intense magnetic field. This field "shoots" its print wire against the page. After the pulse passes, the solenoid's magnetic field collapses. A return spring pulls the wire back to a rest position. In actual practice, DMI solenoids and print wires are very small assemblies. A typical print wire might only travel about 0.5 mm. This distance is known as *wire stroke*.

Not all DMI heads hold their print wire directly within the solenoid's coil. While this approach might work well for smaller, general-purpose heads, heavy-duty heads need larger coils than can be stacked vertically. Each solenoid is mounted offset from one another, then connected to their respective print wires using a mechanical linkage as shown in figure 3-4. Due to the additional mechanical components, heavy-duty print heads operate somewhat slower than "direct-drive" heads, and they usually do not last as long because of the additional mechanical wear.

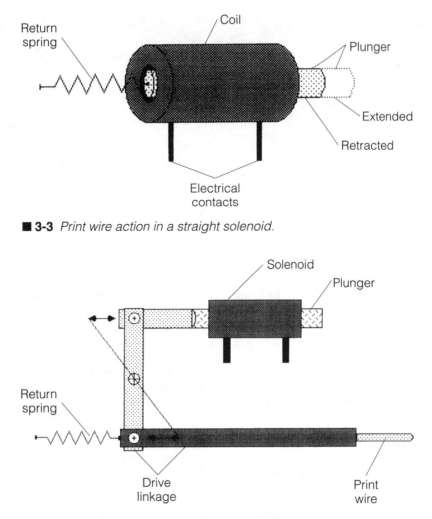

■ 3-3 *Print wire action in a straight solenoid.*

■ 3-4 *Print wire action in an "offset" solenoid.*

Driving the dot

Solenoids require a substantial amount of electrical energy very quickly to develop a magnetic field strong enough to move a print wire. Figure 3-5 shows a driver circuit similar to one used in everyday DMI printers. A printer's ECU produces a narrow logic pulse, which is sent along to a transistor driver. In figure 3-5, a gate array ASIC (marked U3) is responsible for generating the logic signal. However, logic circuits alone cannot handle enough power to operate a solenoid directly, so a transistor driver is used. A high-power driver transistor (such as the 2SD1929) acts as a switch that turns the solenoid on and off. A driver voltage (typically 24 V,

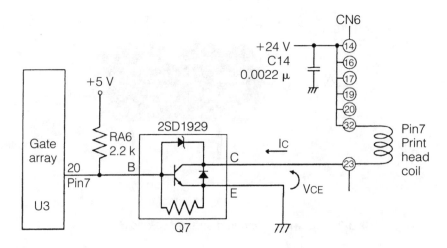

■ 3-5 *A typical print wire driver circuit.* Tandy Corporation

much higher than the +5-V logic supply voltage) provides the energy required. A relatively small nonpolarized capacitor (marked C14) provides supplemental power supply filtering. One such driver circuit is required for every print wire.

To understand more about dynamics of the driver circuit, refer to the oscilloscope diagram of figure 3-6. The upper trace shows what happens to the transistor's output voltage when a logic pulse

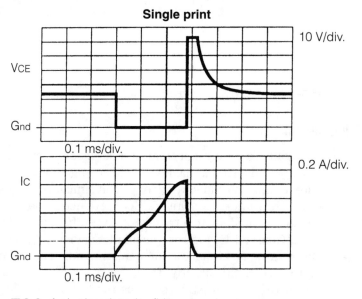

■ 3-6 *A single print wire firing signal.* Tandy Corporation

arrives. The lower trace illustrates the corresponding effects on solenoid current. As you can see, current is not instantaneous; there is a rapid period of ramp-up (about 0.3 ms or 300 µs), followed by a sudden drop. The print wire will overcome the restriction of its return spring and fire at roughly the current peak. You can see the effects of continuous print in figure 3-7.

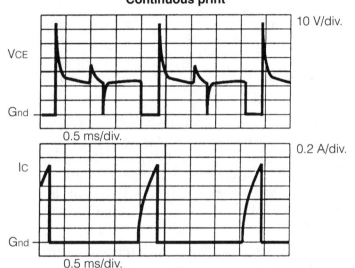

■ 3-7 *Continuous print wire firing signals.* Tandy Corporation

Electrically speaking, solenoids are not very efficient devices. Only 1% or 2% of the energy provided is actually converted to force. The remaining energy is wasted as heat. Heating can have severe effects that include print wire jamming, coil burnout, and even a potential burn hazard. Solenoids usually require anywhere from +12 to +24 Vdc at currents greater than 1.5 A, so the "on-time" for a typical solenoid must be kept very short (well below 0.5 ms, as in figure 3-6) to prevent excessive heating. Metal heat sinks are often die-cast into the head housing to dissipate excess heat as quickly as possible. Short pulses also allow extremely fast firing cycles, and most current DMI print heads can fire at greater than 300 Hz. Some models can be fired as fast as 600 Hz.

Dot specifications

DMI print heads use an array of 9 or 24 print wires arranged in vertical columns as shown in figure 3-8. There are three major mechanical specifications that you should be familiar with. Wire

57

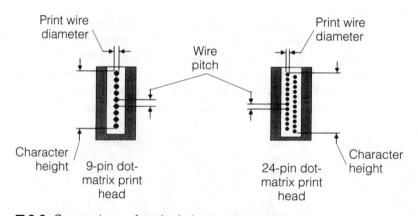

■ 3-8 *Comparison of typical dot-matrix impact print head arrangements.*

diameter specifies the diameter of each print wire (normally expressed in millimeters). This tells you how large each dot will be. The distance between the center of each dot is known as *wire pitch* (also expressed in millimeters). Finally, the height of each fully formed character is specified (in millimeters) as character height (CH). Wire diameter and pitch are much smaller for 24-pin heads than for 9-pin heads.

Not all wires can be used to form every character. For example, a 9-pin print head with a 2.5-mm CH might only use seven wires to form most characters. Wires 8 and 9 could be used to form characters with descenders. The concept of true descenders is illustrated in figure 3-9. When all nine wires are used to form characters, there is no room left for descenders, so characters can be printed with false descenders, also shown in figure 3-9. Overall character size might appear larger when all nine wires are used, but many people find false descenders awkward.

The technique of DMI printing is every bit as straightforward as character printing, but the actual formation of each letter, number, or symbol is a bit more involved. Data sent from a host computer is interpreted by the printer's main logic and converted to a series of vertical dot patterns. Motor commands start the carriage (and print head) moving across the platen. Simultaneously, printer circuits will send each dot pattern to the print head in series. Each dot pattern fires the corresponding print wires through an inked ribbon to leave a permanent mark on the page. This is also called *serial* or moving-head operation. Figure 3-10 is a photo of a heavy-duty DMI print head.

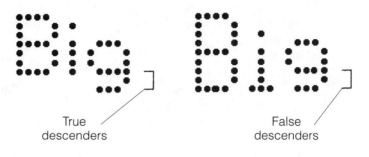

True descenders

False descenders

■ **3-9** *Comparison of "true" and "false" descenders.*

Singer Data Products, Inc.

■ **3-10** *A commercial dot-matrix print head.*

Early DMI printers lacked the dot resolution and electronic sophistication to produce letter-quality print as character printers could. Characters were printed uppercase in a 5 × 7 dot matrix. Such printing was clear and highly legible, but not visually pleasing. Advances in electronics and printer communication allowed both upper- and lowercase characters to be printed, while improvements in materials and construction techniques resulted in smaller, more reliable print heads. This made DMI printing easier to look at, but still could not approach letter quality. With the introduction and widespread acceptance of 9-pin print heads, enhanced features became possible.

The additional dots offered by a 9-pin head improved the appearance of each character even more, and provided enough detail to make several different font styles practical, but individual dots were still visible. Near-letter quality (NLQ) was finally accomplished through multipass printing of the same line. Paper position is shifted just a fraction, then a subsequent series of dots would "fill-in" some of the spaces left by a previous pass. Forming a line of NLQ characters might take two, three, and four passes (depending on the particular printer). A single pass per line became known as *draft mode*.

NLQ printing is limited by the printer's speed. The more passes needed to form a line of print, the more time that is required. For example, suppose a single pass (draft) takes one second. If two passes are required for an NLQ line, print time would be twice as long (two seconds). If four passes are required, print time would be four times as long (four seconds). For long documents, this additional time can really add up. As a general rule, you can save substantial time (and your ribbon) by using the draft mode until you are ready to print the final version of your document.

It was not until the introduction of the 24-pin DMI print head that NLQ print could be accomplished in only one pass. Two vertical columns of 12 wires are offset from one another. As a 24-pin head moves across the page, the leading column places a first "pass" of dots. This forms a basic character image, which is already superior to the 9-pin equivalent. Immediately after the leading column fires, the lagging column places a second "pass" of dots to fill in each character. In this way, two passes are made effectively at the same time.

Advantages & disadvantages of DMI printing

To this day, DMI print heads remain a cornerstone of commercial printing technology. They are flexible and inexpensive devices, capable of a wide variety of fonts and enhancements, as well as draft or NLQ performance and bit-mapped graphics. DMI heads are reasonably fast, so they can achieve speeds easily exceeding 160 CPS. They are reliable devices. Heavy-duty print heads can last through more than 30 million characters. Smaller, general-purpose heads can last for more than 100 million characters. Impact printing is mandatory for printing multicopy forms. Finally, they require very little maintenance except for periodic routine cleaning.

However, impact printing is very noisy. The continuous drone of print wires striking paper can become quite annoying. Although DMI printers are now made with plastic coverings that baffle much of its noise, they do not quiet the printer completely. Limited dot resolution is another concern. You might not notice this for NLQ text, but you can see individual dots in draft or bit graphics modes. You can only achieve just so many dots per inch. Finally, head overheating can be a problem during long printouts, especially when printing graphics where many wires must fire repeatedly.

Thermal printing

Thermal printing technology eliminates the need for impact altogether by using heat to form characters and graphic images on paper. Instead of forming dots through a matrix of electromechanical solenoids, thermal print heads employ an array of resistive semiconductor heating elements (called *dot heaters*). Electrical pulses from the printer's driver circuits transfer energy to each desired dot heater causing them to heat up rapidly. This in turn will discolor the appropriate points on temperature-sensitive paper or a thermal transfer ribbon. Thermal dot-matrix (TDM) print heads are available in serial (moving head) and line-head versions, which can be chosen depending on the particular application. Before you learn about the operation of thermal printing in detail, however, you must understand the way in which TDM heads are constructed.

Thermal head fabrication

TDM heads are not built as an assembly of other parts the way other heads are. They contain no wires, solenoids, linkages, or other individual parts. Instead, they are fabricated much like an integrated circuit; one layer of material at a time is deposited onto previous layers in masked patterns that form resistive elements, their interconnections, electrical insulators, and protective coatings. Three popular fabrication methods are discussed as follows.

Thermal head thick-film fabrication is shown in figure 3-11. A semiconductor resistance element is deposited between two conductors and covered with a protective glass film. Everything is fabricated onto a ceramic substrate (or support structure) insulated by a glass insulating layer.

Thick-film technology can achieve resolutions of 12 dots per millimeter. Each dot heater exhibits a resistance that can range from 160 to 3,000 Ω depending on the number of dots and the print head's intended application. Each dot is roughly round and faintly defined. The working temperature of each dot can easily exceed 350°C at the resistive element itself, yet typical thick-film heater elements will last for more than 30 million pulses, and they are

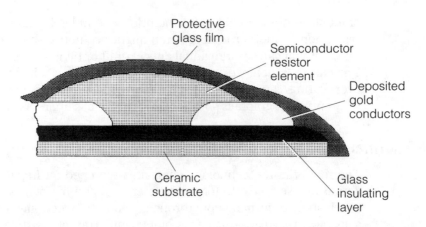

Protective
glass film

Semiconductor
resistor
element

Deposited
gold
conductors

Ceramic
substrate

Glass
insulating
layer

■ **3-11** *Cross-sectional view of a thick-film thermal printing element.*

highly resilient to power overload conditions. Line-print head versions are often manufactured using thick-film technology. Another issue to consider with TDM print heads is wear; after all, the print head must be in physical contact with the page surface (or thermal transfer ribbon). Over time, the friction encountered at the page surface will wear away the TDM head. Today, protective film coverings can endure more than 30 kilometers of wear against the page surface before failure.

Figures 3-12 and 3-13 illustrate typical thin-film fabrication technologies. Both figures are variants of the same approach. A ceramic substrate and glass insulating layer form the foundation of a thin-film heater. Unlike the thick-film technique, a thin-film resistor is much thinner, and is fabricated underneath its electrical conductors. Finally, two separate protective layers are added.

Thin-film technology offers several performance differences versus thick-film devices. Thin-film devices can achieve resolutions as high as 16 dots per millimeter. Each dot heater can range from 1.5 to 50 Ω depending on the number of dots and intended application of the head. Dots appear square and are sharply defined. Thin-film dot heaters can easily survive more than 50 million pulses, but they are not very tolerant of driver overloads. The hard tantalum covering can undergo more than 70 kilometers of surface wear, more than twice that of a thick-film covering. Thin-film technology can also be used to manufacture either serial or line-print heads.

Although TDM heads are considered rugged and reliable devices, the hard glass and ceramic materials they contain are also very fragile. Impact or physical abuse can fracture the entire structure

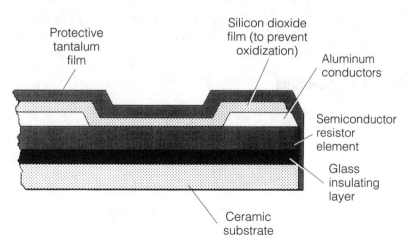

■ 3-12 *Cross-sectional view of a thin-film (total glaze) thermal printing element.*

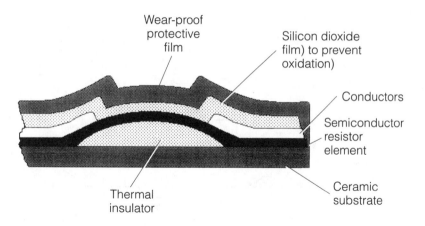

■ 3-13 *Cross-sectional view of a thin-film (partial glaze) thermal printing element.*

and ruin the assembly. Because they are fabricated as a single assembly, there are no serviceable parts inside. When one or more dot heaters fail, the whole TDM head must be replaced.

Contact pressure is surprisingly important in thermal printing systems. The print head must present enough pressure to ensure good thermal transfer to its heat-sensitive paper or thermal ribbon, but not so much pressure where it interferes with the paper transport mechanism. Excessive pressure will also decrease the head's wear resistance by breaking down its protective covering layer faster. As a general rule, contact force should be between

100 to 250 grams. When you examine thermal print heads closely, you might note that some of their faces have a slight curve to match the platen.

Serial head operation

A thermal heater can be fired much the same way as an impact solenoid as shown in figure 3-14. Data for such a dot driver circuit is applied to a tri-state buffer. Note the extra input marked "print." This is an enable input that controls the tri-state buffer. While "print" is brought logic high, the buffer's output state will equal its input state (i.e., a logic 1 input will yield a logic 1 output, and vice versa). This enable line is very handy because it allows data to be set up in advance on any number of dot heaters. The "print" signal can then be strobed on all drivers simultaneously. For the simple driver circuit of figure 3-14, a logic 1 will fire a dot heater, and a logic 0 will not. Each dot heater will have its own driver circuit.

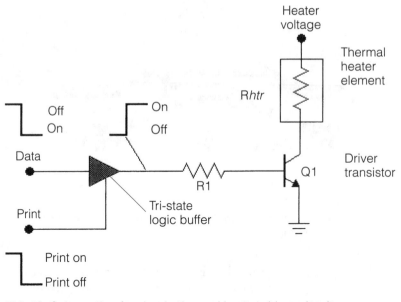

■ **3-14** *Schematic of a simple thermal heater driver circuit.*

Timing is very important to the successful operation of a thermal print head as you can see in figure 3-15. A dot heater cannot work instantaneously; it takes a certain amount of time for it to reach a working temperature. It takes even more time for it to cool off before firing again. This entire cycle is called the *pulse period* (or SLT). When a dot heater is fired by a logic 1, it begins to heat. It

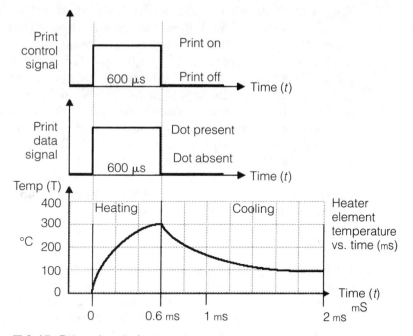

■ 3-15 *Drive signals for a typical thermal heater element.*

only takes a pulse width of about 0.6 to 1.0 milliseconds (ms) to reach a marking temperature of 350°C. For figure 3-15, 0.6 ms is used as an example. At that point, the heater has formed its dot and must be allowed to cool. Cooling can require more than 1.5 ms in addition to heating time. If the head (or paper) moves before sufficient cooling can occur, printed dots will tend to appear "pulled" or smudged. Once the heater has cooled enough, it might be moved and fired again.

It is possible to coax even more heat from TDM heaters by lengthening the driving pulse width. Because energy is applied for a longer time, heaters will reach higher temperatures. However, this must be matched by even longer cooling times. A ROHM NS0907-D1 thermal print head makes an excellent example. Suppose a pulse width of 1.0 ms requires a cool-down time of another 1.0 ms (SLT would be 2.0 ms). If pulse width were extended to 1.2 ms, SLT would have to be 4.0 ms to allow sufficient cooling time. A pulse width of 1.6 ms would require an SLT of 8.0 ms (a much longer cool-down time). It is helpful to understand the relationship of time and temperature, but for the purposes of troubleshooting and repair, you should not be concerned with pulse width and SLT. Those values will be fixed by software in the printer's ECU.

Thermal printing

TDM print heads require much less operating power than DMI devices because only a small amount of current is needed to fire each dot heater. A typical impact solenoid might require a pulse of 1.8 A at 12 Vdc for proper firing. That equates to [1.8 A × 12 Vdc] 21.6 watts (W) per solenoid. For a 9-pin printer, the power supply must be able to provide [21.6 W × 9] about 195 W under "worst case" conditions. Such a power demand makes DMI printers unsuitable for mobile or battery-powered operation. On the other hand, a typical TDM print head requires only about 2 W (or less) per dot. For a 9-pin serial TDM print head, it would take [2 W × 9] 18 W to fire all 9 dot heaters together. As you might imagine, this represents a terrific power savings. Thermal printers are often found in mobile or battery-powered instrumentation.

Line head operation

TDM line-print heads take a bit more complicated approach. Instead of assembling images as a series of vertical dots scanned from side to side, a line-print head uses a single horizontal row of dot heaters to assemble complete images as paper advances. Figure 3-16 illustrates a ROHM TDM line-print head. One dot heater is available for every possible point in a horizontal line. Depending on the width of a head (which will probably be as wide as the paper it is printing), there might be as few as 160 or as many as 7,000 dot heaters.

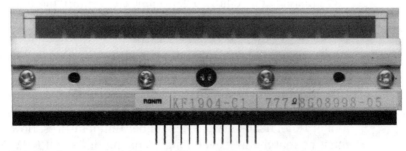

■ **3-16** *A commercial thermal line-print head assembly.* ROHM, Inc.

Because the print head now spans the entire page width, there is no longer any need for a carriage mechanism. Data is sent from the host computer and interpreted by the printer's ECU one FULL line at a time, then translated into a series of horizontal dot patterns. In this way, the line head forms entire lines one horizontal row at a time (as opposed to serial heads that form lines as one vertical column at a time). Line-print heads are available for use with heat-sensitive paper or thermal transfer applications.

Sending the actual print data to a line-print head, however, is a more complex and involved process than for a serial head. After all, nine discrete driver circuits to fire a serial print head (one driver for each dot heater) are simple enough to follow and troubleshoot, but when you deal with a line-print head that can contain thousands of dot heaters, discrete drivers are totally inadequate. In response to this, TDM line head manufacturers take advantage of sophisticated VLSI (Very Large-Scale Integration) technology to fabricate data handling and dot driver circuitry right onto the print head assembly itself. Figure 3-17 shows a simplified internal diagram for a typical line-print head.

After the printer's ECU translates a line of characters into rows of horizontal dots, data for each row is sent in series (one bit at a time) into the DI input of the line head. Remember that data is made up of digital logic conditions corresponding to a heater's output; a logic 1 is on and a logic 0 is off. When a complete row of data is fed into the print head, the dot heaters can be fired to produce a row of finished print. Ideally, all dot heaters should be able to fire together; it would print one full row at a time. Unfortunately, the cumulative current required by every heater firing together would put a strain on even the best power supply. It would also create a strong heat surge. For the example head of figure 3-17, each of the 2,048 dot heaters need 0.19 W (190 mW) of power. If every dot fires together, the head will dissipate a stunning [0.19 W × 2,048] 389 W! A 24-V power supply would have to provide [389 W / 24 Vdc] about 16 A.

In order to limit line head power dissipation, power surges are limited by firing the line-print head in segments. The 2,048 dots of figure 3-17 are divided into four segments of 512 dots. A segment is enabled using the trigger (or "strobe") lines (marked STR1 through STR4). There might be as many as four or more strobe lines depending on the density of the particular head. One segment of 512 dots firing simultaneously would dissipate [0.19 W × 512] about 98 W; a much more manageable power requirement for a power supply. It will take four times longer to print with this "segmented" approach, but it places much less stress on printer electronics, and allows extra cooling time for unused segments.

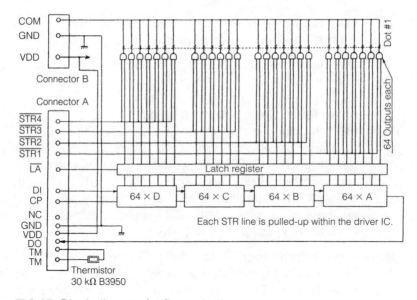

■ 3-17 *Block diagram for figure 3-16.* ROHM, Inc.

Thermal printing media

Heat generated by each dot must be transferred to a page surface as required in order to form permanent images. This is accomplished in either of two ways: direct contact or thermal transfer. As the name implies, direct contact requires that a print head be in direct contact with the paper surface. Heat is used to activate heat-sensitive chemicals in the paper, which cause those points to discolor and leave permanent marks. Most heat-sensitive papers generate either blue or black marks.

This offers a simple, reliable printing approach, but it is not without its disadvantages. Perhaps the most notable disadvantage is that standard paper (such as 20-lb bond xerography-grade paper) cannot be used. Temperatures developed by a thermal print head are insufficient to cause discoloration on normal paper without burning it. Heat-sensitive chemicals cause thermal paper to discolor at much lower temperatures. Such sensitized paper is usually delicate and sensitive material. It only has a limited shelf life (typically a few years). Age, sunlight, humidity, and a variety of chemical vapors will ruin the paper. Finally, thermal paper is manufactured on long, continuous rolls. Its tendency to curl makes it difficult to handle.

Thermal transfer uses heat generated by each dot to deposit solid ink onto a page surface. A narrow plastic ribbon is coated with a heat-sensitive solid ink made of waxes, oils, and dies. By altering the ratio of these materials, ink viscosity and melting point can be optimized to suit a print head. As with DMI print heads, ribbons are inserted between the print head and paper. When dot heaters fire, they melt corresponding points of plastic ink, which remain on the paper to form images. Because it is an ink ribbon (not the paper) that must be heated, thermal transfer printing will work with just about any kind of smooth, standard-bond paper. Print color is determined by the color of the plastic ink. TDM print heads used for thermal transfer printing require slightly more power than direct contact heads in order to melt plastic ink reliably.

Advantages & disadvantages of TDM printing

Thermal dot-matrix technology enjoys several advantages over other printing methods. First, TDM is very quiet. Even though contact is required, there are no moving parts to generate noise. Power consumption is also very low in TDM heads. Thermal dot heaters in serial heads often require less than 2 W per dot. Line-print heads need even less (in the range of 0.2 W per dot). Compare this to a DMI head that usually takes more than 12 W per solenoid. Low power consumption makes TDM heads ideal for mobile and battery-powered applications. TDM heads are simple and reliable devices. They are totally self-contained; there are no moving parts or apertures that can jam or clog. About the only preventive maintenance required for TDM heads is an occasional wipe-down to remove accumulations of dust and dirt. Line-print heads simplify the printer by eliminating the need for a carriage mechanism. They also incorporate on-board circuitry for data handling and dot driving. This too can greatly simplify the printer's corresponding ECU. Print is clear and crisp, with resolutions that can rival that of an electrophotographic printer.

TDM print heads have a few disadvantages. Print time is slightly slower than other printing methods because of cooling time added to each printing cycle. This is just a slight difference. Thermal head life is shorter than most DMI heads, largely due to the surface wear caused by paper friction. TDM heads are usually good for 10 million characters or less, where DMI heads can support 50 million characters or more. Finally, TDM heads cannot be serviced; if a dot heater burns out or its internal circuitry fails, the entire head assembly must be replaced.

Ink jet printing

While impact and thermal technologies require a print head to actually contact the page surface in order to print, ink jet dot-matrix printing is a "noncontact" technology; that is, ink jet heads never come into contact with the printing surface. Instead, liquid ink is literally spray-painted onto a page. Commercial ink jet printers use one of two approaches to spray ink: drop-on-demand, and intermittent jet. These techniques are both very similar. The drop-on-demand technique requires an individual command for every dot that is ejected from an ink jet print head, much like DMI print wires that are fired by discrete pulses. The intermittent jet approach fires streams of ink (a continuous series of dots) rather than individual dots. By timing the start and duration of each ink stream very precisely, it is possible to achieve extremely fine detail at resolutions that are now exceeding 360×360 dpi. For the purposes of this book, ink jet print heads are assumed to use intermittent jet technology.

Inside the ink jet print head

When compared to other types of print heads, ink jet heads are perhaps the simplest and most straightforward design. The ink reservoir is provided in a prepackaged container. The ink container might come in a replaceable cassette, or it might be integrated into a throw-away print head assembly. When the ink is installed, gravity and capillary action feed the ink along narrow channels, and the ink comes to rest at each nozzle, the exit aperture, where the ink is ejected. There might be 12, 24, 50, or more depending on the vintage and complexity of the particular ink jet head. The nozzles themselves are really nothing more than fine holes (each about ⅛ the diameter of a human hair) drilled into a metal plate with a laser. You might wonder how liquid will stay at the nozzles without spilling out all over the place. The answer is that the ink's viscosity and surface tension hold the ink in place until it is pumped out. Figure 3-18 shows you the internal components of a throw-away ink jet print head.

Finally, every ink channel is fitted with an ink pump. In actual practice, an ink pump is almost microscopic, but it is the key element in any ink jet head. Because each pump requires electrical pulses in order to run, a series of contacts will run to a connector on the print head. Like other dot-matrix technologies, each ink pump can be fired independently for maximum flexibility in image formation. The circuitry needed to time and fire the ink pumps is

3-18 *A disposable ink jet cartridge assembly.*

contained in the printer's ECU, or on a supplemental driver board located close to the carriage transport system. There are two basic types of ink pumps: piezoelectric and bubble.

Piezoelectric pumps

In a piezoelectric pump, a ring of piezoelectric ceramic material is built into an ink channel as in figure 3-19. When a high-energy electrical pulse is applied across the ceramic ring, its piezoelectric quality causes it to constrict the channel. This causes a sudden displacement of volume that pushes out a single droplet of ink. After the electrical driver pulse passes, ceramic returns to its original shape, and more ink is drawn into the channel to make up the expelled droplet. Piezoelectric ceramic requires short pulses (in the 5- to 10-μs range) at high energy levels. Pulse amplitudes can be anywhere from 70 to 200 V depending on the particular design of the channel and the type of ceramic used. One pump is required per channel, and can fire at rates approaching 5 kHz (5,000 dots per second). This is one droplet every 200 μs.

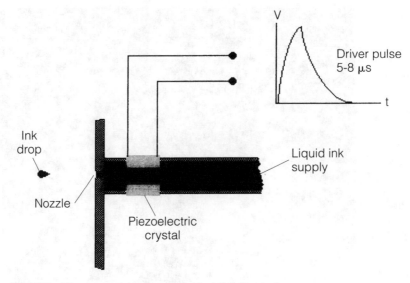

■ **3-19** *Diagram of a piezoelectric ink jet nozzle.*

Bubble pumps

Bubble pumps (used in "thermal ink jet" or "bubble jet" printers) are also widely used to generate ink droplets; they are now perhaps the single most popular technology for disposable ink jet print heads. As you see in figure 3-20, nozzles and channel construction are very similar to piezoelectric heads, but ceramic rings are replaced by ring heaters. An electrical driver pulse fires a ring heater. In turn, this heats ink in the immediate vicinity. As ink heats, a bubble forms and expands in the channel. When the bubble finally bursts, its force ejects an ink droplet, and more ink is drawn in to fill the void. Heated ink droplets also dry faster on paper. Although bubble pumps are fast-working devices, they are limited to firing rates of 1,000 dots per second. Ring heaters (like the dot heaters found in thermal print heads) require a finite amount of time to cool after firing. If there is not enough cooling time, ink might actually dry out and clog inside the channel. However, bubble pumps do not require nearly as much energy to operate. Early units typically used 24- to 50-V pulses, but the new generation of mobile lap-top bubble jet printers uses even lower signals. You can see the proliferation of nozzles and electrical contacts in the HP ink jet print head of figure 3-21.

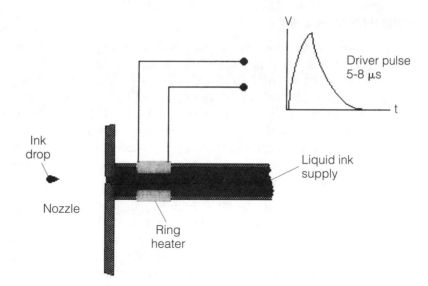

Driver pulse
5-8 μs

Ink
drop

Liquid ink
supply

Nozzle

Ring
heater

■ **3-20** *Diagram of a bubble jet nozzle.*

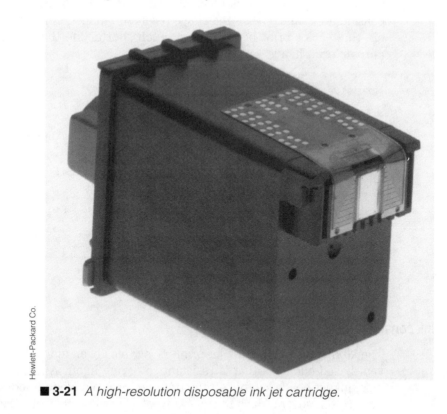

Hewlett-Packard Co.

■ **3-21** *A high-resolution disposable ink jet cartridge.*

Monochrome vs. color

Because the color of an ink jet's output depends entirely on the color of its ink, the ink jet printer has become a preeminent platform for low-cost, high-quality color printing. The principles involved in a color ink jet design are virtually identical to those of monochrome printers, but where a monochrome printer merely prints black and white (the presence or absence of dots), a color printer must interpret the colors of each dot, and provide a more sophisticated signal stream that will not only drive the print head, but will drive the proper color nozzles.

This marginal increase in complexity also demands a slightly more complex print head as shown in figure 3-22. Essentially, a color ink jet head can have three or four separate ink systems integrated into the same head; three to handle the three key colors (yellow, magenta, and light blue), and a fourth color (black) might be included. As you might imagine, cramming three or four ink systems into a single print head is no small feat, and each ink reservoir is considerably smaller than the single-reservoir (monochrome) models; thus, color print heads have a much shorter life than monochrome print heads.

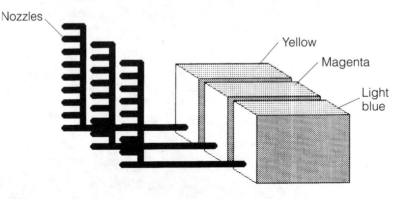

■ **3-22** *Diagram of a color ink jet cartridge.*

Ink considerations

The kind of ink used in ink jet printers is typically an indelible, solvent-based chemical that is resistant to drying in air. As a result, most ink jet heads can be left unattended for prolonged periods of time (often several days to several weeks) without fear of nozzle clogs due to drying. Most ink jet printers also have a type of "ink cap" sponge in the printer's carriage home position that wipes each nozzle whenever the head reaches its home position, and covers them whenever the printer is turned off. This can be an important factor in older ink jet printers because sponges become dried and hardened with ink (as well as age), so they are less effective at keeping the print head clean. Cleaning sponge replacement should be a routine procedure when servicing older ink jet printers.

Sooner or later, solvent will evaporate into the air. Evaporation begins to increase the ink viscosity inside each channel; it becomes thicker. In early stages, this can cause ink to sputter or travel off course to the page. In advanced stages, solvent might evaporate entirely, or enough to allow ink to dry and harden in the channel. This is a clog. The afflicted nozzle(s) might still fire electrically, but no ink will flow until the clog is cleared. The ink jet head will have to be cleaned or purged. Clogs might also be dislodged through normal use. Once any viscous ink is forced out through normal use, proper operation will return automatically. Of course, disposable ink jet print heads (which do not support purging) can simply be replaced.

Recycling considerations

One of the side effects of disposable ink cartridges or ink jet print heads is that a huge amount of waste material is generated. When coupled with the relatively high cost of original manufacturer's ink cartridges, the expended ink units open a wide market for recycling. If you stroll through your favorite computer store, you will probably notice "ink cartridge refill kits" sitting near the ink cartridges, so the drive to recycle has made it to the retail level. Unfortunately, the refill kits are not compatible with all print heads, are not much less expensive than new print cartridges, and generate just about as much waste material as throwing the old ink cartridge/print head away. Chapter 7 discusses ink jet recycling in more detail, but if you plan to handle recycling on a regular basis, be sure to obtain the proper tools and materials for the job, preferably not from retail vendors. If you do not have the inclination to recycle ink jet cartridges yourself, there are a number of companies that accept and rework cartridges as a specialty.

Paper considerations

When ink droplets leave a nozzle, they are still in a liquid form. Once a droplet reaches paper, it must dry almost immediately so that the finished page can be handled. This is not always easy to accomplish if you are using the wrong paper type. Paper must accept ink into its fibers just the right way to dry it quickly, yet leave droplets on the surface for a crisp image. If paper absorbs ink too readily, the dried image might appear light or faint (lacking contrast). This is a typical problem with standard-weight xerography-grade paper. If paper does not absorb ink quickly enough, ink might remain a liquid, which can smear and smudge when touched. Although this is less common, gloss or specially coated papers might act this way.

To guarantee just the right drying characteristics, there is a specially made ink jet paper impregnated with clay or solvent-absorbing chemicals that cause ink to dry quickly while leaving a clear, dark image. The best way to determine the compatibility between paper and ink is to test the printer in actual operation. Either ink or paper (usually the paper) might have to be changed to optimize the printer's performance.

Advantages & disadvantages of ink jet printing

Ink jet dot-matrix technology offers a method of "noncontact" printing that can mark a wide variety of surfaces and paper types. Printing speeds rival any DMI printer, yet operation is very quiet. Nozzles and ink channels are incredibly small, so dot resolution can be extremely high (the head in figure 3-21 is used in a 300×300 dpi ink jet printer). Ink jet heads have no mechanical parts, so they are exceptionally inexpensive to manufacture, enjoy high reliability, and a long working life—some piezoelectric heads are rated for more than 1 billion dots. The low power requirements have made ink jet technology the forerunner of mobile, battery-powered printers.

Unfortunately, ink jet heads are sealed devices. If one ink pump fails, the entire head must be replaced. Costs can also add up; even though ink cartridges can be made inexpensively, they can still be expensive items by the time they reach store shelves. Given the fairly limited number of pages handled by a single ink cartridge, the per-page cost of ink jet printing can still be high. The ink itself can be a frustrating problem. Ruptured print heads or leaking cartridges can spill thick, indelible ink everywhere. Fabrics and other porous materials are particularly susceptible to permanent stains; even your skin can be stained. While contemporary commercial "disposable head" printers have become very clean, there is still a risk of ink spillage.

4

Electrophotographic printing technology

ELECTROPHOTOGRAPHIC (EP) PRINTERS ARE DIFFERENT fundamentally from the other types of printers discussed in Chapter 3. Those "conventional" printers develop dots as a one-step process using impact, heat, or ink. EP printers are not nearly as simple (figure 4-1). Images are formed by a complex and delicate interaction of light, static electricity, chemistry, pressure, and heat, all guided by a sophisticated ECU. This chapter shows you the operation and intricacies of EP operation.

Hewlett-Packard Co.

■ **4-1** *Hewlett-Packard LaserJet printers.*

The classical electrophotographic approach

Electrophotographic printing is accomplished through a "process" rather than a "print head." The collection of components that performs the EP printing process is called an *Image Formation System* (or IFS). An IFS is made up of eight distinctive areas: a photosensitive drum, cleaning blade, erasure lamp, primary corona, writing mechanism, toner, transfer corona, and fusing rollers. Each

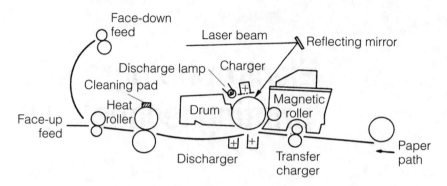

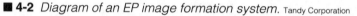

■ 4-2 *Diagram of an EP image formation system.* Tandy Corporation

of these parts, as shown in figure 4-2, play an important role in the proper operation of an IFS. Trouble in any of these areas will adversely effect the printed output.

A photosensitive drum is generally considered to be the heart of any IFS. An extruded aluminum cylinder is coated with a nontoxic organic compound that exhibits photoconductive properties. That is, the coating will conduct electricity when exposed to light. You might see this referred to as *organic photoconductive chemicals* (or OPC). The photosensitive compound gives the EP drum a bright green appearance. The aluminum base cylinder is connected to ground of the high-voltage power supply; this is an important point, serious printing problems can occur if the ground becomes loose.

It is the drum that actually receives an image from a writing mechanism, develops the image with toner, then transfers the developed image to paper. Although you might think that this constitutes a print head because it delivers an image to paper, the image is not yet permanent; other operations must be performed by the IFS. Complete image development is a six-step process that involves all eight IFS components: cleaning, charging, writing, developing, transfer, and fusing. To really understand the IFS, you should know each of these steps in detail. The following sections of this chapter show you how each part of the IFS works together.

Cleaning

Before a new printing cycle can begin, the photosensitive drum must be physically cleaned and electrically erased (typically referred to as *conditioning*). Cleaning might sound like a rather unimportant step, but not even the best drum will transfer every microscopic granule of toner to a page every time. A rubber cleaning blade is applied across the entire length of the drum to gently scrape away any residual toner that might remain from a previous image (figure 4-3). If residual toner were not cleaned, it could adhere to subsequent pages and appear as random black speckles. Toner that is removed from the drum is deposited into a debris cavity. Keep in mind that cleaning must be accomplished without scratching or nicking the drum. Any damage to the photosensitive surface would become a permanent mark that appears on every subsequent page. Some EP printer designs actually return scrap toner back to the supply for reuse. This kind of recycling technique can substantially extend the life of your electrophotographic cartridge, and eliminate the need for a large debris cavity.

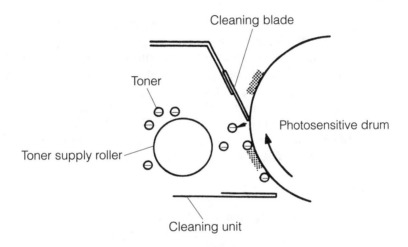

■ **4-3** *Cleaning residual toner from the drum.* Tandy Corporation.

Images are written to a drum's surface as horizontal rows of electrical charges, which correspond to the image being printed. A dot of light causes a relatively positive charge at that point. This corresponds to a visual dot in the completed image. Absence of light allows a relatively negative charge to remain and no visible dots are generated. All drum charges caused by light MUST be removed (or discharged) before any new images can be written; otherwise images would overwrite and superimpose on one another. A series of erasure lamps are placed in close proximity to the drum's surface as shown in figure 4-4. Their light is carefully filtered to allow only effective wavelengths to pass. Erase light bleeds away any charges along the drum. Charges are carried to ground through the aluminum cylinder. After erasure, the drum's surface is completely neutral; it contains no charges at all.

The classical electrophotographic approach

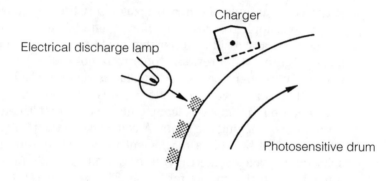

■ **4-4** *Erasing latent charges from the drum.* Tandy Corporation

Charging

A neutral drum surface is no longer receptive to light from the writing mechanism. New images cannot be written until the drum is charged again. In order to condition the drum, a uniform electrical charge must be applied evenly across its entire surface. Surface charging is accomplished by applying a tremendous negative voltage (often about –6,000V) to a solid wire called a primary corona located close to the drum. Since the drum and high-voltage power supply share the same ground, a powerful electrical field is established between the corona wire and drum, as in figure 4-5.

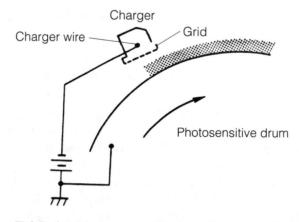

■ **4-5** *Applying a conditioning charge to the drum.* Tandy Corporation

For low voltages, the air gap between a corona wire and drum would act as an insulator. With thousands of volts of potential, however, the insulating strength of air breaks down and an electric corona forms. A corona ionizes any air molecules surrounding the wire, so negative charges migrate to the drum's surface. A corona also breaks down air into ozone, which must be filtered and exhausted from the printer assembly.

The trouble with ionized gas is that it exhibits a very low resistance to current flow. Once a corona field is established, there is essentially a "short-circuit" between the wire and drum. This is not good for a high-voltage power supply. A primary grid (part of the primary corona assembly) is added between the wire and drum. By applying a negative voltage to the grid, charging voltage and current to the drum can be carefully regulated. This "regulating grid voltage" (often –600 to –1,000V) sets the charge level actually applied to the drum. Image intensity can be adjusted to an extent by varying the grid voltage generated by the high-voltage power supply. The drum is now ready to receive a new image.

Writing

In order to form a latent image on a drum surface, the uniform charge that has conditioned the drum must now be discharged in the precise areas where images are to be produced. Images are written to the drum using light as shown in figure 4-6. Any points on the drum exposed to light will discharge to a very low level (about –100V), while any areas left unexposed retain their conditioning charge. The device that produces and directs light to the drum surface is called a *writing mechanism*.

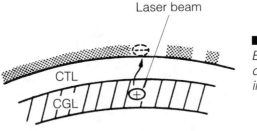

Laser beam

CTL

CGL

■ 4-6
Exposing the charged drum to create a latent image. Tandy Corporation

Because images are formed as a series of individual dots, a larger number of dots (per unit area) will allow finer resolution of the image, and generally higher quality. For example, suppose a writing mechanism can place 300 dots per inch along a single horizontal line on the drum, and the drum can rotate in increments of ¹⁄₃₀₀ of an inch. This means your printer can develop images with a resolution of 300 × 300 dots per inch (DPI). Current EP printers are providing resolutions of 600 × 600 dpi and 1,200 × 1,200 dpi. Lasers have been traditionally used as writing mechanisms, and are still used in many EP printer designs, but new printers are replacing lasers with bars of light-emitting diodes (LEDs) to direct light as needed. Writing mechanisms are covered more extensively later in this chapter.

Developing

Images written to the drum are initially invisible, merely an array of electrostatic charges. The latent image must be "developed" into a visible one before it can be transferred to paper. Toner is used for this purpose. Toner itself is an extremely fine powder of plastic resin and organic compounds bonded to iron particles. Individual granules can be seen under extreme magnification of a microscope.

Toner is applied using a transfer roller in a developer unit as shown in figure 4-7. A transfer roller is basically a long metal sleeve containing a permanent magnet. It is mounted inside the toner supply trough. When the cylinder turns, iron in the toner attracts it to the cylinder. Once attracted, toner acquires a negative static charge provided by the high-voltage power supply. This static charge level is typically equal to the drum's charge level (about –600V). A restricting blade (called a *doctor blade*) limits toner on the cylinder to just a single layer.

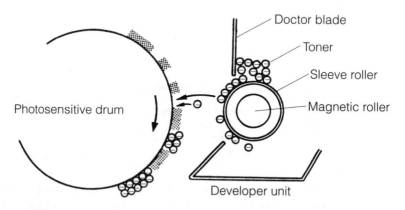

■ **4-7** *Developing the latent image.* Tandy Corporation

Charged toner on the transfer roller now rotates into close proximity with the exposed drum. Any points on the drum that are not exposed will retain a strong negative charge. This repels toner, which remains on the toner cylinder and is returned to the supply. Any points on the drum that are exposed now have a much lower charge than the toner particles (about –100V). This attracts toner from the cylinder to corresponding points on the drum. Toner "fills-in" the latent image to form a visible (or *developed*) image.

In actual practice, an ac booster bias is added in series to the dc intensity bias on the transfer roller. The ac causes fluctuations in the toner's charge level. As the ac signal goes positive, the intensity level increases to help toner particles overcome attraction of the cylinder's permanent magnet. As the ac signal goes negative, intensity levels decrease to pull back any toner particles that might have falsely jumped to unexposed areas. This technique greatly improves print density and image contrast. The developed image can now be applied to paper.

Transfer & discharge

At this point, the developed toner image on the drum must be transferred onto paper. Because toner is now strongly attracted to the drum, it must be pried away by applying an even larger attractive charge to the page. A transfer corona wire charges the page as shown in figure 4-8. The theory behind the operation of a transfer corona is exactly the same as that for a primary corona, except that the potential is now positive. This places a powerful positive charge onto paper, which attracts the negatively charged toner

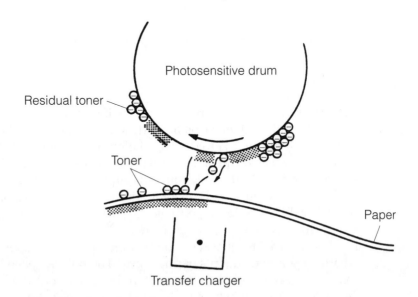

■ 4-8 *Transferring the developed image to paper.* Tandy Corporation

particles. Remember that transfer is not a perfect process. Not all toner is transferred to paper, which is why a cleaning process is needed.

Caution is needed here. Because the negatively charged drum and positively charged paper tend to attract each other, it is possible that paper could wrap around the drum. Even though the small-diameter drum and natural stiffness of paper tend to prevent wrapping, a static charge eliminator (or *static eliminator comb*) is included to counteract positive charges and remove the attractive force between paper and drum as in figure 4-9. Paper now has no net charge. Another good reason for discharging the paper is to prevent charge irregularities from shifting the toner now applied on paper before fusing can take place. Also, discharging the paper will keep subsequent sheets from repelling each other when the sheets are ejected. The drum can be cleaned and prepared for a new image.

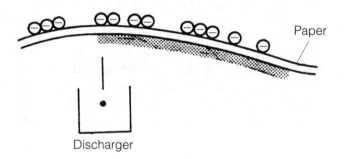

Paper

Discharger

■ **4-9** *Discharging the charged page.* Tandy Corporation

Fusing

Once the toner image has reached paper, it is only held to the page by gravity and weak electrostatic attraction. Toner must be fixed permanently (or fused) to the page before it can be handled. Fusing is accomplished with a heat and pressure assembly like the one shown in figure 4-10. A high-intensity quartz lamp heats a nonstick roller to about 180°C. Pressure is applied with a pliable rubber roller. When a developed page is passed between these two rollers, heat from the top roller melts the toner, and pressure from the bottom roller squeezes molten toner into the paper fibers where it cools and adheres permanently. The finished page is then fed to an output tray. Note that both rollers are referred to as *fusing rollers*, even though only the top roller actually "fuses."

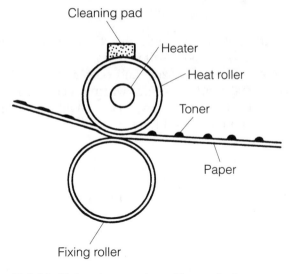

Cleaning pad

Heater

Heat roller

Toner

Paper

Fixing roller

■ **4-10** *Fixing the transferred image to the page.*
Tandy Corporation

To prevent toner particles from sticking to a fusing roller, it is coated with a nonstick resin such as Teflon. A cleaning pad is added to scrape away any toner that might yet adhere. The pad also applies a thin coating of silicon oil to prevent further sticking. Fusing temperature must be carefully controlled. Often a thermistor is used to regulate current through the quartz lamp in order to maintain a constant temperature. A snap-action thermal switch is also included as a safety interlock in the event that lamp temperature should rise out of control. If temperature is not controlled carefully, a failure could result in damage to the fusing assembly, or even a fire hazard.

Writing mechanisms

As discussed earlier in this chapter, the newly charged photosensitive drum contains a uniform electrostatic charge across its surface. To form a latent image, the drum must be discharged at any and all points that comprise the image. Directed light is used to discharge the drum as needed. Images are scanned onto the drum one horizontal line at a time. A single pass across the drum is called a *trace* or *scan line*. Light is directed to any points along the scan line where dots are required. When a scan line is completed, the drum increments in preparation for another scan line. It is up to the printer's control circuits to break down an image into individual scan lines, then direct the writing mechanism accordingly.

Lasers

Lasers have been around since the early 1960s, and they have developed to the point where they can be manufactured in a great variety of shapes, sizes, and power output. To understand why lasers make such a useful writing mechanism, you must understand the difference between laser light, and ordinary "white" light as shown in figure 4-11.

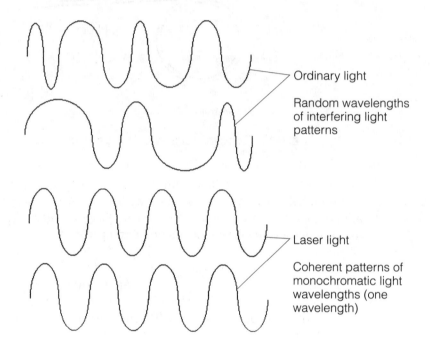

Ordinary light

Random wavelengths of interfering light patterns

Laser light

Coherent patterns of monochromatic light wavelengths (one wavelength)

■ **4-11** *Ordinary light vs. laser light.*

Ordinary white light is actually not white. The light you see is composed of many different wavelengths, each traveling in their own directions. When these various wavelengths combine, they do so virtually at random. This makes everyday light very difficult to direct and almost impossible to control as a fine beam. As an example, take a flashlight and direct it at a far wall. You will see just how much white light can scatter and disperse over a relatively short distance.

The nature of laser light, however, is much different. A laser beam contains only one major wavelength of light (it is *monochromatic*). Each ray travels in the same direction and combines in an additive fashion (known as *coherence*). These characteristics make laser light easy to direct at a target as a hair-thin beam, with almost no scatter (or *divergence*). Older EP printers used Helium-Neon (HeNe) gas lasers, but semiconductor laser diodes have essentially replaced gas lasers in just about all laser printing applications.

Laser diodes are very similar to ordinary light-emitting diodes as in figure 4-12. When the appropriate amount of voltage and current is applied to a laser diode, photons of light will be liberated that have the characteristics of laser light (coherent, monochromatic, and low divergence). A small lens window (or laser aperture) allows light to escape, and helps to focus the beam. Laser diodes are not very efficient devices; a great deal of power is required to generate a much smaller amount of light power, but this trade-off is usually worthwhile for the small size, light weight, and high reliability of a semiconductor laser.

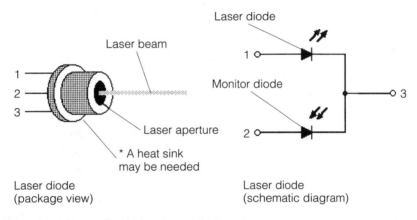

■ **4-12** *Views of a typical laser diode.*

Generating a laser beam is only the beginning. The beam must be modulated (turned on and off) while being swept across the drum's surface. Beam modulation can be accomplished by turning the laser on and off as needed (easily accomplished with fast semiconductor diodes), as shown in figure 4-13, or by interrupting a continuous beam with an electro-optical switch (typically used with gas lasers that are difficult to switch on and off rapidly). Mirrors are used to alter the direction of the laser beam, while lenses are used to focus the beam and maintain a low divergence at all points along the beam path. Figure 4-13 is just one illustration of a laser writing mechanism, but it shows some of the complexity that

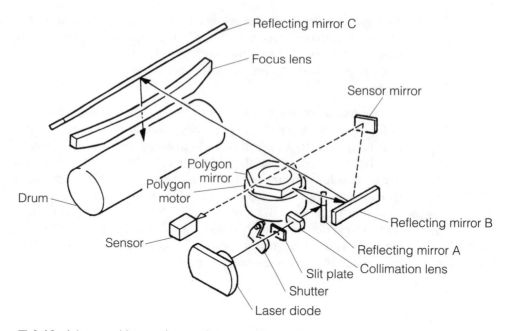

■ **4-13** *A laser writing and scanning arrangement.* Tandy Corporation

is involved. The weight of glass lenses, mirrors, and their shock mountings have kept EP laser printers bulky and expensive.

Alignment has always been an unavoidable problem in complex optical systems such as figure 4-13. Consider what might happen to the beam if any optical component should become damaged or fall out of alignment; focus and direction problems could render a drum image unintelligible. Unfortunately, the realignment of optical systems is virtually impossible without special alignment tools, and is beyond the scope of this book. Finally, printing speed is limited by the speed of moving parts, and the rate at which the laser beam can be modulated.

LEDs

Fortunately, a photosensitive drum is receptive to light from many different sources. Even light from light-emitting diodes (LEDs) can expose the drum. By fabricating a series of microscopic LEDs into a single scan line, as shown in figure 4-14, an individual LED can be provided for every possible dot in a scan line. For example, the ROHM JE3008SS02 is an LED print bar containing 2,560 microscopic LEDs over 8.53 inches. This equates to 300 dots per

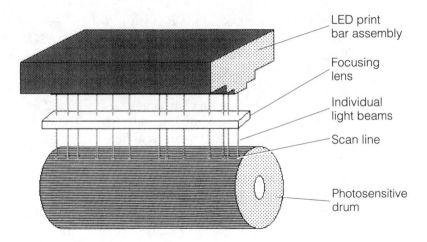

LED print
bar assembly

Focusing
lens

Individual
light beams

Scan line

Photosensitive
drum

■ **4-14** *LED print bar operation.*

inch. Each LED is just 50 × 65 micrometers (μm) and they are spaced 84.6 μm apart.

The operation of an LED print bar such as the one shown in figure 4-15 is remarkably similar to that of a thermal line-print head discussed in Chapter 3. An entire series of data bits corresponding to each possible dot in a horizontal line is shifted into internal digital circuitry within the print bar. Dots that will be visible are represented by a logic "1," and dots that are not visible will remain at logic "0." For a device such as the JE3008SS02, 2,560 bits must be entered for each scan line.

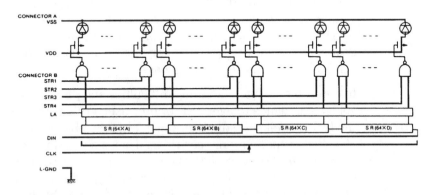

■ **4-15** *Block diagram of an LED print bar.* ROHM, Inc.

After a complete line of data has been loaded through the DIN pin, the LEDs must be fired. This is performed in segments to reduce the power surges that would be generated if every LED were fired together. The JE3008SS02 is divided into four segments of 640 dots. A trigger signal (or *strobe*) can be applied to STR1 through STR4. This passes data to each segment's driver circuits. LEDs that illuminate will leave latent points on the drum's surface. LEDs that do not light will have no effect. Each strobe is fired sequentially until all four segments have been strobed. All 2,560 dots can be scanned in under 2.5 milliseconds (ms). The drum is incremented $\frac{1}{300}$ of an inch, and a new scan line can be loaded into the print bar.

You can probably see the advantages of an LED print bar system over a laser approach. There are no moving parts involved in light delivery, no mirror motor to jam or wear out. The printer can operate at much higher speeds because it does not have to overcome the dynamic limitations of moving parts. There is only one focusing lens between the print bar and drum. This greatly simplifies the optics assembly, and removes substantial weight and bulk from the printer. An LED system overcomes almost all alignment problems, so a defective assembly can be replaced or aligned quickly and easily.

The electrophotographic cartridge

Electrophotographic printers mandate the use of extremely tight manufacturing tolerances to ensure precise, consistent operation. A defect of only a few thousandths of an inch could cause unacceptable image formation. Even the effects of normal mechanical wear can have an adverse effect on print quality. Many key IFS components would have to be replaced every 5,000 to 10,000 pages to maintain acceptable performance. Clearly it would be undesirable to send your printer away for a complete (and time-consuming) overhaul every 10,000 pages.

In order to ease manufacturing difficulties and provide fast, affordable maintenance to every EP printer user, critical components of the IFS, as well as a supply of toner, are assembled into a replaceable electrophotographic cartridge. This is sometimes referred to as an *engine*. As figure 4-16 shows, a typical EP cartridge contains the toner roller, toner supply, debris cavity, primary corona (and primary grid), photosensitive drum, and cleaning blade assembly. All necessary electrical connectors and drive gears are included. By assembling sensitive components into a single replaceable cartridge, printer reliability is substantially improved by preventing problems before they ever become noticeable.

Hewlett-Packard, Inc.

■ **4-16** *A HP electrophotographic cartridge.*

The complexity of an EP cartridge varies with the particular printer's design. Some printers use a stand-alone engine, while providing simple, low-cost toner cartridges (usually with enough toner for several hundred prints). In these designs, the toner is cheap, but the replacement engine can be rather expensive (anywhere from $200 to $300 U.S.) and rated to last for tens of thousands of pages. An alternative to this approach, the more popular approach, is to integrate the toner cartridge and engine components into a single assembly. Hewlett-Packard's engines (e.g., the 95A, 75A, 91A, and 98A) use this strategy. The major difference here is that the engine contains enough toner for 5,000 to 10,000 pages or more. So by the time toner runs out, the major engine components need to be replaced anyway.

You should also understand that the life expectancy of a toner/cartridge and engine are not absolute, but it will vary greatly depending on the way in which the printer is used. The amount of toner used will depend on the relative darkness of each page; pages with more black space use more toner, and vice versa. As an example, printers that produce complex graphics will tend to use toner faster. Because toner is largely an organic compound, it also suffers from a limited shelf life. Most toner containers are marked with an expiration date, but once the toner/engine assembly is unsealed to the air, the toner should be used within six months. Another consideration is the engine components. Life expectancy (in pages) is typically defined around standard 20-lb. xerography-grade (photocopier) stock. Thicker or specially coated papers will tend to place more stress on the engine mechanics, which translates into shorter engine life.

Protecting an EP cartridge

As you might imagine, the precision components in an EP cartridge are sensitive and delicate. The photosensitive drum and toner supply are particularly sensitive to light and extreme environmental conditions, so it is important to follow several handling and storage guidelines. First, the photosensitive drum is coated with an organic material that is extremely sensitive to light. Although a metal shroud covers the drum when the cartridge is exposed, light might still penetrate the shroud and cause unwanted exposure (also known as *fogging*). Deactivating the printer for a time will often eliminate mild fogging. Do not defeat the shroud in open light unless absolutely necessary, and then only for short periods. This will certainly fog the drum. A seriously fogged cartridge might have to be placed in a dark area for several days. Also, never expose the EP drum to direct sunlight; direct sunlight can permanently damage the drum's coating.

Next, avoid extremes of temperature and humidity. Temperatures exceeding 40°C can permanently damage an EP cartridge. Extreme humidity is just about as dangerous. Do not allow the cartridge to become exposed to ammonia vapors or other organic solvent vapors; they break down the drum's photosensitive coating very quickly. Finally, keep a cartridge secure and level. Never allow it to be dropped or abused in any way.

Finally, as the toner supply diminishes, it might be necessary to re-distribute remaining toner so that it reaches the toner roller. Because toner is available along the entire cartridge, it must be redistributed by rocking the cartridge back and forth along its long axis. If you tip a cartridge upright, remaining toner will fall to one end and cause uneven distribution. You should check the notes provides with your toner/engine cartridges for specific redistribution hints and precautions.

Refinements to the EP process

This chapter has focused largely on SX-type EP architecture used extensively in laser printers such as the Hewlett-Packard LaserJet II and III from 1986 through 1990. Since 1990, however, continuing advances in EP design have simplified the electrophotographic process, while improving reliability. All of the basic principles have remain unchanged, but the process is more refined, as illustrated in figure 4-17.

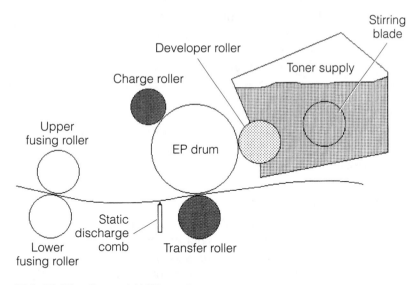

■ **4-17** *The Canon LX EP engine.*

Charge rollers

One of the first things you should notice is that the primary and secondary coronas have been eliminated and replaced with charge rollers. Introduced with the Canon LX EP engine in 1990, the charge rollers have replaced corona wires as a means of applying electrical charges to the EP drum and paper. The charge rollers themselves are little more than a specialized formulation of foam rubber, which accepts an even charge across its entire surface area (a key attribute for EP printers).

The main advantage to the charge rollers is the lower high-voltage requirement and the elimination of primary grid circuitry. While the primary corona demands about −6,000 V, the primary charging roller requires only about −1,000 V to achieve the required −600-V charge on the drum. This behavior also holds true for the transfer roller working at +1,000 V and delivering a charge of +600 V to the page. A handy side effect to these lower charging voltages is a significant reduction in the amount of ozone developed in the printer. Lower ozone eases environmental concerns about ozone health hazards.

Unfortunately, a noted disadvantage to the charge roller scheme is that direct contact is required with the EP drum and page. In theory, this is not a problem. But in practical applications, you cannot escape the dust, hair, pollen, and other airborne debris that will invariably find its way into the printer. Debris that collects on charge rollers has a chance of being transferred to the page (or interfering with the charge uniformity along the roller). Accumulations of foreign matter on the primary charge roller also stand a much greater chance of damaging the EP drum's photosensitive coating. As it turns out, the transfer roller is more likely to attract foreign matter because of its lower physical position in the printer assembly. To help keep the transfer roller clean, its charge is reversed when the printer is idle. Charge reversal tends to repel foreign matter off the roller.

Erase lamps

Another thing you will notice about figure 4-17 is that the erase lamps are missing. This is not an accident. With the introduction of charge rollers, the need for erase lamps disappeared. EP designers discovered that adding an ac signal to the dc charging voltage on the primary charge roller effectively recharged the drum without any light exposure. Reports from the field indicate that the absence of erase lamps has made no visible difference in image quality.

Doctor blade

As discussed, a precise leveling blade (called a *doctor blade*) in the development unit serves to keep toner on the development roller limited to a single layer. As you might expect, any damage or misadjustment to the blade will have a pronounced effect on the image. Recent advances in materials have resulted in a rubber blade that actually rests on the development roller's surface. This streamlines toner charging while still providing even toner distribution on the roller.

New EP coatings

Traditionally, the photosensitive coating applied to an EP drum has been notoriously delicate; even the slightest nick or scratch left marks permanently in the drum. Over the last few years, EP coatings have improved dramatically in response to changing market forces. One major reason for the improvements coincides with the introduction of charge rollers. Even a foam rubber roller in direct contact with the EP drum demands a more durable photosensitive coating. Another factor fueling the drum improvements is attributed to the growing remanufacturing market for toner cartridges and engines; drums are being expected to serve much longer lives as they are tested and recycled into remanufactured assemblies.

Toner paddle

One of the great complaints about toner supplies is that the toner is not always distributed evenly along the development roller. While this is not an issue while the toner is relatively new, a cartridge that is approaching exhaustion might experience faint streaks in the print where toner is particularly low. Newer toner/engine designs include a toner paddle that circulates the toner supply. This serves two important purposes: it keeps the toner from clumping, and it keeps the toner level steady across the development roller.

New toner

Even the toner itself has undergone some substantial improvements over the last few years. Remember that toner is a microfine powder composed of plastic, iron, and coloring. As the resolution of an EP printer improves, the size of a toner granule becomes a gating issue. After all, a grain of toner must be smaller than a single "dot," otherwise, fine detail will be lost. Toner is continuing to improve in the remanufacturing market as well, but its overall quality still lags the original manufacturer's toner.

Using test equipment

5

THIS CHAPTER INTRODUCES YOU TO THE TOOLS AND TEST equipment needed for printer troubleshooting. If you are unfamiliar with test equipment, read this chapter carefully. If you are an experienced troubleshooter, you might want to skip this chapter.

Small tools & materials

It might sound strange, but hand tools can often make or break your repair effort. If you have ever started a repair, then been delayed by one missing screwdriver, pair of pliers, or wrench ("you knew it was there last week"), then you know how much time and frustration can be saved by gathering the proper tools before you begin. Check your toolbox! If you do not have a toolbox just yet, now is a good time to consider what you need for electronics repairs.

Hand tools

Screwdrivers are always a good place to start. Figure 5-1 illustrates six general types of screws and rings that you will likely encounter. A healthy variety of both regular and Phillips-type screwdrivers goes a long way for basic assembly, disassembly, and adjustment tasks. Avoid excessively large or unusually small screwdrivers unless there is a specific need for them. Several short-shaft screwdrivers can come in handy when working in confined areas; printer assemblies can be very densely packaged. Allen-type (hexagonal hole) screws are also common, so include a set of Allen keys in your toolbox. Keep an eye out for specialty screws. Spline-type and torx-type screws are growing in popularity as manufacturers seek to keep untrained personnel out of their equipment. Fortunately, large hardware stores usually stock spline and torx drivers.

Silver washer screws are highly conductive fasteners, and are frequently used to ensure a reliable electrical contact between assemblies. The silver machine screws are generic fasteners used almost exclusively to mount modules and printer subassemblies.

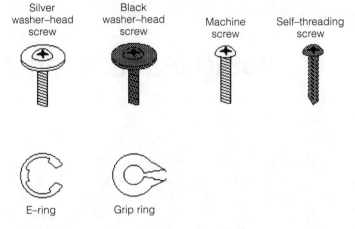

■ 5-1 *A comparison of basic printer fasteners.*

Black washer screws are anodized, which makes their surfaces nonconductive. Such fasteners are used to secure plastic parts to metal or other plastic parts. Black machine screws are typically found securing the printer's structural parts such as the chassis. Black self-threading screws are almost always used to secure a part to a plastic mounting hole. Be very careful to avoid stripping the plastic mounting hole; use just enough force to seat the screw, but no more. E-rings are slipped over grooves machined into metal axles, and are typically used to hold rotating parts (such as gears and rollers) in place on their shafts. A grip ring is similar to an e-ring, but the grip ring is used when there are no machined grooves on the axle or shaft.

Consider three types of pliers. A garden variety pair of mechanic's pliers is useful for keeping tight hold on any nut, bolt, or other pesky part. Two pairs of needlenose pliers (one short nose, one long nose) round out the collection. Needlenose pliers are great for grabbing and holding parts in tight spaces. They also come in handy as heat sinks when soldering.

A set of small, electronics-grade, open-end wrenches work well to hold small nuts during assembly and disassembly. Wrenches should have thin bodies (for tight spaces), and should be below $\frac{5}{16}$" (or 8 mm for metric sizes). If you have the choice between metric or U.S. wrenches, get metric. Many printers are made in Asia and the Pacific rim where metric parts are standard. A small adjustable wrench is always a good addition to a toolbox.

Because you are working with electronic systems, two sizes of wire cutters (one small and one medium) should be on hand to cut jumper wires or trim replacement component leads. A separate wire stripper should be included to remove insulation from wires or components. Resist the temptation to strip insulation using wire cutters. Even if insulation should be removed successfully, cutting blades often leave a nick or pinch in the conductor, which later might fatigue and break. If you plan to be making or repairing connectors, you might need special crimp or insertion tools to do the job. Those will vary depending on exactly what types of connectors you need to make.

Circuit work requires a soldering iron. Invest in a good-quality soldering pencil, something in the 25- to 30-W range. Avoid the heavy-duty, high-power soldering guns. They might work fine for plumbers, but that much power can easily destroy delicate printed circuit boards and components. Always be sure to have one or two spare tips on hand. Irons exceed 500°F at the tip. This temperature is a serious hazard to your personal safety, and a tremendous fire hazard as well. Be sure to park a hot soldering iron in a strong, wire frame holder. Never leave an iron unattended on a bench or table top.

Materials

To use a soldering iron, you are also going to need an ample supply of solder. Use a 60% tin, 40% lead (60/40) solder containing a resin cleaning agent. Under no circumstances should you ever use paste flux containing acids or solvents, or use solder containing acid flux (sometimes called *acid core* or *plumber's solder*). Harsh solvents destroy delicate component leads and circuit traces.

A spool of hook-up wire can serve a variety of uses ranging from printed circuit repair and wire splices, to makeshift test leads. Solid wire (18 to 20 gauge) is often the easiest to work with, but stranded wire can be used just as readily. You might find it easier to keep several smaller spools of different colored wire so that wire can be color coded for different purposes (e.g., red for dc voltage, blue for dc ground, yellow for signal wires, and so on).

Include a set of alligator leads. They are available in a selection of lengths, colors, and wire gauges. Alligator leads are handy for temporary jumping of signals during a repair, or to test the reaction of another signal in a circuit. A set of small alligator leads and a set of large alligator leads will cover you under most circumstances. Individual alligator clips make excellent heat sinks for components during soldering or desoldering.

Heat shrink tubing provides quick, clean, and effective insulation for exposed wiring or splices. Most general-purpose electronics supply stores sell heat shrink tubing in three-foot rolls, or in packages of assorted sizes. You will need a heat gun to shrink the tubing, although a blow drier usually works just as well. Heat shrink tubing is available in a selection of colors.

These are only a few of the more common tools and materials that you will need to get started. The list is by no means complete. There are literally hundreds of general- and special-purpose tools that you can use to aid your repairs, far too many to cover completely. Experience will be your best guide in deciding which tools and materials are best for you.

Soldering

Soldering is the most commonly used method of connecting wires and components within an electrical or electronic circuit. Metal surfaces (e.g., component leads, wires, or printed circuit traces) are heated to high temperatures, then joined together with a layer of compatible molten metal. When done correctly, soldering forms a lasting, corrosion-proof, inter-molecular bond that is mechanically strong and electrically sound. All that is needed is an appropriate soldering iron and electronics-grade (60/40) solder. This section of the chapter looks at both regular soldering and surface-mount soldering.

Soldering background

By strict definition, soldering is a process of bonding metals together. There are three distinct types of soldering: brazing, silver soldering, and soft soldering. Brazing and silver soldering are used when working with hard or precious metals, but soft soldering is the technique of choice for electronics work.

In order to bond wire and component leads (typically made of copper), a third metal must be added while in its molten state. The bonding metal is known simply as *solder*. Several different types of solder are available to handle each soldering technique, but the chosen solder must be molecularly compatible with the metals to be bonded; otherwise, a bond will not form. Lead and tin are two common, inexpensive metals that adhere very well to copper. Unfortunately, neither metal by itself has the strength, hardness, and melting point characteristics to make them useful. Therefore, lead and tin are combined into an alloy. A ratio of approximately 60%

tin and 40% lead yields an alloy that offers reasonable hardness, good pliability, and a relatively low melting point that is ideal for electronics work. This is the solder that you should use.

Although solder adheres very well to copper, it does not adhere well at all to the oxides that form naturally on its surface. Even though conductors might "look" clean and clear, some amount of oxidization is always present. Oxides must be removed before a good bond can be achieved. A resin cleaning agent (called *flux*) can be applied to conductors before soldering. While resin is chemically inactive at room temperature, it becomes extremely active when heated to soldering temperatures. Active flux combines with oxide and strips it away, leaving clean copper surfaces for molten solder. As a completed solder joint cools, any residual resin cools and returns to an inactive state. Never use acid or solvent-based flux to prepare conductors. They can clean away oxides as well as resin, but acids and solvents remain active after the joint cools. Over time, this will dissolve copper wires and connections, and eventually cause a circuit failure. Resin flux can be purchased as a paste that can be brushed into conductors before soldering, but most electronic solders have a core of resin manufactured right into the solder strand itself. This is much cleaner and more convenient because resin-core solder cleans the joint as solder is applied. Such resin-core solder is much more convenient than working with flux paste.

Irons & tips

A soldering iron is little more than a resistive heating element built into the end of a long steel tube as shown in the cross-sectional diagram of figure 5-2. When 120 Vac is applied to the heater, it warms the base of a tip. Any heat conducted down the cool-down tube is dissipated harmlessly to the surrounding air. This keeps the handle temperature low enough to hold comfortably.

Most of the heat is channeled into a soldering tip similar to the one shown in figure 5-3. Tips often have a core of copper that is plated with iron. It is coated with a layer of nickel to stop high-temperature corrosion, then plated with chromium. A chromium coating renders the tip nonwettable; solder will not stick. Because solder must stick at the tip's end, that end is plated with tin. A tin coating makes the end wettable. Tips can be manufactured in a wide range of shapes and sizes. Before you select the best tip for the job, you must understand ideal soldering conditions.

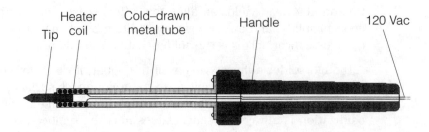

Tip | Heater coil | Cold-drawn metal tube | Handle | 120 Vac

■ **5-2** *A simple soldering pencil.*

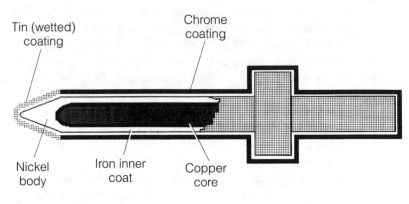

Tin (wetted) coating | Chrome coating | Nickel body | Iron inner coat | Copper core

■ **5-3** *A soldering tip.*

The very best electronic soldering connections are made within only a narrow window of time and temperature. A solder joint heated between 500 to 550°F for one to two seconds will make the best connections. You must select your soldering iron wattage and tip to achieve these conditions. Too many amateur technicians are content with a single soldering iron and tip, but this can present serious problems. The entire purpose of soldering irons is not to melt solder. Instead, a soldering iron is supposed to deliver heat to a joint; the joint will melt solder. A larger joint (with more numerous or larger conductors) requires a larger iron and tip than a small joint (with fewer or smaller conductors). If you use a small iron to heat a large joint, the joint might dissipate heat faster than the iron can deliver it, so the joint might not reach an acceptable soldering temperature. Conversely, using a large iron to heat a small joint will overheat the joint. Overheating can melt wire insulation and damage printed circuit traces. The skilled technician will match wattage to the situation. Most general-purpose electronics work can be done using an iron below 30 W.

Because the end of a tip actually contacts the joint to be soldered, its shape and size can assist heat transfer greatly. When heat must be applied across a wide area (such as a wire splice), a wide area tip should be used. A screwdriver (or flat-blade) tip such as shown in figure 5-4 is a good choice. If heat must be directed with pinpoint accuracy for small, tight joints or printed circuits, a narrow blade or conical tip is the best. Two tips for surface mount desoldering are also shown in figure 5-4.

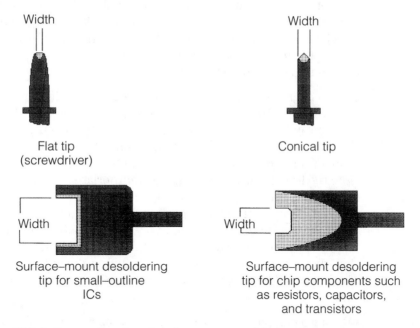

Width

Flat tip
(screwdriver)

Width

Conical tip

Width

Surface–mount desoldering
tip for small–outline
ICs

Width

Surface–mount desoldering
tip for chip components such
as resistors, capacitors,
and transistors

■ **5-4** *Conventional and surface-mount soldering tips.*

Soldering technique

First, consider these important safety points:

- ☐ Always park your soldering iron in a secure holder while it is on.
- ☐ Never allow a hot iron to sit freely for any length of time on a table top or anything flammable.
- ☐ Make it a point to always wear safety glasses when soldering. Active resin or molten solder can easily flick off the iron or joint and do permanent damage to the tissue in your eyes.

Give your soldering iron plenty of time to warm up (five minutes is usually adequate). Once the iron is at its working temperature, you should coat the wettable portion of the tip with fresh solder (this is known as *tinning* the iron). Rub the tip into a sponge soaked in clean water to wipe away any accumulations of debris or carbon that might have formed, then apply a thin coating of fresh solder to the tip's end. Solder penetrates the tip to a molecular level and forms a cushion of molten solder that aids heat transfer. Re-tin the iron any time its tip becomes blackened, perhaps every few minutes or after several joints.

It also might be helpful to tin each individual conductor before actually making the complete joint. To tin a wire, prepare it by stripping away ³⁄₁₆ to ¼ of an inch of insulation. As you strip insulation, be sure not to nick or damage the conductor. Heat the exposed copper for about one second, then apply solder into the wire, not into the iron. If the iron and tip are appropriate, solder should flow evenly and smoothly into the conductor. Apply enough solder to bond each of a stranded wire's exposed strands. When tinning a solid wire or component lead, apply just enough solder to lightly coat the conductor's surface. Park the soldering iron safely and allow the tinned wire(s) to cool. You will find that conductors heat faster and solder flows better when all parts of a joint are tinned in advance.

Making a complete solder joint is just as easy. Bring together each of your conductors as necessary to form the joint. For example, if you are soldering a component into a printed circuit board, insert the component leads into their appropriate locations. Place the iron against all conductors to be heated, as shown in figure 5-5. For a printed circuit board, heat the printed trace and component lead together. After about one second, flow solder gently into the conductors, not the iron. Be sure that solder flows cleanly and evenly into the joint. Apply solder for another one or two seconds,

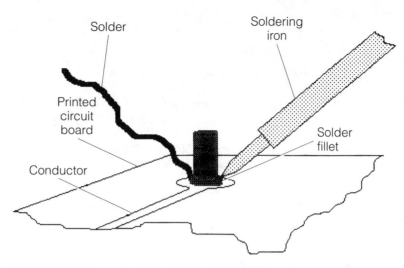

■ 5-5 *Soldering a typical printed circuit junction.*

then remove both solder and iron. Do not attempt to touch or move the joint until molten solder has set for several seconds (and cooled for at least 30 seconds). If the joint requires additional solder, reheat the joint and flow in a bit more solder.

You can identify a good solder joint by its smooth, even, silvery-gray appearance. Any charred or carbonized flux on the joint indicates that your soldering temperature is too high (or that heat is being applied for too long). Remember that solder cannot flow unless the joint is hot. If it is not, solder will cool before it bonds. The result is a rough, built-up, dull-gray or blackish mound that does not adhere very well. This is known as a *cold* solder joint. Cold joints can cause poor connections that translate to intermittent operation. Fortunately, a cold joint can easily be corrected by reheating the joint properly and applying fresh solder; the hard part is finding the cold joint in the first place.

Surface mount soldering

Classic printed circuit boards use *through-hole* components. This is the tried-and-true fabrication technique where parts are inserted on one side of the PC board, and their leads are soldered to printed traces on the other side. Surface-mounted components do not penetrate a PC board. Instead, they rest on only one side of the board as shown in figure 5-6. Metal tabs replace component lead wires. Surface mount components range from discrete components, such as resistors and capacitors, to active parts, such as

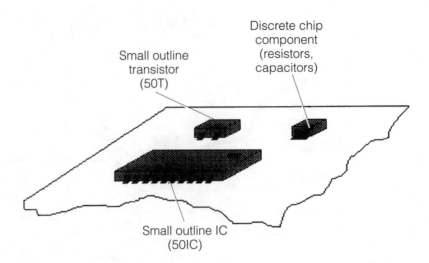

Small outline
transistor
(50T)

Discrete chip
component
(resistors,
capacitors)

Small outline IC
(50IC)

■ **5-6** *Partial view of a simple surface-mount PC board.*

transistors and integrated circuits. Even sophisticated ICs like microprocessors and ASICs are frequently found in surface mount packages.

During manufacture, surface mount parts are glued into place on a PC board, then the board is brought quickly up to soldering temperature in a special chamber. A wave of molten solder is passed over the board where it adheres to heated component leads and PC traces. The remainder of the board is chemically and physically masked prior to soldering to prevent molten solder from sticking elsewhere. The finished board is then cooled slowly to prevent thermal shock to the components, masks are stripped away, and the board can be tested (or any through-hole parts can be added). This type of fabrication is called *flow soldering*, and it is similar to the principle used to mass-solder through-hole PC boards.

A close variation of the surface-mount manufacturing technique applies a layer of solder paste to a masked PC board before components are applied, then the board is heated to flow solder into PC trace. After components are glued into place, the board is quickly reheated so solder will adhere to each component lead. The finished board is then cooled slowly. This is known as *reflow soldering*. Although the specific methods of surface mount soldering will have little impact on your troubleshooting, you should understand how surface mount components are assembled in order to disassemble them properly during your repair.

Desoldering

Ideally, desoldering a connection involves removing the intermolecular bond that has been formed during soldering. In reality, however, this is virtually impossible. The best that you can hope for is to remove enough solder to gently break the connection apart without destroying the joint. Desoldering is basically a game of removing as much solder as possible. Some connections are very easy to remove. For instance, a wire inserted into a printed circuit board might be removed just by reheating the joint and gently withdrawing the wire from its hole once solder is molten. You can use desoldering tools to clear away the solder itself after the connection is broken.

Surface mount components present a special problem, because it is impossible to move the part until it is desoldered completely. By using special desoldering tips, as shown in figure 5-4, all leads can be heated simultaneously so the part can be separated in one quick motion. There are also special tips for desoldering a selection of IC packages. Once a part is clear, excess solder can be removed with conventional desoldering tools, such as a solder vacuum or solder wick.

Desoldering through-hole components is not as easy as it looks. You must heat each solder joint in turn, and use a desoldering tool to remove as much solder as possible. Once each lead is clear, you will probably have to break each lead free as shown in figure 5-7. Grab hold of each lead and wiggle it back and forth gently until it breaks free. An alternate method is to heat each joint while withdrawing the lead with a pair of needlenose pliers, then clean up any excess solder later. Unfortunately, this is not possible with all components. Experience is the best teacher when it comes to desoldering.

Test equipment

Much of today's printer troubleshooting is performed symptomatically; that is, subassemblies are typically replaced based on the symptoms exhibited in the printed output. This is a well-understood technique, and generally does not require the use of test equipment. On the down side, symptomatic troubleshooting is not always 100% reliable. Because new assemblies and circuits represent an investment of your time and money, it would benefit you to be as certain as possible of the fault before proceeding. Diagnostics (such as PRINTERS) and test equipment can help narrow

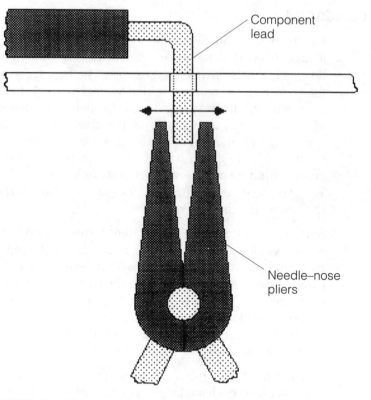

Component
lead

Needle–nose
pliers

■ **5-7** *Breaking a stubborn through-hole solder joint.*

down the problem. This section introduces you to the background, operation, and testing techniques for three major test instruments: a multimeter, logic probe, and oscilloscope.

Multimeters

Test meters can go by many names. Some people call them multimeters or just meters, while others might refer to them as volt-ohm meters (VOMs) or multitesters. Regardless of what name you choose to call them, multimeters are the handiest and most versatile piece of test equipment that you will ever use. If your toolbox does not contain a good-quality multimeter yet, now is a good time to consider purchasing one. Even the most basic multimeters are capable of measuring ac and dc voltage, current, and resistance. For less than $150, you can buy a good digital multimeter that also includes features like a capacitance checker, continuity checker, diode checker, and transistor checker. Digital multimeters are easier to read, more tolerant of operator error, and more precise than analog multimeters. Figure 5-8 shows a B+K digital multimeter.

■ **5-8** *A B+K Model 2912 digital multimeter.* B+K Precision

There are usually just two considerations when using a multimeter. First, the meter must be set to the desired function (voltage, current, capacitance, etc.). Second, the range must be set properly for that function. If you are unsure what range to use, start by choosing the highest possible range. Once you have a better idea of what readings to expect, the range might be reduced to achieve a more precise reading. If your signal exceeds the meter's range, an "overrange" warning will be displayed. Many digital multimeters are capable of selecting the proper range automatically (autoranging).

A multimeter can be used for two types of testing: static and dynamic. Dynamic tests are made with power applied to a circuit, while static tests are made on unpowered circuits or components. Measurements like voltage, current, and frequency are dynamic tests, but most other tests such as resistance/continuity, capacitance, diode and transistor junction quality are static tests. The following is a review of basic multimeter measurement techniques.

Measuring voltage

Multimeters can measure both dc voltages (marked dcV or Vdc) and ac voltages (marked acV or Vac). It is important to remember that all voltages (either ac or dc) must be measured in parallel with the desired circuit or component. Never interrupt a circuit and attempt to measure voltage in series with other components. Any such reading would be meaningless, and your circuit might not even function.

Set your multimeter to its appropriate function (dcV or acV), then select the proper range. If you are unsure about what range to use, start at the largest range to prevent possible damage to the meter. An autoranging multimeter will select its own range. Place your test leads across (in parallel) with the part under test, as shown in figure 5-9, and read voltage directly from the digital display. The dc voltage readings are polarity sensitive, so if you read +5 Vdc and then reverse the test leads, you will see –5 Vdc. The ac voltage readings are not polarity sensitive.

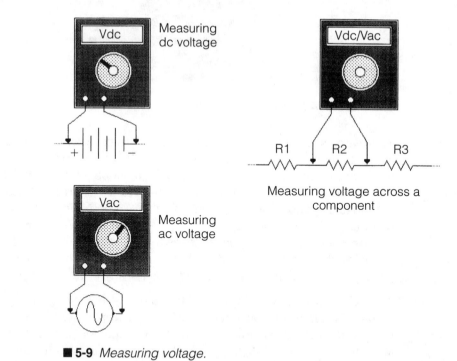

Measuring dc voltage

Measuring ac voltage

Measuring voltage across a component

■ **5-9** *Measuring voltage.*

Measuring current

Most general-purpose multimeters allow you to measure ac current (acA or Iac) and dc current (dcA or Idc) in a circuit, although there are often few ranges to choose from. As with voltage measurements, current is measured in a working circuit with power applied, but current must be measured in series with the circuit or component under test. Inserting a meter in series, however, is not always an easy task. In many cases, you must physically interrupt a circuit at the point you wish to measure, then connect test leads across the break. While it might be easy to interrupt a circuit, keep in mind that you must also put the circuit back together, so use care when choosing a point to break. Never try to read current in parallel. Current meters, by their nature, exhibit very low resistance across their leads (sometimes below $0.1\ \Omega$). Placing a current meter in parallel can cause a short circuit across a component that can damage the part, the circuit under test, or your multimeter.

Set your multimeter to the desired function (dcA or acA) and select the appropriate range. If you are unsure about the proper range, set the meter to its largest range. It might be necessary to plug one of your test leads into a different "current input" jack on the meter. Unless your multimeter is protected by an internal fuse, it can be damaged by excessive current. Make sure that the meter can handle the amount of current you are expecting.

Turn off all power to a circuit before inserting a current meter. This prevents unpredictable circuit operation when it is interrupted. If you wish to measure power supply current feeding a circuit, such as in figure 5-10, break the power supply line at any convenient point (often at the supply or circuit board connectors). Insert the meter and reapply power. Read current directly from the display. This procedure can also be used for measuring current within a circuit.

Measuring resistance

Resistance (ohms) is the most common static measurement that your multimeter is capable of. This is a handy function, not only for checking resistors, but for checking other resistive elements such as wires, connectors, motors, solenoids, and some semiconductor components. Resistance is measured in parallel across components with all circuit power off, as shown in figure 5-11. It might be necessary to remove at least one component lead from its circuit to prevent interconnections with other components from causing false readings.

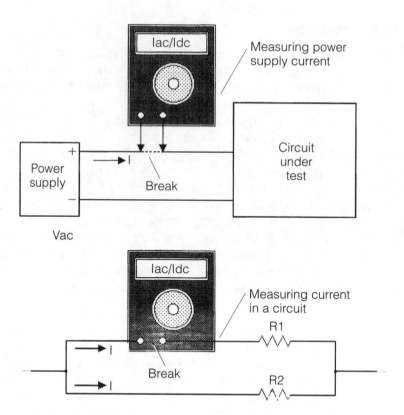

Measuring power supply current

Measuring current in a circuit

■ **5-10** *Measuring current.*

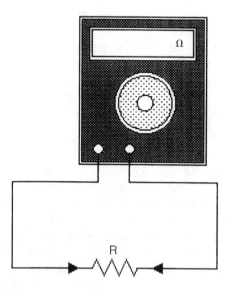

■ **5-11**
Measuring resistance.

Ordinary resistors can be checked simply by switching to a resistance function and selecting the proper range. Many multimeters can reliably measure resistance up to 20 MΩ. Place your test leads across the component and read resistance directly from the display. If resistance exceeds the selected range, the display will indicate an overrange or infinite resistance condition. Continuity checks are made to ensure a reliable, low-resistance connection between two points. For example, you could check the continuity of a cable between two connectors to ensure that both ends are connected properly. Set your multimeter to a low resistance scale, then place your test leads across both points to measure, as shown in figure 5-12. Ideally, good continuity should be about 0 Ω.

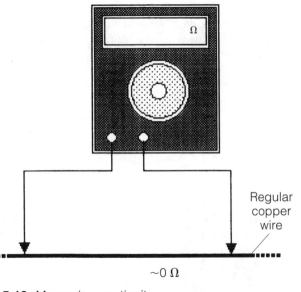

~0 Ω

■ **5-12** *Measuring continuity.*

Checking a capacitor

There are two methods of checking a capacitor using your multimeter. If your meter is equipped with a built-in capacitance checker, all you need to do is select the capacitance function and set the desired range. You might have to place test probes in parallel across the capacitor under test, or you might have to remove the capacitor and insert it into a special fixture on the meter's face. A capacitance checker will usually display capacitance directly in microfarads (μF) or picofarads (pF). As long as your reading is within the tolerance of the capacitor's marked value, you know the part is good.

If your multimeter is not equipped with an internal capacitor checker, you could use the resistance ranges to approximate the quality of a capacitor. This type of check provides a "quick and dirty" judgment of whether the capacitor is good or bad. The principle behind this is simple. All ohmmeter ranges use an internal battery to supply current for the component under test. When that current is supplied to a working capacitor, as shown in figure 5-13, it will charge the capacitor. Charge accumulates as the ohmmeter is left connected, and can be seen as changing resistance on the ohmmeter display. When first connected across an ohmmeter, the capacitor will draw a relatively large amount of current; this reads as low resistance. As the capacitor charges, it draws less and less current, so resistance appears to increase. Ideally, a fully charged capacitor draws no current, so your resistance reading should climb to infinity. When a capacitor behaves this way, it is probably good.

114

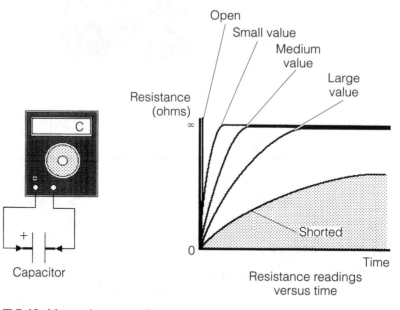

■ **5-13** *Measuring capacitance.*

You are not actually measuring resistance or capacitance here, but only the profile of a capacitor's charging characteristic. If the capacitor is extremely small, or is open-circuited, it will not charge substantially, so it will instantly read infinity. If a capacitor is partially (or totally) short-circuited, it will not hold a charge, so you might read 0 Ω (or resistance might climb to some level below infinity and remain there). In either case, the capacitor is probably defective. If you doubt your readings, check several other capacitors of the same value and compare readings. Be sure to make this test on a moderate to high resistance scale. A low resistance scale might charge to infinity too quickly for a clear reading.

Semiconductor checks

Many multimeters offer a semiconductor junction checker for diodes and transistors. Meters equipped with a "diode" range in their resistance function can be used to measure the static resistance of most common diodes in their forward or reverse-biased conditions, as shown in figure 5-14. Select the diode range from your meter's resistance function and place test leads across the diode in the "forward" direction. A working silicon diode should exhibit a resistance between about 450 and 700 Ω, which will read directly on your meter. Reverse your test leads to reverse-bias the diode. Because a working diode should not conduct at all in the reverse direction, you should read infinite resistance.

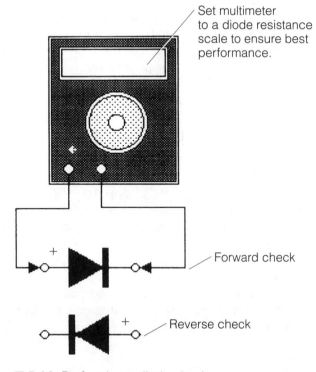

Set multimeter to a diode resistance scale to ensure best performance.

Forward check

Reverse check

■ **5-14** *Performing a diode check.*

A shorted diode will exhibit a very low resistance in the forward and reverse-bias directions. This indicates a shorted semiconductor junction. Be certain that at least one of the diode's two leads is removed from the circuit before testing. This will prevent its interconnections with other components from causing a faulty reading. An opened diode will exhibit very high resistance (usually infinity) in both its forward and reverse directions. In this case, the semiconductor junction is open-circuited. If you feel unsure how to interpret your measurements, test several other comparable diodes and compare readings.

Transistors can be checked in several ways. Some multimeters feature a built-in transistor checker that measures a transistor's gain (or hfe) directly. If your meter offers a transistor checker, insert your transistor into the test fixture on the meter's face in its correct lead orientation (emitter, base, and collector). Manufacturer's specifications can tell you whether a gain reading is correct for a particular part. A low (or zero) reading indicates a shorted transistor, while a high (or infinite) reading suggests an open-circuited transistor.

Your meter's diode checking feature can also be used to check a bipolar transistor's base-emitter and base-collector junctions as shown in figure 5-15. Each junction acts just like a diode junction. Test one junction at a time. Set your multimeter to its diode range, then place its test leads across the base-collector junction. If your transistor is npn, place the positive test lead at the base. This should forward-bias the base-collector junction and cause a normal amount of diode resistance. Reverse your test leads across the base-collector junction. It should now be reverse-biased and show infinite resistance. Repeat this procedure for the base-emitter junction.

If your transistor is pnp, your test lead placement must be reversed. For example, a forward-biased junction in an npn transistor is reverse-biased in a pnp transistor. You can refer to manufacturer's specification sheets to determine which leads in the transistor are the base, emitter, and collector. As a final check, measure the resistance from emitter to collector. You should read infinite resistance in both directions. Although this is not a diode junction, short circuits can develop during a transistor failure, which might not appear across normal junctions. Replace any diode with an open or shorted junction, or a short from emitter to collector.

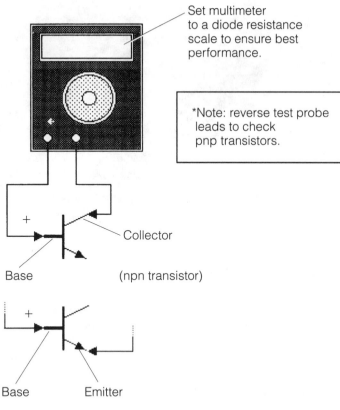

Set multimeter
to a diode resistance
scale to ensure best
performance.

*Note: reverse test probe
leads to check
pnp transistors.

Collector

Base   (npn transistor)

Base   Emitter

■ **5-15** *Performing a transistor check.*

Logic probes

The problem with most multimeters is that they do not relate well to digital logic circuits. A multimeter can certainly measure whether a logic voltage is on or off, but if that logic level changes quickly, a dc voltmeter function cannot track it properly. Logic probes provide a fast and easy means of detecting steady-state or alternating logic levels. Some logic probes can detect logic pulses faster than 50 MHz.

Logic probes are rather simple-looking devices, as shown in figure 5-16. A probe might be powered from its own internal battery, or from the circuit under test. Connect the probe's ground lead to a convenient circuit ground. If a probe is powered from the circuit under test, attach its power lead to a logic supply voltage in the circuit. A small panel on the probe's body holds several LED indicators and a switch that allows the probe to work with two common logic families (TTL and CMOS). You might find TTL and CMOS devices mixed into the same circuit, but one family will usually dominate.

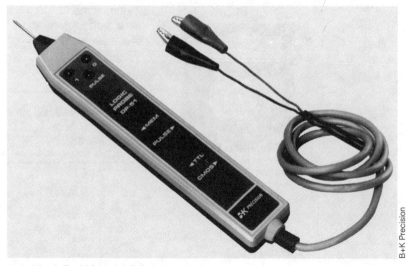

■ **5-16** *A B+K Model DP-51 logic probe.*

When the metal probe tip is touched to an IC lead, its logic state is displayed on one of the three LED indicators, as shown in Table 5-1. Typical choices are "logic 0," "logic 1," and "pulse" (or "clock"). Logic probes are most useful for troubleshooting working logic circuits where logic levels and clock signals must be determined quickly and accurately.

■ **Table 5-1 Typical logic probe display patterns.**

Input signal	Hi LED	Low LED	Pulse LED
Logic "1" (TTL or CMOS)	On	Off	Off
Logic "0" (TTL or CMOS)	Off	On	Off
Bad logic level or open circuit	Off	Off	Off
Square wave (<200 kHz)	On	On	Blink
Square wave (>200 kHz)	On/Off	On/Off	Blink
Narrow "high" pulse	Off	On/Off	Blink
Narrow "low" pulse	On/Off	Off	Blink

Oscilloscopes

Oscilloscopes offer a tremendous advantage over multimeters and logic probes. Instead of reading signals in terms of numbers or lighted indicators, an oscilloscope will show voltage versus time on a visual display. Not only can you observe ac and dc voltages, but it enables you to watch digital voltage levels, high-energy motor pulses, or other unusual signals occur in real time. If you have used

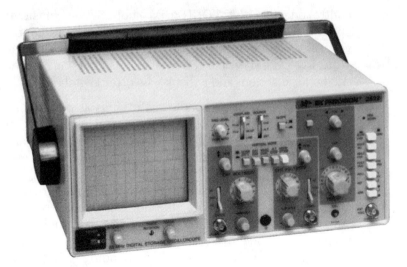

■ **5-17** *A B+K Model 2522 digital oscilloscope.* B+K Precision

an oscilloscope in the past, you know just how useful it can be. Oscilloscopes such as the one shown in figure 5-17 may appear somewhat overwhelming at first, but many of their operations work the same way regardless of what model you are working with.

Controls

In spite of the wide variations of features and complexity between models, most controls are common to the operation of every oscilloscope. Controls fall into four categories: horizontal (time base) signal controls, vertical (amplitude) signal controls, "housekeeping" controls, and optional (enhanced) controls.

Since an oscilloscope displays voltage versus time, adjusting either voltage or time settings will alter the display. Horizontal controls manipulate the left-to-right time appearance (sweep) of the voltage signal. Your oscilloscope's master time base is adjusted using a TIME/DIV knob or button. This sets the rate at which voltage signals are swept onto the screen. Smaller settings allow shorter events to be displayed more clearly, and vice versa. Remaining horizontal controls include a horizontal display mode selector, sweep trigger selection and sensitivity, trigger coupling selection, and trigger source selection. Your particular oscilloscope may offer additional controls.

An adjustment to an oscilloscope's voltage sensitivity will also alter your display. Vertical controls affect the deflection (up-to-down) appearance of your signal. An oscilloscope's vertical sensitivity is controlled with the VOLTS/DIV knob. When sensitivity is increased (VOLTS/DIV becomes smaller), signals will appear larger vertically. Reducing sensitivity will make signals appear smaller vertically. Other vertical controls include coupling selection, vertical mode selection, and a display inverter switch.

Housekeeping controls handle such things as oscilloscope power, trace intensity, graticule intensity, trace magnification, horizontal trace offset, vertical trace offset, and a trace finder—any control that affects the quality and visibility of a display. Your oscilloscope may have any number of optional controls depending on its cost and complexity, but cursor and storage controls are some of the most common. Many scopes offer horizontal and vertical on-screen cursors to aid in the evaluation of waveforms. Panel controls allow each cursor to be moved around the screen. The distance between cursors is then converted to a corresponding voltage, time, or frequency value, and that number is displayed on the screen in appropriate units. Storage oscilloscopes allow a screen display to be held right on-screen, or in memory within the scope to be recalled on demand.

Oscilloscope specifications

Oscilloscopes have a variety of important specifications that you should be familiar with when choosing and using an oscilloscope. The first specification to know is bandwidth. It represents the range of frequencies that the scope can work with. This does not necessarily mean that all signals within that bandwidth can be displayed accurately. Bandwidth is usually rated from dc to some maximum frequency (often in megahertz or MHz). For example, an inexpensive oscilloscope might cover dc to 20 MHz, while a more expensive model might work up to 150 MHz or more. Good bandwidth is very expensive, more so than any other feature.

The vertical deflection (or vertical sensitivity) is another important specification. It is listed as the minimum to maximum volts/DIV settings that are offered, and the number of steps that are available within that range. A typical model might provide vertical sensitivity from 5 mV/DIV to 5 V/DIV broken down into 10 steps. A time base (or sweep range) specification represents the minimum to maximum time base rates that an oscilloscope can produce, and the number of increments that are available. A range of 0.1 μs/DIV to 0.2 s/DIV in 20 steps is not unusual. You will typically find a greater number of time base increments than sensitivity increments.

There is a maximum voltage input that can be applied to an oscilloscope input. A maximum voltage input of 400 V (dc or peak ac) is common for most basic models, but more sophisticated models can accept inputs greater than 1,000 V. An oscilloscope's input will present a load to whatever circuit or component it is placed across. This is called *input impedance*, and is usually expressed as a value of resistance and capacitance. To guarantee proper operation over a model's entire bandwidth, select a probe with load characteristics similar to those of the oscilloscope. Most oscilloscopes have an input impedance of 1 MΩ with 10 to 50 pF of capacitance.

The accuracy of an oscilloscope represents the vertical and horizontal accuracy of the final CRT display. In general, oscilloscopes are not as accurate as dedicated voltage or frequency meters. A typical model can provide ±3% accuracy, so a 1-V measurement can be displayed between 0.97 V to 1.03 V. Keep in mind that this does not consider human errors in reading the CRT marks (or graticule). However, because the strength of an oscilloscope is its ability to graphically display complex and fast signals, 3% accuracy is usually adequate.

Oscilloscope start-up procedures

Before you begin taking measurements, a clear, stable trace must be obtained (if not already visible). If a trace is not visible, make sure that any CRT screen storage modes are off, and that intensity is turned up at least 50%. Set triggering to its "automatic" mode and adjust the horizontal and vertical offset controls to the center of their ranges. Be sure to select an "internal" trigger source, then adjust the trigger "level" until a trace is visible. Vary your vertical offset if necessary to center the trace across the CRT.

If a trace is not yet visible, use the "beam finder" to reveal its location. A beam finder simply compresses the vertical and horizontal ranges. This forces a trace onto the display and gives you a rough idea of its relative position. After your trace is moved into position, adjust your focus and intensity controls to obtain a crisp, sharp trace. Keep intensity as low as possible to improve display accuracy, and preserve phosphors in the CRT.

Your oscilloscope probe must be calibrated before use. Calibration is a quick and straightforward operation that requires only a low-amplitude, low-frequency square wave. Many models have a built-in "calibration" signal generator (a 1-kHz, 300-mV square wave with a 50% duty cycle). Attach your probe to the desired input jack, then place it across the calibration signal. Adjust your horizontal (Time/DIV) and vertical (Volts/DIV) controls so that one or two complete cycles are clearly shown on the CRT.

Observe the characteristics of your test signal, as shown in figure 5-18. If the square wave's corners appear rounded, there might not be enough probe capacitance (Cprobe). Spiked square wave corners suggest too much capacitance in the probe. Either way, the scope and probe are not matched properly. You must adjust the probe capacitance to establish a good electrical match; otherwise, signal distortion might result. Slowly adjust the variable capacitance on your probe until the corners of your calibration signal are as square as possible. If you cannot achieve a clean square wave, try a different probe.

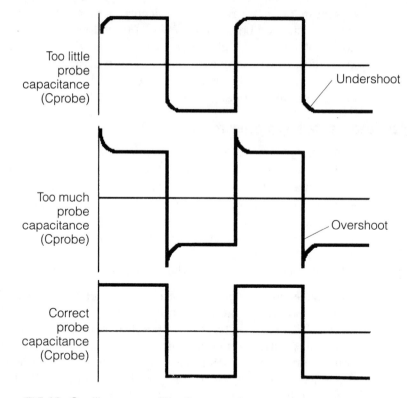

■ 5-18 *Oscilloscope calibration waveforms.*

Voltage measurements

The first step in any voltage measurement is to set your normal trace (or *baseline*) where you want it. Normally, a baseline is placed along the center of the graticule during start-up, but it can be placed anywhere so long as it is visible. To establish a baseline, switch your input coupling control to its "ground" position. This disconnects the input signal and grounds the channel to ensure a zero reading. Adjust the vertical offset control to shift the baseline wherever the zero reading is to be. If you have no particular preference, simply center it in the CRT.

To measure dc, set your input coupling switch to its "dc" position, then adjust the Volts/DIV control to provide the desired amount of sensitivity. If you are unsure about what sensitivity is appropriate, start with a very low sensitivity (a large Volts/DIV setting), then carefully increase the sensitivity (reduce the Volts/DIV setting) after your signal is connected. This prevents a trace from simply jumping off the display when an unknown signal is first applied. If your signal does happen to leave the visible display, you could reduce sensitivity (increase the Volts/DIV setting) to make the trace visible again.

For example, suppose you were measuring a +5-Vdc power supply output. If Volts/DIV is set to 5 V/DIV, each major vertical division represents 5 V, so your +5-Vdc signal should appear one full division above your baseline (5 V/DIV × 1 division = 5 V), as shown in figure 5-19. At a Volts/DIV setting of 2 V/DIV, the same +5-V signal would now appear 2.5 divisions above your baseline (2 V/DIV × 2.5 divisions = 5 V). If your input signal were a negative voltage, the trace would appear below the baseline, but it would read the same way.

The ac signals can also be read directly from the oscilloscope. Switch your input coupling control to the "ac" position, then set a baseline just as you would for dc measurements. If you are unsure about how to set the vertical sensitivity, start with a low sensitivity (a large Volts/DIV setting), then slowly increase the sensitivity (reduce the Volts/DIV scale) after a signal is connected. Keep in mind that ac voltage measurements on an oscilloscope will not match ac voltage readings on a multimeter. An oscilloscope displays instantaneous peak values for a waveform, while ac voltmeters measure in terms of rms (root mean square) values. To convert an rms value to peak, multiply rms × 1.414. To convert a peak voltage reading to rms, divide peak/1.414.

123

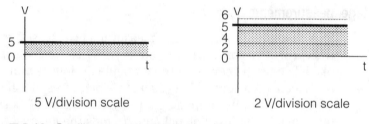

5 V/division scale   2 V/division scale

■ **5-19** *Oscilloscope dc voltage readings.*

When actually measuring an ac signal, it might be necessary to adjust the oscilloscope's trigger level control to obtain a stable (still) trace. As figure 5-20 shows, signal voltages can be measured directly from the display. For example, the sinusoidal waveform of figure 5-20 varies from −10 to +10 V. If scope sensitivity were set to 5 V/DIV, its peaks would be two divisions above and below the baseline. Because this is a peak measurement, an ac voltmeter would show the signal as peak/1.414, or [10/1.414] 7.07 V_{rms}.

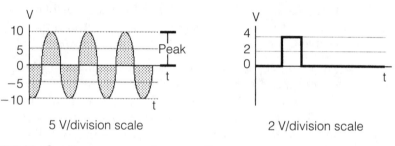

5 V/division scale   2 V/division scale

■ **5-20** *Oscilloscope ac voltage readings.*

Time & frequency measurements

An oscilloscope is perfect for measuring critical signal parameters such as pulse width, duty cycle, and frequency. It is the horizontal sensitivity (Time/DIV) control that comes into play with time and frequency measurements. Before making any measurements, you must first obtain a clear baseline as you would for voltage measurements. When a baseline is established and a signal is connected, adjust the Time/DIV control to display one or two complete cycles of the signal.

Figure 5-21 illustrates two typical period measurements. With Volts/DIV set to 5 ms/DIV, the sinusoidal waveform repeats every two divisions. This represents a period of [5 ms/DIV × 2 divisions] 10 ms. Because frequency is the simple reciprocal of the period,

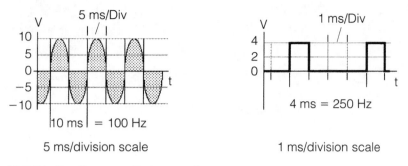

■ 5-21 *Oscilloscope timing readings.*

frequency can be calculated directly from period. A period of 10 ms would represent a frequency of (⅒ ms) 100 Hz. This also works for square waves and other nonsinusoidal waveforms. The square wave in figure 5-21 repeats every four divisions. At a TIME/DIV setting of 1 ms/DIV, its period would be 4 ms. This corresponds to a frequency of (¼ ms) 250 Hz.

Instead of measuring the entire period of a pulse cycle, you can also read the time between any two points of interest. For the square wave in figure 5-21, you could read its pulse width to be 1 ms. You could also read the low portion of the cycle as a pulse width of 3 ms (added together for its total period of 4 ms). A signal's duty cycle is simply the ratio of a signal's ON time to its total period expressed as a percentage. For example, a square wave on for 2 ms and off for 2 ms would have a duty cycle of [2 ms/(2 ms + 2 ms) × 100%] 50%. For an on time of 1 ms and an off time of 3 ms, its duty cycle would be [1 ms /(1 ms + 3 ms) × 100%] 25%. Use caution in duty cycle measurements.

125

Troubleshooting guidelines

6

ELECTRONIC TROUBLESHOOTING IS A STRANGE PURSUIT, an activity that falls somewhere between an art and a science. Success is typically influenced by experience, access to technical information, and the availability of replacement parts and sub-assemblies. However, troubleshooting success also depends on a thorough, logical, troubleshooting approach (along with a selection of useful utilities such as PRINTERS). So often, novice troubleshooters are overwhelmed by the perceived complexity of their printer, but as you have seen in earlier chapters, the printer is really just a series of simple assemblies. This chapter shows you how to approach any troubleshooting situation, locate technical data, and present a series of printer service guidelines that can ease your work.

127

The troubleshooting cycle

Regardless of how complex your particular circuit or system might be, a reliable troubleshooting procedure can be broken down into four basic steps, as shown in figure 6-1:

1. Define your symptoms.
2. Identify and isolate the potential source (or location) of the problem.
3. Replace or repair the suspected component or subassembly.
4. Retest the system thoroughly to be sure that you have solved the problem.

If you have not solved the problem, start again from step one. This is a universal procedure that you can use for any sort of trouble-shooting, not just for printers.

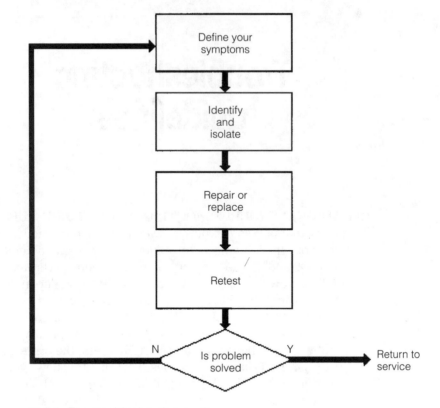

■ **6-1** *The troubleshooting cycle.*

Define your symptoms

Sooner or later, your printer is going to break down. It might be as simple as a sticky gear, or as complicated as an extensive electronic failure. However, before you open the toolbox, you must have a firm understanding of the symptoms. It is not enough to simply say "it's broken." Think about the symptoms carefully. Ask yourself what is (or is not) happening. Consider when it is happening. If this is a new installation, ask yourself if the computer is set up properly, or if the right cables are being used, or if DIP switches are set up correctly. If you have used your printer for a while, do you remember the last time you cleaned and lubricated it? Is the print just light or is it completely missing? Is the paper or print head advancing freely? By recognizing and understanding your symptoms from the start, it can be much easier to trace a problem to the appropriate subsection or components. Of course, you should reference the many different symptom descriptions used throughout this book.

Use your senses and write down as many symptoms as you can, whatever you smell, see, or hear. This might sound tedious now, but when you are up to your elbows in repair work, a written record of symptoms and circumstances will keep you focused on the task at hand. This is even more important if you are a novice troubleshooter.

Identify & isolate

Before you try to isolate a problem within the printer, you must first be sure that the printer is actually causing the problem in the first place. In many circumstances, this will be fairly obvious, but there are some situations that might appear ambiguous (e.g., no print with power on, erratic printing, missing characters, and so on). Always remember that a printer itself is just a subsection of a larger system made up of your computer, printer, and interconnecting cable. It is possible (especially in new installations) that a computer failure, software incompatibility, or cable problem might be causing your symptoms. Chapter 12 illustrates many of the problems you will encounter under Windows and Windows 95.

This is an easy application of the universal troubleshooting procedure. Once you have carefully identified your symptoms, isolate the printer. This can be done by removing it from its communication cable. You can replace it by testing it on another computer system with a known-good printer. A friend or colleague might let you test your printer on their computer system. Because various computers might be set up to communicate in different fashions, you might have to alter the internal settings of your printer to match those of the known-good one. If your printer exhibits the same symptoms on another computer, there is an excellent chance that the problem is within the printer. You can then proceed with specific troubleshooting procedures outlined in this book. If, however, those symptoms disappear and your printer works properly, you should suspect a problem in your computer, software configuration, dip switch settings, or interconnecting cable (i.e., the printer "setup").

Another test is to try a known-good printer on your computer system. If another printer works properly, it verifies that your computer, software configuration, and cable are intact. If a known-good printer fails to work on your system, check the computer's communication interface, software settings, and interconnecting cable. This can be done in addition to testing your questionable printer on another system.

When you are confident that the printer is at fault, you can begin to identify any possible problem areas. Start at the subassembly level. You might recall from Chapter 1 that a printer consists of five major subsections (depending on the particular type of printer technology employed): a paper feed system, a print head (or Image Formation System), a print head carriage system (not used in EP printers), a power supply, and an electronic control unit. Your printer's fault will be located in at least one of these five subsections. The troubleshooting procedures shown throughout this book will aid you in deciding which subassemblies are at fault. Once you have identified a potential problem, you can begin the actual repair process (and sometimes even track the fault to a component level).

Repair or replace

Once you have an understanding of what is wrong and where to look, you might begin the actual repair procedures that you feel will correct the symptoms. Some procedures require only simple adjustments or cleaning, while others might require the exchange of electrical or mechanical parts, but all procedures are important and should be followed very carefully.

Parts are usually classified as components or subassemblies. A component part is the smallest possible individual part that you can work with. Components can serve many different purposes in a printer. Resistors, capacitors, gears, belts, motors, and integrated circuits are just a few types of component parts. As a general rule, components contain no serviceable parts; they must be replaced. A subassembly is composed of a variety of individual components. Unlike components, a complete subassembly serves a single, specific purpose in a printer, but it can be repaired by locating and replacing any faulty components. It is certainly acceptable to repair a defective subassembly simply by installing a new one in the printer. They are generally easier to obtain, but complete subassemblies can be very expensive (compared to the original cost of your printer).

Replacement electronic components can be purchased from several different sources, but keep in mind that many mechanical parts, assemblies, and fittings might only be available through the manufacturer or distributor. Many mail-order companies will send you their complete catalogs or product listings at your request. Going to the manufacturer for subassemblies or components is often somewhat of a calculated risk; they might do business only with their affiliated service centers, or refuse to sell parts directly to consumers. If you find a manufacturer willing to sell you parts, you must often know the manufacturer's exact part number or code. Remember that many manufacturers are not equipped to deal with consumers directly, so be patient and be prepared to make several different calls.

During a repair, you might reach a roadblock that requires you to leave the printer for a day or two (or longer). Make it a point to reassemble the printer as much as possible before leaving it. Place any loose parts into plastic bags and seal them shut. Reassembly will prevent a playful pet, curious child, or well-meaning spouse from accidentally misplacing or discarding parts while the printer sits on your workbench. This is much more important if your workspace is in a well-traveled area. You will also not forget how to put it back together later on.

Retest

When your repair is complete, the printer must be carefully reassembled and tested before connecting it to a computer. Run a thorough suite of tests with PRINTERS to check the printer's operation. PRINTERS tests the print head, carriage, paper advance, power supply, and much of the ECU. If symptoms persist, you will have to reevaluate them and narrow the problem to another part. If normal operation is restored (or significantly improved), the printer might be returned to service. Do not be discouraged if the printer still malfunctions. Simply walk away, clear your head, and start again by defining your symptoms. Never continue with a repair if you are tired or frustrated; tomorrow is another day. You should also realize that there might be more than one bad component to deal with. Remember that a computer printer is just a collection of assemblies, and each assembly is a collection of components. Normally, everything works together, but when one part fails, it might cause one or more interconnected parts to fail as well. Be prepared to make several repair attempts before the printer is repaired completely.

Gathering technical data

Information is perhaps your most valuable tool in tackling a printer repair. Just how much information you actually need will depend on the particular problems you are facing. Simple adjustments and cleaning might be accomplished with little or no technical information (except for your own observations and common sense judgment), but complex electronic troubleshooting might require a complete set of schematics and parts lists. More intricate repair procedures generally need more comprehensive technical literature. Luckily, there are some avenues of information.

The user's manual

A user's manual is always a good place for basic printer information. A user's manual describes how to set up and operate the printer, outlines its important specifications and communication interface, and points out its major assemblies and controls. If you are unfamiliar with the printer or unaccustomed to changing its configurations, a user's manual can keep you out of trouble. Some user's manuals also present a short selection of very basic troubleshooting and maintenance procedures, but these are almost always related to the printer's setup and operation, not to its internal circuitry or mechanics.

Fax back information

Along with telephone technical support, many manufacturers are providing setup instructions, user details, and frequently asked questions (FAQs) through a "fax back" service. By calling into an automated request system, you can order one or more documents to be returned to your fax machine. Fax back resources offer some unique advantages such as the ability to add or correct documents quickly, as well as promote upgrades, suggest bug fixes, and answer user questions.

On-line resources

The proliferation and popularity of modems has resulted in the tremendous growth of on-line services, which can be priceless sources for sales information and technical support. Where the demand for "live" telephone support results in long, wasteful periods on hold, on-line services can provide immediate and interactive answers. On-line resources also offer an effective media for providing new software drivers, patches, updates, documents, forms, electronic catalogs, and a whole host of other resources that can be downloaded right to your PC. For the purpose of this book, on-line resources embrace three areas: private services, commercial services, and Internet services.

The private service is basically a BBS (Bulletin Board System). A BBS is basically a PC (or several PCs networked together) fitted with one or more modems. Although the individual user runs a simple communication package (such as Smartcom for Windows or Procomm Plus), the PCs on the BBS end run special "BBS applications" that allow users to call the BBS and interact with it without human attention. As a consequence, the BBS is largely an automated system (except for periodic file updates). Many manufacturers employ a BBS to supply software and electronic documents, technical support notes, and exchange support messages (e-mail). Try TechNet BBS at 508-366-7683, which specializes in PC diagnostics and utility shareware.

While most BBS facilities are relatively easy to use and maintain, the direct modem-to-modem connection usually means that you are making a toll call. Although the value of the information available often justifies the expense, a 15- to 45-minute call can still result in a serious cost. The commercial service is a national or international network with "nodes" in major cities or key areas. This allows users to make less costly (often local) calls. The architecture of a network also provides additional access points, so more users can be accommodated at the same time. This approach has given rise to major services like CompuServe and America Online. Manufacturers frequently open forums on commercial services. Like bulletin boards, however, commercial services are generally limited to the distribution of software and electronic documents, as well as the exchange of e-mail, though live chats and other features are appearing.

The Internet has grown significantly as an on-line resource for PC technicians and enthusiasts. Rather than a single network, the Internet is actually a "network of networks" all working together. With Internet service providers popping up all over the country (and around the world), people and organizations are embracing the Internet in record numbers. From a practical standpoint, the Internet serves three major functions: the exchange of e-mail, transferring files, and web browsing. Many manufacturers have added direct connections to the Internet, and technical support questions can often be answered via e-mail. The file transfer protocol (ftp) allows companies on the Internet to provide file libraries. Though ftp sites are often private (i.e., a corporate network server), many more are open and can be accessed anonymously. An ftp transfer lacks the refined look of a BBS upload or download, but the principles are the same. New (beta) software, patches, updates, electronic documents, and other files can be obtained through Internet ftp.

The Internet's World Wide Web (WWW) is a relatively new development for the Internet that allows users to browse hypertext documents right on-line. Product information (and graphics), technical data and FAQs, news briefs, and other kinds of information can be accessed through the WWW. In fact, a well-designed web site can allow you to download files and send e-mail right from the web site itself, making the WWW an efficient, highly integrated resource. Finally, the WWW provides the capacity for forms, which supports ordering parts, materials, subscriptions, and so on, right on-line. If you find yourself on the Internet, try the Dynamic Learning Systems site at http://www.dlspubs.com/home.htm. Finally, the Internet supports an incredible selection of newsgroups and mailing lists (many of which are PC-related) that can keep you in touch with new troubleshooting ideas and techniques.

Technical data & schematics

Technical information down to a component level can be obtained from data sheets published by the component's manufacturer. For example, if you want a pin diagram of an IC manufactured by Motorola, you could refer to a Motorola data book containing information on that particular component. This will tell you what the part is, what it does, what purpose each pin performs, and what its electrical specifications are. Some suppliers sell a selection of up-to-date component data books. Although data books bear no direct relationship to your particular printer, they can give you tremendous insight on the purpose and functions of individual components.

If you intend to jump into a detailed electrical repair, a complete set of schematics can quickly and efficiently guide you through even the most complicated printer. If you are working on an older printer, there might be a complete documentation package published by Howard W. Sams & Co. Their comprehensive Sams Photofact series has long been an indispensable part of the electronic service industry.

Your printer's manufacturer can be a key source of technical information, but not all manufacturers are willing to sell technical information to individuals or private organizations. Start by checking directly with the manufacturer. Their phone number is usually listed somewhere in the user's manual. You can try to contact their technical literature, parts order, or service departments to order a service or repair manual. Service information can be expensive (as much as $50 or more), so be prepared.

134

If you cannot get satisfaction from the manufacturer, check with a local dealer (not a retail store) that sells for that manufacturer. The yellow pages of your local telephone book can give you good leads. A reputable dealer has access to parts and technical information that you do not. Finally, try contacting a service organization that repairs your type of printer. They might be willing to order a copy for you.

Electricity hazards

No matter how harmless your printer might appear, always remember that potential shock hazards do exist. Once the printer is disassembled, there can be several locations where live ac voltage is exposed and easily accessible. Domestic U.S. electronic equipment operates from 120 Vac at 60 Hz. Many European countries use 240 Vac at 50 Hz. When this kind of voltage potential establishes a path through your body, it causes a flow of current that might be large enough to stop your heart. Because it only takes about 100 mA to trigger a cardiac arrest, and a typical printer fuse is rated for 1 or 2 A, fuses and circuit breakers will not protect you.

Understanding power supply dangers

It is your skin's resistance that limits the flow of current through the body. Ohm's law states that for any voltage, current flow increases as resistance drops (and vice versa). Dry skin exhibits a high resistance of several hundred thousand ohms, while moist, cut, or wet skin can drop to only several hundred ohms. This means that even comparatively low voltages can produce a shock if your skin resistance is low enough. Some examples will help to demonstrate this action.

Suppose a worker's hands come across a live 120-Vac circuit. If their skin is dry (say 120 kΩ), they would experience an electrical shock of [120 Vac/120,000 Ω] 1 mA. The result would be harmless, probably a brief, tingling sensation. After a hard day's work, perspiration could decrease skin resistance (perhaps 12 kΩ). This would allow a far more substantial shock of [120 Vac/12,000 Ω] 10 mA. At that level, the shock can paralyze the victim and make it difficult or impossible to let go of the "live" conductors. A burn (perhaps serious) could result at the points of contact, but it probably would not be fatal. Consider a worker whose hands or clothing are wet. Their effective skin resistance can drop very low (1.2 kΩ for example). At 120 V, the resulting shock of [120 Vac/1,200 Ω] 100 mA could be fatal without immediate CPR.

Electrophotographic printers use power supplies that are every bit as dangerous. Most can produce voltages easily exceeding 2,000 Vdc. Based on the examples just described, even dry skin at 200,000 Ω could receive a paralyzing shock. Not only is there a risk of injury, but normal test probes (such as multimeter test leads) only provide insulation to about 600 V. Testing high voltages with standard test leads could electrocute you right through the lead's insulation! Avoid taking direct measurements on high-voltage power supplies whenever possible. Fortunately, high-voltage supplies are rarely capable of supplying enough current to inflict real injury, but extreme caution is recommended.

The dangers of ozone

Electrophotographic printers also pose another, more subtle, danger for technicians: the presence of ozone. High-voltage operation ionizes the nearby air. This results in the development of ozone gas. While ozone is rarely dangerous in the small concentrations found with EP printers, prolonged exposure in confined areas can result in respiratory irritation. As a rule, run EP printers in well-ventilated areas, and be sure to replace exhausted ozone filters.

Steps for protection

Take the following steps to protect yourself from injury:

1. Keep the printer unplugged (not just turned off) as much as possible during disassembly and repair. When you must perform a service procedure that requires power to be applied, plug in the printer just long enough to perform your procedure, then unplug it again. This makes the printer safer for you, as well as a spouse or child that might happen by.

2. Whenever you must work on a power supply, try to wear rubber gloves. Gloves will insulate your hands just like insulation on a wire. You might think that rubber gloves are inconvenient and uncomfortable, but they are far better than the inconvenience and discomfort of an electric shock. Make it a point to wear a long-sleeved shirt with sleeves rolled down to insulate your forearms.

3. If rubber gloves are absolutely out of the question for one reason or another, remove all metal jewelry and work with one hand behind your back. The metals in jewelry are excellent conductors. Should your ring or watchband hook onto a "live" ac line, it can conduct current directly to your skin. By

keeping one hand behind your back, you cannot grasp both ends of a live ac line to complete a strong current path through your heart.

4. Inspect your test probes carefully before testing high-voltage circuitry. Standard "off-the-shelf" probes do not necessarily have the insulating properties (or dielectric strength) to protect you. If you must make powered tests on a high-voltage circuit, be sure to use test leads that offer sufficient protection.

5. Work dry! Do not work with wet hands or clothing. Do not work in wet or damp environments. Make sure that nearby fire extinguishing equipment is suitable for electrical fires.

6. Treat electricity with tremendous respect. Whenever electronic circuitry is exposed (especially power supply circuitry), a shock hazard does exist. Remember that it is the flow of current through your body, not the voltage potential, that can injure you. Insulate yourself as much as possible from any exposed wiring.

Static electricity

Another troubleshooting hazard can come from static voltages accumulated on your body or tools. If you have ever walked across a carpeted floor on a cold, dry, winter day, you have probably experienced the effects of electrostatic discharge (ESD) while reaching for a metal object. Under the right conditions, your body can accumulate static charge potentials greater than 20,000 V. When you provide a conductive path for electrons to flow, that built-up charge rushes away from your body at the point closest to the object. The result is often a brief, stinging shock. Such a jolt can be startling and annoying, but it is generally harmless to people. Semiconductor devices, however, are highly susceptible to damage from ESD while you handle or replace circuit boards and components. This section will introduce you to static electricity, and show you how to prevent ESD damage during your repairs.

Static formation

When two dissimilar materials are rubbed together, the force of friction causes electrons to move from one material to another. The excess (or lack) of electrons cause a charge to develop on each material. Because electrons are not flowing, there is no current, so the charge is said to be static. However, the charge does exhibit a voltage potential. As materials continue to rub together, their charges increase, sometimes to potentials of thousands of volts.

In a human, static charges can be developed by normal everyday activities such as walking on a carpet. Friction between the carpet and shoe soles cause opposing charges to be developed. The shoe's charge induces an equal (but opposite) charge in your body, which acts as a capacitor. Sliding across a vinyl car seat, pulling a sweater on or off, or taking clothes out of a dryer are just some of the ways that a static charge can appear in the body.

Device damage

ESD poses a serious threat to many modern semiconductor devices. Huge static voltages that build up in the environment (or in your body) can find their way into all types of advanced ICs. If that happens, the result for the component can be catastrophic. Static discharge can damage bipolar transistors, transistor-transistor logic (TTL), emitter-coupled logic (ECL), operational amplifiers, SCRs, and junction field-effect transistors (JFETs), but certainly the most susceptible components are those fabricated using metal-oxide semiconductor (MOS) technology.

MOS devices (PMOS, NMOS, HMOS, CMOS, etc.) have become the cornerstone of high-performance ICs such as memories, high-speed logic, microprocessors, and other advanced digital components. It offers high-speed, high component density, and low power consumption. Typical MOS ICs can easily cram over one million transistors onto a single IC. Every part of the transistor must be made continually smaller to keep pace with the demands for higher levels of integration. As each part of the transistor shrinks, however, breakdown voltages drop, and ESD damage problems escalate.

A typical MOS transistor breakdown is illustrated in figure 6-2. Notice the areas of positive and negative semiconductor material that forms its three terminals: source, gate, and drain. The gate is isolated from other parts of the transistor by a thin film of silicon dioxide (sometimes called the oxide layer). Unfortunately, this layer is extremely thin, and it can be overcome easily by high voltages like those from static discharges. Once this happens, the oxide layer is punctured. This renders the entire transistor (and the whole IC) defective.

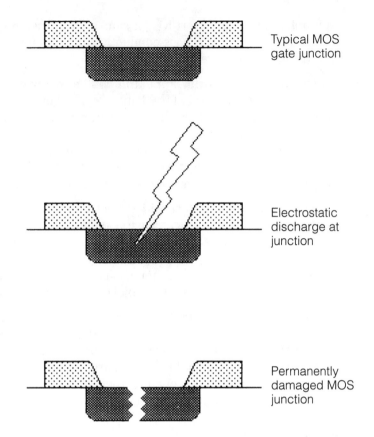

Typical MOS gate junction

Electrostatic discharge at junction

Permanently damaged MOS junction

■ **6-2** *Action of electrostatic discharge on a MOS gate.*

Controlling static electricity

Do not underestimate the importance of static control during your printer repairs. Without realizing it, you could destroy a new IC or circuit assembly before you even have a chance to install it, and you would never even know that static damage has occurred. All it takes is the careless touch of a charged hand, tool, or piece of clothing. Take the necessary steps to ensure the safe handling and replacement of your sensitive (and expensive) electronics.

One way to control static is to keep charges away from boards and ICs. This is often accomplished as part of a device's packaging and shipping container. ICs are usually packed in a specially made conductive foam. Carbon granules are compounded right into the polyethylene foam to achieve conductivity (about 3,000 Ω/cm). Foam prevents bending of IC leads, absorbs vibrations and shocks, and its conductivity helps to keep every IC lead at the same potential (also called *equipotential bonding*). Conductive foam is reusable, so you can insert ICs for safekeeping, then remove them as needed. You can purchase conductive foam from just about any electronics retail store.

Circuit boards are normally held in conductive plastic bags that dissipate static charges before damage can occur. Anti-static bags are made up of different layers, each with varying amounts of conductivity. The bag acts as a "Faraday cage" for the device it contains. Electrons from an ESD will dissipate along the bag's surface layers, instead of passing through the bag to its contents. Bags are also available through many electronics retail outlets.

Whenever you work with sensitive electronics, it is a good idea to dissipate charges that might have accumulated on your body. A conductive fabric wrist strap that is soundly connected to an earth ground will bleed away all charges from your skin. Avoid grabbing hold of a ground directly. Although this will discharge you, it can result in a sizable jolt if you have picked up a large charge.

Remember to make careful use of your static controls. Keep ICs and circuit boards in their anti-static containers at all times. Never place parts onto synthetic materials (such as plastic cabinets or fabric coverings) that could hold a charge. Handle static-sensitive parts carefully and avoid touching their metal pins if possible. Be sure to use a wrist strap connected to a reliable earth ground.

Other EP printer hazards

EP printer designers have taken significant time and effort to develop safeguards in and around the printer. When disassembling and working inside the printer, however, there are some inherent risks. You have already seen the risks of electrocution and static damage, but there are two other attributes of the printer that can also cause injury: the laser and the fusing assembly.

Scanning laser

The lasers used in today's laser printers are typically low-power laser diodes. Most laser diodes generate well under 1 W of light energy, hardly enough to burn skin. Still, laser light poses a potential danger to your eyes (especially with regular or prolonged exposure). As a result, you should avoid exposing yourself to the laser beam. Fortunately, virtually all modern laser printers incorporate the laser into a sealed laser/scanning (L/S) assembly. The L/S assembly is interlocked with mechanical shutters that cut off the laser beam while the printer enclosure is opened. Do not tamper with or defeat the safety interlocks on the L/S assembly.

Fusing assembly

In order to melt toner, a fusing roller must reach at least 180° C. This temperature is hot enough to cook meat, even you. After running a laser/LED printer for any period of time, you should always unplug the printer and allow at least 15 minutes for the fusing assembly to cool before attempting service, even if your service objectives have nothing to do with the fusing assembly. This waiting period also allows an opportunity for the dc and high-voltage power supplies to discharge somewhat.

Ozone hazards

When the surrounding air is ionized by high-voltage electricity, air is broken down into ozone gas. Unfortunately, ozone is not breathable, and can actually cause respiratory distress in high concentrations. EP printers try to reduce the amount of ozone generated by the printer by exhausting the air through a catalytic filter, which neutralizes the ozone. When a technician works on an opened printer, however, chances are that ozone will escape into the surrounding air. To prevent a buildup of ozone in the work area, try running the EP printer closed as much as possible, and try to work in a well-ventilated area, which can exhaust the ozone (especially in the winter when doors and windows are typically sealed).

Disassembly hints

Sooner or later, you will have to disassemble your printer to some extent in order to perform your repair. While the actual process of disassembly and reassembly is usually pretty straightforward, there are some important points for you to keep in mind during your procedures.

Housing disassembly

Most printer enclosures are designed in two halves, as shown in the exploded diagram of figure 6-3. By removing the top cover (marked 1-4-1), major subassemblies should be exposed. Removing the top cover, however, is not always as easy as it might seem. Examine your enclosure very carefully before beginning the disassembly. Some enclosures are held together with simple screws in an obvious, easily accessible fashion. Other types of enclosures use unusual screw patterns and types, such as spline or torx. They might also incorporate cleverly hidden internal clips that latch the enclosures together and provide a "seamless" appearance. Seam-

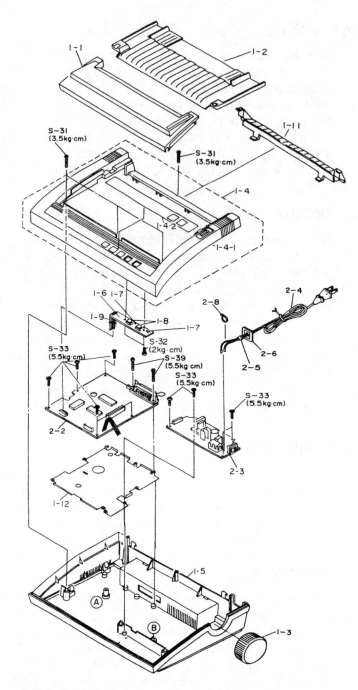

■ 6-3 *Exploded diagram of a dot-matrix printer.*
Tandy Corporation

less housings might need special tools to disengage these internal latches before housings can be separated. You might like to get another person's assistance when disassembling this type of housing.

Note: Electrical safety is critically important whenever you are working with electronic circuitry, so be sure to unplug the printer before starting any work. Plug the printer in only long enough to follow your particular troubleshooting and testing procedures, then unplug it again.

Electromechanical disassembly

Once the external cover has been removed, you can examine the other key subassemblies, such as the ECU (2-2), the mechanical assembly (not shown in figure 6-3), the control panel (1-6), and so on. Before attempting to disassemble the printer any further, note the location of each subassembly carefully. When proceeding with your disassembly, pay particular attention to electrical connectors and mechanical part locations and alignments.

Printers contain a wide array of electrical connectors handling everything from ac line voltage to print head signals. During disassembly, you might have to remove one or more connectors to free a circuit board or other subassembly. Never remove a connector by yanking on its wires; many connector shells use keys or latches to hold them in place. Always remove a connector by holding its shell. Take careful note of each connector's location and orientation. Some connectors are keyed so they can only be reinserted in their proper orientation, but other types of connectors might not be this foolproof.

Take careful note of mechanical parts as well, especially when you must disassemble complex drive trains of gears or pulleys. It will help you tremendously when it comes time to reassemble the system. Mark your parts before disassembly with an indelible felt-tip marker. Feel free to use any kind of markings that are clear to you, but marks should show how each part is mounted in relation to its adjacent parts. Figure 6-4 is just one simple example.

Reassembly hints

The reassembly process can offer even more challenges than disassembly. Although most printers are rather straightforward in their construction, putting those pieces back together in the right order can prove difficult unless you pay close attention. Soldering, connector replacement, and shielding are the most important concerns.

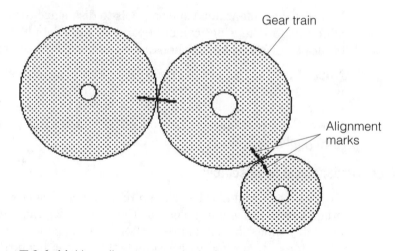

Gear train

Alignment marks

■ **6-4** *Making alignment marks before disassembly.*

Whenever you must replace ICs on a through-hole printed circuit board, always solder an IC holder in its place, then plug the replacement IC into its holder. Printed circuit boards (especially complex boards) are very delicate. Printed traces can be damaged by excessive or repeated heating. If you install an IC holder in its place, you will never have to desolder those points again. In order to replace that IC in the future, just unplug it and install a new one.

Next, always double-check your connector locations and orientations before reapplying printer power. If a connector is engaged backward or is skipping pins, your circuits can be seriously damaged. If you have made orientation marks on the connectors before disassembly, they should be a snap to install properly. Further, see that no cable is crimped or crushed between two or more assemblies that can damage the cable.

Finally, metal shields or shrouds are often added to limit RF (radio-frequency) interference between circuits. Switching-regulated power supplies and high-speed devices, such as microprocessors, are often shielded thoroughly. This prevents noise generated in one circuit from causing false signals in another circuit. If you have ever seen or heard radio or television reception in close proximity to a computer, then you have probably witnessed the effects of RF noise. Because a printer uses many of the same electronic components that a computer does, it too can generate noise. Be certain that all noise shielding is installed and secured properly. Guards (metal or plastic) can also be added to protect physical parts, such as drive trains. Be certain that any protective covers are replaced.

Conventional print head service techniques

7

A PRINT HEAD IS THAT PART OF A PRINTER THAT DELIVERS a permanent image to the page surface. The image might be a letter, number, or symbol (in the case of a character printer). But it might also be a pattern of dots that come together to form any type of characters or graphic images. As discussed in Chapter 3, there are three conventional technologies used to deliver permanent images: impact, thermal, and ink jet. The remainder of the printer, from its ECU to its paper feed system, is designed to accommodate the particular print head technology (figure 7-1). This chapter identifies and offers solutions for common print head problems. Chapter 11 is entirely devoted to EP printer service.

145

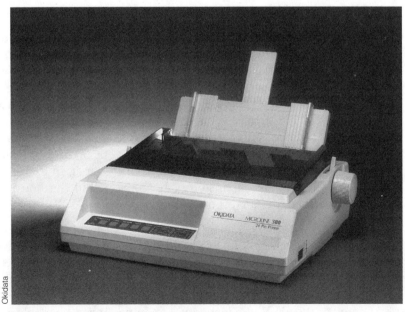

Okidata

■ **7-1** *An Okidata 24-pin impact printer.*

Dot-matrix impact print heads

A complete dot-matrix impact (DMI) printer assembly is illustrated in figure 7-2. Individual print wires are assembled into a die cast metal housing (the print head itself is marked 4-2). There might be 7, 9, or 24 print wires in the head depending on the sophistication of the particular printer. The print head is mounted on a carriage (shown as 8-2) that is carried back and forth along a rail (marked 7-3) by a belt (marked 8-1), which is driven by a motor (shown as 9-2). The print head is connected to the ECU (marked 2-2) through a long, flat, flexible cable, called the *print head cable*.

When the host computer sends a character to be printed, a series of vertical dot patterns representing that character (in its selected font and size) are recalled from the printer's permanent memory. The ECU sends each dot pattern in turn through a series of print wire driver circuits. A typical print wire driver circuit is shown in figure 7-3. It is the driver circuits that amplify digital logic signals from main logic into the fast, high-energy pulses needed to fire a print wire. As a pulse reaches the firing solenoid, it creates an intense magnetic field, which shoots its print wire forward against the page. After the pulse is complete, a spring (not shown in figure 7-3) pulls the print wire back to its rest position. Figure 7-4 shows a typical printing sequence for a letter "A."

Impact heating problems

One of the major problems with impact printing is the eventual buildup of heat. The substantial amount of current needed to fire a solenoid is mostly given up as heat that must be dissipated by the print head housing. Under average use, the metal housing will dissipate heat quickly enough to prevent problems. Heavy use, however, can cause heat to build faster than it dissipates. This happens most often when printing bit-image graphics where many print wires might fire continuously. Excessive or prolonged heating can cause unusual friction and wear in print wires. In extreme cases, uneven thermal expansion of hot pins within the housing might cause them to jam or bend.

To combat the buildup of heat, impact print heads are cast with a series of heat sink fins. If you look at the print head in figure 7-2 closely, you can see the heat sink fins depicted. In order for heat sink fins to be effective, they must be exposed to the open air; buildup of dust, dirt, and paper debris will prevent heat from venting to the air. When inspecting an impact print head, always make sure that the heat sink fins are clear.

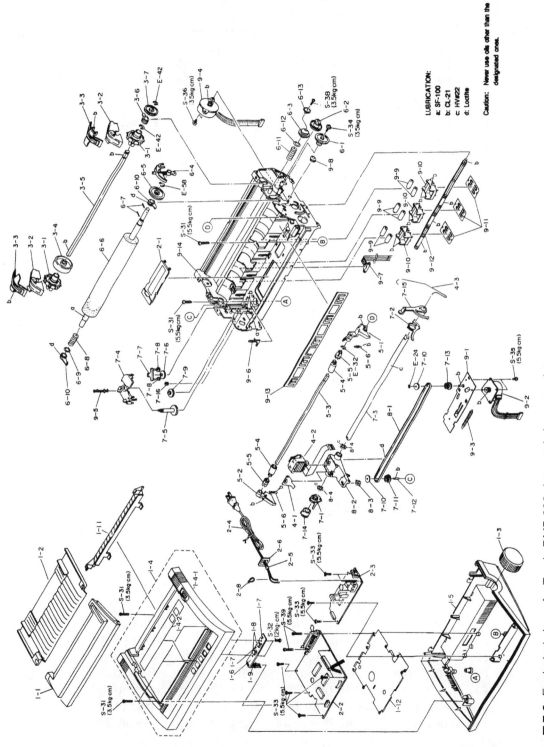

7-2 *Exploded view of a Tandy DMP 203 dot-matrix impact printer.* Tandy Corporation

Dot-matrix impact print heads

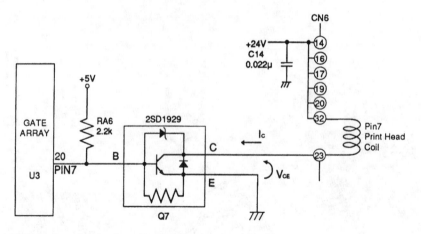

■ 7-3 *A typical print wire driver circuit.* Tandy Corporation

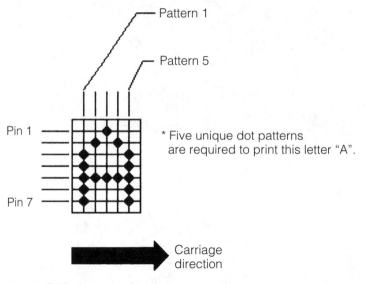

Pattern 1

Pattern 5

Pin 1

Pin 7

* Five unique dot patterns
are required to print this letter "A".

Carriage
direction

■ 7-4 *DMI character formation.*

Evaluating the print head

The first step in all diagnosis is examination. In the case of print head testing, you should print a sample page that will place the print head under a certain amount of stress. Now, it is certainly possible to discern glaring problems while printing ordinary text, but to really push the print head's capabilities, the printer should generate a demanding graphic. You can use the companion utility PRINTERS to generate a print head test pattern, the large black rectangle shown in figure 7-5. By observing inconsistencies in the

■ 7-5 *The PRINTERS print head test pattern.*

resulting image, you can make an accurate determination of the print problems. Of course, if you have other utilities that will generate similar results, feel free to use them as well.

Troubleshooting impact print heads

Generally speaking, impact print heads are some of the most reliable electromechanical devices available. With life expectancies easily exceeding 30 million characters, each print wire is rated for hundreds of millions of operations. In spite of their reliability, however, operating practices and environments will affect the print head's service life. Reliable print head performance is also related to routine maintenance; neglected maintenance will typically result in print problems.

Symptoms

Warning: The impact print head can become extremely hot during the printing process. Be sure to allow at least 10 minutes of idle time for the print head to cool before attempting any service.

Symptom 1 Print quality is poor. Dots appear faded or indistinct. All other operations appear normal. Refer to Chart 7-1. The Print Head Test pattern should yield a solid black rectangle. In actual practice, the rectangle drawn by an impact print head will not be

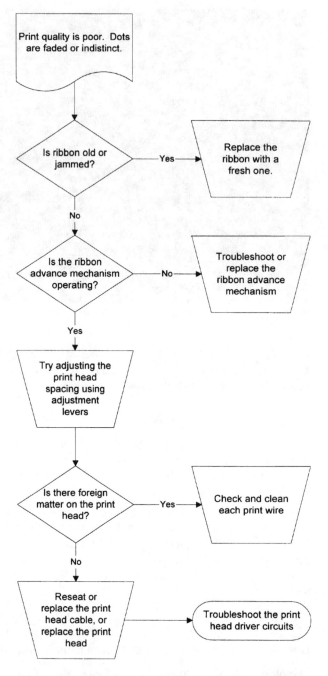

Chart 7-1 *Flowchart for Impact Print Head Service Symptom 1.*

absolutely homogeneous, you will be able to discern slight tone differences in each pass, but the print should be sharp and the color tone should be consistent. If the print is excessively light, or there are patches of light and dark areas, you should begin by carefully examining the ribbon. It should be reasonably fresh and it should advance normally while the carriage moves back and forth. A ribbon that is not advancing properly (if at all) might be caught or jammed internally, so install a fresh ribbon and retest the printer. If the ribbon still does not advance properly, troubleshoot the printer's ribbon advance mechanics as specified in Chapter 10. A fresh ribbon might improve image quality, but that will fade again quickly if the ribbon does not advance.

Examine your print head spacing next. Most printers are designed with one or two small mechanical lever adjustments that can alter the distance between print head and platen by several thousandths of an inch (such as the level shown in figure 7-6). This adjustment allows print intensity to be optimized for various paper thicknesses, and can be adjusted to keep the print head perfectly parallel to the platen. If the print head is too far away from the platen for your current paper thickness, the resulting print might appear light or faded. When there are two adjustments (as in fig-

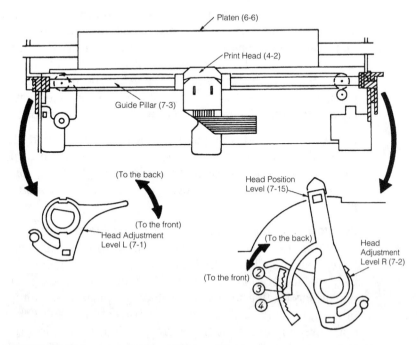

■ **7-6** *Print head adjustment levers.* Tandy Corporation

ure 7-6), adjust the right lever so that there is no smudging or missed dots along the right side of the image, then adjust the left lever so that there is no smudging or missed dots along the left side of the image. If spacing is already close or nonadjustable, turn your attention to the print head itself.

Turn off and unplug the printer, then check each print wire in the head assembly. Print wires should all be free to move, sliding in and out without restriction, except for mechanical tension from the return spring. Keep in mind that you will probably have to remove the print head from its carriage assembly. Over an extended period of time, a combination of paper dust and ink form a sticky glue that can work its way into each print wire. As this substance dries, it can easily restrict a wire's movement or jam it all together. If you find a tremendous buildup of foreign matter, wipe off each wire as gently as possible. Use a stiff cotton swab dipped lightly in alcohol or light-duty household oil. Do not use harsh chemical solvents! Finally, wipe down the front face of the print head with a soft, clean cloth. Once all wires are moving freely again (and any loosened glue has been removed), replace the print head and retest the printer. If you do remove the print head for cleaning, be sure to readjust the head spacing lever(s) to keep the head parallel to the platen.

Another possible problem area might be in the print head cable itself. If the cable (especially the ground) connections are loose or marginal, the increased resistance in the print wire circuits might lighten the print. Try reseating the print head cable one or more times. If this does not help, try replacing the print head cable. If problems continue (or the cable cannot be replaced without exchanging the print head), try a new print head assembly.

Finally, you should remember that it is the print head driver circuitry that supplies energy necessary for print wire operation. There could be a loss of solenoid driving voltage or some other defect in your drivers. Chapter 9 details driver troubleshooting procedures.

Symptom 2 Print has one or more missing dots that resemble "white line(s)." This also takes place during a self test or Test Page. Refer to Chart 7-2. This type of symptom generally resembles the illustration in figure 7-7. The white line(s) should be equally noticeable in text as well as graphics. Assuming that all other operations of the printer are correct, a loss of one or more dot rows suggests that the corresponding print wire(s) will not fire. In most cases, this is due to a fault in the print head, the print

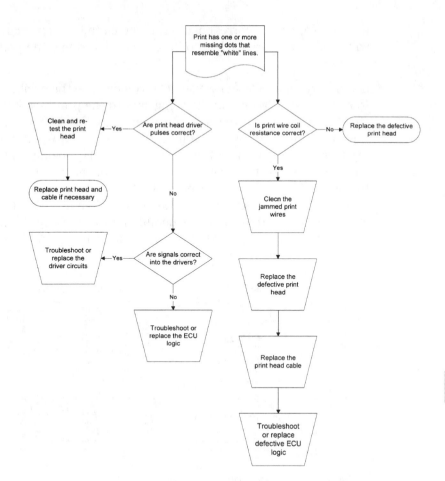

Chart 7-2 *Flowchart for Impact Print Head Service Symptom 2.*

■ **7-7** *Missing (white) lines in the print.*

Dot-matrix impact print heads

head cable, or the corresponding driver circuitry. However, you will need to explore each step in order to determine the point of failure.

If you have an oscilloscope available, you can use the oscilloscope to measure the driving pulses being sent to the print head, which should appear similar to those shown in figure 7-8. The schematic fragment shown in figure 7-9 illustrates a typical ECU driver-to-print head configuration. If you see driver pulses being sent to a pin, but the corresponding pin is not firing, chances are good that the print wire is jammed (or has failed completely). You can then proceed to shut off and unplug the printer, then clean each print wire carefully to remove any accumulations of buildup. Reinstall the cleaned print head and retest the printer. If the problem persists, replace the print head and print head cable.

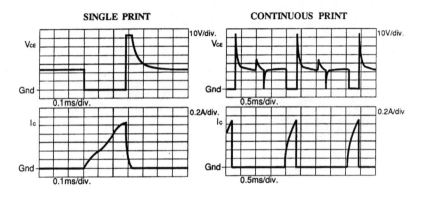

■ **7-8** *Pin driver waveforms.* Tandy Corporation

If the expected pulse signal(s) are missing, it means that the offending print wires are not receiving driver signals; this is a sign of driver circuit failure in the ECU. As you see in figure 7-9, each print wire is operated by its own discrete driving transistor. You can then use your oscilloscope to check the logic signal driving the suspect transistor. If the logic signal is present on the transistor's base, but the output signal on the transistor's collector is missing, the transistor is faulty and should be replaced. If your inspection reveals that the logic signal itself is missing, then the ASIC in the ECU that generates the logic outputs is defective and should be replaced. From a practical standpoint, it might not be possible for you to obtain replacement components or replace those components with the soldering tools you have available. In that case, you should simply replace the ECU outright because it typically contains all of the logic and driver circuitry.

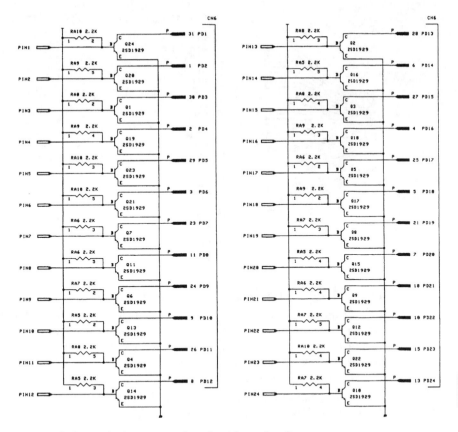

■ **7-9** *Schematic fragment of a pin-driver circuit.* Tandy Corporation

So what if you don't happen to have an oscilloscope handy? Well, you can shut off and unplug the printer, then use your multimeter to measure the resistance between the 24-V ground and each pin terminal, as shown in figure 7-10. You should read approximately 19.5 Ω between ground and any pin (the resistance of a solenoid coil). If you read an open or short circuit instead, the solenoid is defective, and the entire print head must be replaced. When replacing a print head, you should also replace the print head cable if possible.

If all of the solenoids check out properly, then you might be facing one or more jammed print wires, so remove the print head and carefully clean the face (and the print wires if possible). If any of the wires appear jammed, or do not extend to their full length (even after a thorough cleaning), the pin(s) might be damaged, in which case you should replace the print head and cable. After any accumulations of buildup have been removed and the unit is dry, reinstall the print head and retest the printer. If problems persist, there might still be a fault in the print wire itself, the print head ca-

Dot-matrix impact print heads

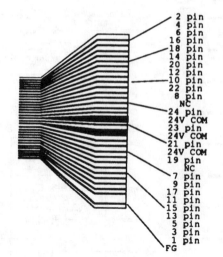

■ 7-10
Print head cable layout for resistance measurement. Tandy Corporation

ble, or the ECU. Because you have already demonstrated that the print solenoids are intact, check the continuity across the print head cable. If any connections are broken or intermittent, replace the print head cable. Otherwise, replace the ECU outright (which will effectively replace the print head logic and driver circuitry). In the unlikely event that you have not yet corrected the problem, the only remaining possibility is that one or more pins in the print head have actually broken. In that case, try a new print head.

Symptom 3   Print appears "smeared" or exceedingly dark. All other operations appear normal. Unfortunately, this type of problem cannot be detected by reviewing a dark graphic. Instead, smeared or extremely dark print is best discovered with a textured graphic or text (such as the printer's self test or Print Test Page feature in PRINTERS), as shown in figure 7-11. However, smeared print is relatively easy to correct.

■ 7-11 *An example of smeared print.*

One of the most likely reasons for smeared print in an impact printer is that the print head is too close to the platen. When the head is too close, the print wires might not have enough time to retract before the carriage moves, thus smearing the print. The answer is to retract the print head a bit using the position adjustment lever(s), shown in figure 7-6.

Another possible problem can occur when buildup accumulates on and in the print head. Foreign matter can affect the dimensions and stroke of the print wires. It can also jam the print wires in an extended position and prevent them from retracting before the carriage moves, again smearing the print. Turn off and unplug the printer. You can then remove the print head and clean the face and print wires. If problems continue at this point, chances are good that your print head is defective, or fouled with build up so badly that cleaning is not effective. Try replacing the print head.

Symptom 4 Printer does not print under computer control. Operation appears correct in self-test mode. Before you disassemble the printer, take a moment to check its "on-line" status. There is almost always an indicator on the control panel that is lit when the printer is selected. If the printer is not selected (on-line), then it will not receive information from the computer, even if everything is working correctly.

A printer can be off-line for several reasons. Paper might have run out, in which case you will have to reselect the printer explicitly after paper is replenished. Even the simplest printers offer a variety of options that are selectable through the keyboard (e.g., font style, character pitch, line width, etc.). However, you must often go off-line in order to manipulate those functions, then reselect the printer when done. You might have selected a function incorrectly, or forgotten to reselect the printer after changing modes. Also consider software compatibility. If you are using a "canned" software package, make sure that its printer driver settings are configured properly for your particular printer.

Check your communication interface cable next. It might have become loose or unattached at either the printer or computer end. If this is a new or untested cable, make sure that it is wired correctly for your particular interface. An interface cable that is prone to bending or flexing might have developed a faulty connection, so disconnect the cable at both ends and use your multimeter to check cable continuity. If this is a new, homemade cable assembly, double check its construction against your printer and computer interface diagrams.

Double check the printer's DIP switch settings or set-up configuration. DIP switches are often included in the printer to select certain optional functions such as serial communication format, character sets, default character pitch, or automatic line feed. If you are installing a new printer, or you have changed the switches to alter an operating mode, it might be a faulty or invalid condition. DIP switches also tend to become unreliable after many switch cycles. If you suspect an intermittent DIP switch, rock it back and forth several times, then retest the printer.

When everything checks out, you will have to disassemble the printer and troubleshoot its interface circuits and main logic. Use the troubleshooting procedures provided in Chapter 9.

Symptom 5 Print head moves back and forth, but does not print (or prints only intermittently). This also takes place during a self-test or test patterns. Refer to Chart 7-3. If the print appears to gradually fade in and out, check your ribbon first. Make sure that it is installed and seated properly between the platen and print head. If the ribbon has dislodged from the head path or is totally exhausted, no ink will be deposited on paper. If the ribbon is in place, make sure that it advances properly as the carriage moves. A ribbon that does not advance properly might be caught or jammed internally, so install a fresh ribbon and retest the printer. If the ribbon still fails to advance, troubleshoot the printer's ribbon advance using procedures in Chapter 10.

If the print appears to cut in and out suddenly, suspect an intermittent connection. Intermittent connections in the print head cable or within the print head itself can lead to highly erratic head operation. A complete cable break can shut down the print head entirely, especially if the break occurs in a common (ground) conductor. Unplug the printer and use your multimeter to check continuity across each conductor in the print head cable. You might have to disconnect the cable at one end to prevent false readings. Replace any print head cable that appears defective.

You should check the printer's power supply outputs next. Apply printer power and measure the print head driver voltage (usually +12 or +24 Vdc). If driver voltage is low or nonexistent, drivers will not produce enough energy to fire a print wire. Marginal supply outputs can result in intermittent operation. Check your circuit connections. If you find that the power supply voltage is low or erratic, troubleshoot the power supply using the procedures of Chapter 8. If the power supply is integrated into the ECU and must be replaced, you will have to replace the entire ECU outright.

Your trouble could exist in the print head itself. Turn off and unplug the printer, then disengage the print head from its carriage. Use your multimeter on a low resistance scale to measure the resistance of each firing solenoid as in figure 7-10. If the head is completely inoperative, the common lead might be intermittent or

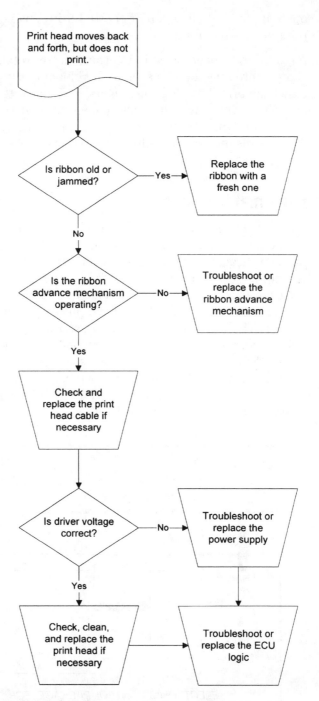

Chart 7-3 *Flowchart for Impact Print Head Service Symptom 5.*

Dot-matrix impact print heads

open. If you find any open or shorted firing solenoid, replace the print head mechanism.

If your print head is still intermittent or totally inoperative, chances are that there is a serious problem in the driver circuits; either the print head logic or drivers are intermittent or has failed entirely. If you have an oscilloscope, you can troubleshoot the print head signals through the ECU; the point at which the signals disappear is the point of failure. If you do not have an oscilloscope handy, replace the ECU.

Thermal dot-matrix print heads

Thermal dot-matrix (TDM) print heads substitute print wires with individual solid-state dot heaters. Much like DMI print heads, a serial TDM head, as shown in figure 7-12, is assembled as a vertical column of seven or nine dot heaters. The head is mounted on a carriage, which carries it back and forth across the page. Every dot can be fired independently by logic pulses amplified with driver circuits. Thermal line-print heads are stationary devices that hold a single horizontal row of dot heaters, one for every possible dot in a row. Figure 7-13 illustrates a block diagram for a typical line-print head. Because each dot heater in a line consumes relatively little power, the driver circuits for a line-print head are usually fabricated into the head assembly itself.

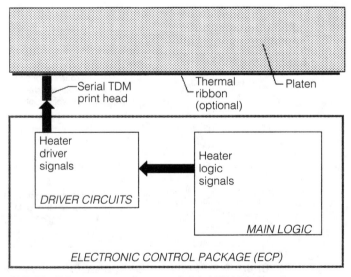

■ **7-12** *Print system diagram (serial TDM printer).*

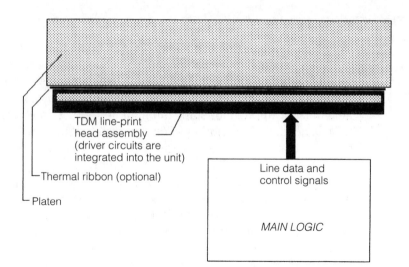

TDM line-print
head assembly
(driver circuits are
integrated into the unit)

Thermal ribbon (optional)

Platen

Line data and
control signals

MAIN LOGIC

■ **7-13** *Print system diagram (line-print TDM printer).*

A "serial" TDM print head works almost identically to an impact print head. When a character is received from the host computer, a series of vertical dot patterns representing that character (including font and size) are recalled from the printer's permanent memory. Each dot pattern is sent out in turn, where they are amplified by driver circuits and finally delivered to the print head. Drivers convert the digital information developed in main logic into the high-energy pulses required to fire dot heaters. As a pulse reaches a dot heater, it causes a sudden, large temperature rise. This localized heat is used to discolor corresponding points on thermally sensitive paper, or melt corresponding points of plastic ink on a thermal ribbon.

"Line-print" heads are more sophisticated, and depend much more on ECU's processing power. After a complete line of characters is sent from the host computer (a "line" ends when the printer sees a "carriage return" or "line feed" character), ECU logic translates the entire line into a series of horizontal rows that span the paper's width. A horizontal row is printed onto thermally sensitive paper (or through a thermal ribbon), paper advances a fraction to the next adjacent row, then another row is printed. This continues until the entire line is complete. If a new row of characters is received, the process will repeat. While the circuitry needed to implement a line head assembly is a bit more complex than that needed for a serial head, the line head architecture eliminates the need for a carriage transport system.

Thermal contact problems

In order for a thermal print head to function at all, it must be in contact with the heat-sensitive paper surface (or the thermal transfer ribbon). The problem here is that contact pressure must be within a fairly narrow margin; to ensure proper printing, adequate heat must be applied for some minimum period of time. If the print head is not applying enough force, the resulting print will likely appear faint or fuzzy where dots form improperly. Make sure that there are no accumulations of dirt or debris that are forcing the print head away from the paper. If the print head is applying too much force, the print head might have trouble moving smoothly. This can result in unusually dark or smudged print. In extreme cases, the paper might not advance properly. When checking print problems on a thermal printer, always wipe the face of the print head to remove dust and foreign matter, then make sure that the print head contact pressure is within acceptable limits. Thermal print heads are also fragile, static-sensitive devices, so use caution when handling them.

Evaluating the print head

The first step in all diagnosis is examination. In the case of thermal print head testing, you should print a sample page, which will place the print head under a certain amount of stress. You can discern glaring problems while printing ordinary text, but to really push the print head's capabilities, the printer should generate a demanding graphic. You can use the companion utility PRINTERS to generate a print head test pattern, such as a test page, or the large black rectangle shown in figure 7-5. By observing inconsistencies in the resulting image, you can make an accurate determination of the print problems. Of course, if you have other utilities that will generate similar results, feel free to use them as well.

Troubleshooting thermal print heads

Thermal print heads are uniquely self-contained devices. Unlike impact and ink jet technologies, there is nothing to maintain; thermal print heads are sealed devices. Even the more complex line heads that incorporate driver circuitry are little more than large integrated circuits. They cannot be recycled, rebuilt, or refurbished. When any part of the print head fails or wears out, the entire print head must be replaced. The only two operations that you can safely perform on a thermal print head are cleaning and contact adjustment.

Symptoms

Symptom 1 Print head will not print at all. All other functions appear correct. Refer to Chart 7-4. Before starting any work on the printer, take special notice of the thermal paper or transfer ribbon. If your thermal printer uses heat-sensitive paper, be sure that the paper supply is fresh and inserted with the heat-sensitive side (usually the shiny side) facing the print head. If the paper is inserted backward (dull side showing), no print will be generated, even if everything is working properly. Keep in mind also that thermal paper has a limited working life, so if it is very old or discolored, print might appear faded, or not appear at all. Transfer ribbons deserve just as much attention. If your printer uses a thermal transfer ribbon, be sure that it is installed and seated properly, and check that there is still fresh ribbon available in the cartridge. Replace the ribbon cartridge if it is exhausted.

Inspect your print head cable next. A loose or broken cable can easily disable a print head, especially if the fault is in a common conductor. Turn off and unplug the printer, then use your multimeter to measure continuity across the cable. You might have to disconnect the cable from at least one end to prevent false readings. Wiggle the cable while checking continuity to stimulate any intermittent connections. Replace any defective print head cable.

Check your serial print head for any obvious faults, such as cracks, or burned or discolored areas. Remove the print head from its carriage and observe its face closely. Clean away any accumulation of dust or debris that might be formed on the print head. Remember that the print head must be in contact with the ribbon or paper for proper heat transfer to take place. Any dust or debris that builds up on the face can push the head just far enough away to interfere with heat transfer. Wipe the face with a clean, soft cloth dampened with ethyl alcohol. Replace the print head assembly and print head cable, especially if they show signs of excessive surface wear. Line-print heads contain a substantial amount of logic and driver circuitry built right into its assembly. If a line-print head fails to work, it might have suffered an internal multiplexer or driver failure, so try replacing the print head as well. If the problem persists, replace the printer's ECU.

Symptom 2 Printer does not print under computer control. Operation appears normal in its self-test mode. Check your printer's "online" status before performing any service procedures. Most control panels provide an indicator that displays the printer's on-line (selected) status. If your printer is not on-line, it will not receive in-

163

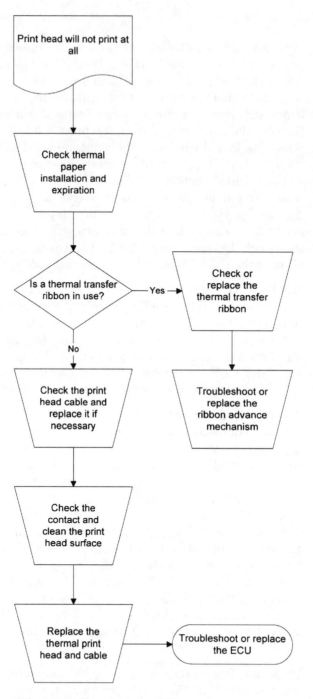

Chart 7-4 *Flowchart for Thermal Print Head Service Symptom 1.*

formation from the computer, even if everything is working properly.

A printer can go off-line for several reasons. A printer will go off-line automatically if paper is exhausted. That will inhibit a printer until paper is replenished and the printer is reselected manually. You must often take a printer off-line in order to set most functions from the control panel (e.g., font style, character pitch, line width, etc.). After functions are manipulated, the printer must be reselected manually. If you forget to reselect the printer, or you have selected an incorrect mode, the printer might remain off-line. Finally, you should consider software compatibility. If you are using a "canned" software package, make sure that the correct printer driver is installed.

Inspect your communication interface cable next. It might have become loose or detached at one end. If this is a new or untested cable, make sure that it is wired correctly for your particular interface. Cables that are prone to bending or flexing might develop faulty connections, so disconnect the cable at both ends and use your multimeter to check cable continuity. Refer to interface information for your printer and computer to verify cable wiring.

Examine any DIP switch configurations in your printer. DIP switches are commonly used to change optional settings, such as serial communication formats, character sets, or automatic line feed. If this is a new installation or you have changed operating modes in any way, the selection might be faulty. DIP switches also have the tendency to become unreliable after many switch cycles. If you suspect a faulty switch, rock it back and forth several times, then retest the printer. If problems persist at this point, the printer might have a defective interface circuit, or trouble in its main logic. Follow the troubleshooting procedures in Chapter 9.

Symptom 3 Print contains one or more missing (white) or consistent black lines through the print. This also takes place during self-test or test pages. When one or more dot heaters fail to fire, you will see lines through the print that are devoid of any dots. This gives the optical illusion of "white" lines as shown in figure 7-14, and can be reviewed even better with a dark graphic, such as a Print Head Test. Although figure 7-14 shows a horizontal line as the result of a print head fault, the missing line(s) would appear vertically in a line-print head. Turn off and unplug the printer, then remove the head from its mounts and carefully observe the head's face. A white or discolored point corresponding to a missing line indicates a burned-out dot heater, so replace the thermal print head.

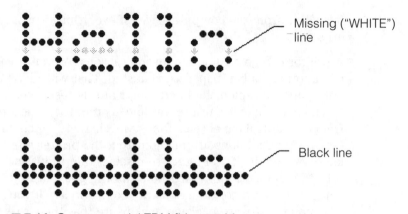

Missing ("WHITE") line

Black line

■ **7-14** *Common serial TDM firing problems.*

If there are no obvious defects on the thermal print head (or there are black lines in the print), the problem might be caused by a fault in your print head cable. Intermittent or broken connections in the cable can disable one or more dot heaters, resulting in missing lines. Shorted connections can cause a heater to fire continuously, resulting in a smeared black line. Unplug the printer and use your multimeter to measure continuity across the print head cable. You might have to disconnect at least one end of the cable to prevent false readings. Also check for shorted connections between individual wires. Replace any print head cable that appears open or shorted.

If both the print head and the print head cable appear to be intact, you might have a fault in one of the printer's driver circuits. A driver transistor usually fails in the open or shorted condition. In the open-circuit condition, a driver will not fire at all, so it appears as a missing line. A short-circuited driver can keep its dot heater turned on continuously to form a black line. If you have an oscilloscope handy, you can check the input and output signals into each driver channel in order to find the defective driver's or logic circuit. If all of the logical signals appear present, a driver is likely at fault. If one or more logic signals are missing at the driver, the printer's logic might have failed. Chapter 9 presents the troubleshooting procedures for driver circuits. If you do not have an oscilloscope handy (or you do not have the parts or tools to troubleshoot the ECU), simply replace the printer's ECU outright.

Symptom 4 The print appears excessively dark or smudged. This particular symptom is best reviewed with reasonably light printing (such as text or a Test Page). Because thermal print heads must be in contact with thermal paper (or a thermal transfer ribbon), contact pressure plays an important role in print quality. Unusually dark or smudged print is typically a sign of excessive pressure at the thermal print head. Check and correct the contact pressure if necessary. Another possibility might be that the thermal print head has simply worn out, and the dot heaters, effectively "closer" to the page, are allowing larger, hotter dots to reach the page. If contact pressure falls within normal limits, it might be necessary to simply replace the thermal print head assembly outright.

Ink jet dot-matrix print heads

Where impact and thermal printing depend on actual contact between the head and paper, ink jet printing provides a "noncontact" printing medium. Noncontact printing is accomplished by spraying liquid ink onto a page from individual ink nozzles. The block diagram for an ink jet print system is shown in figure 7-15. In many ways, this is the same approach used for serial impact and thermal printers, with the notable exception of a liquid ink supply. Depending on the particular print head design, ink might be located in a large reservoir away from the head, or in a small reservoir integrated into the print head assembly. Older ink jet print heads usually hold 9, 12, or 24 nozzles, but contemporary print heads can carry 50 nozzles or more. Each nozzle can be fired independently, so the ink jet system is capable of producing high-quality characters and graphics on plain paper.

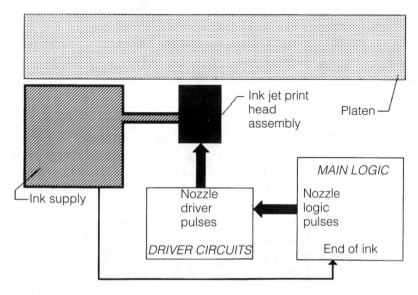

■ **7-15** *Print system diagram, ink jet printer.*

Like other serial print heads, ink jet images are formed one character line at a time while the print head sweeps back and forth across the page. Characters that are received from a host computer are converted into a series of vertical dot patterns recalled from the printer's permanent memory. Each dot pattern is sent in turn to the head as it moves. Driver circuits amplify and condition signals produced by main logic into the short, high-energy pulses that are needed to operate each ink nozzle. As a pulse reaches an ink nozzle, it causes a small piezoelectric ring or heating element around the nozzle to constrict. This sudden constriction literally launches a droplet, as shown in figure 7-16.

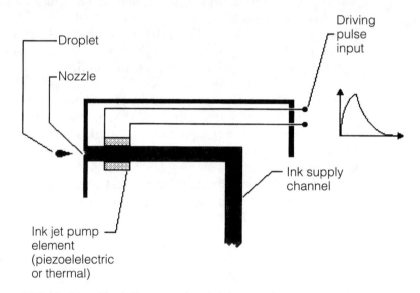

■ **7-16** *Simplified diagram of an ink jet nozzle.*

Cleaning ink jet heads

The proper operation of an ink jet head depends largely on the viscosity of the ink, and its surface tension at each nozzle. During normal use, it is perfectly natural for ink to evaporate. Unfortunately, this raises the ink's viscosity and its surface tension, which ultimately clogs the nozzles. The other enemy of ink nozzles is foreign matter, ink mixed with everyday dust and dirt, which can work into nozzles and also cause clogs. Virtually all contemporary ink jet printer designs incorporate some type of "ink cap" that will seal off the nozzles while the printer is idle. Such caps often take the form of small sponges. As printers age, however, the ink sponges dry out and lose their pliability, so older printers tend to have more serious clogging problems than newly built printers. In actual practice, you will probably encounter clogging problems with most ink jet printers from time to time. Fortunately, they are easy devices to clean.

Some printers are equipped with a PURGE (sometimes called PRIME) function, which is especially intended to clear clogs automatically. You need only press the PRIME button, then recheck the printer. If the problem persists, you should try cleaning the nozzles with a soft cloth or cotton swab dampened lightly in isopropyl alcohol. Note that you will need to turn off and unplug the printer, then remove the print head cartridge for this operation. Wipe the nozzles gently, and make sure that the cartridge face is dry before reinstalling. If the clog still won't clear, check the ink cartridge for a manual PRIME port, which can usually be actuated manually with a paper clip. Be sure to rest the print cartridge on a clean, highly absorbent material; purging will push ink out of the cartridge, which can easily cause a mess. When you press on the PRIME port, press gently; you do not want to rupture the ink bladder. If neither cleaning nor purging will clear the clog, you will simply have to replace the ink jet print head.

Refilling ink jet print heads

If you deal with a large number of ink jet cartridges and you find yourself throwing a large number of them away, you might consider recycling (refilling) the exhausted ink cartridges yourself. You can refill a cartridge with bulk ink, a syringe, and a small drill (#55 or so). The refill process typically involves using the drill to create or enlarge a hole in the ink cartridge (such as shown in figure 7-17). Draw some ink into the syringe (12 to 15 cc), push the syringe into the ink head as far is it will go, then inject ink slowly and lift up slightly as you go to distribute the ink. The process takes several minutes, and costs a fraction of a new ink cartridge. If you do choose to refill ink cartridges as a sideline, you can find an excellent reference in *The Ink Jet Cartridge Book*, published by The Tech Press.

Limited cartridge life

Another important attribute to keep in mind when working with ink jet cartridges is their limited life. Ink does not remain viable indefinitely; the unavoidable evaporation of solvent in the ink will eventually clog the print head, or reduce the cartridge's life. There are two expiration dates to keep in mind when selecting and pur-

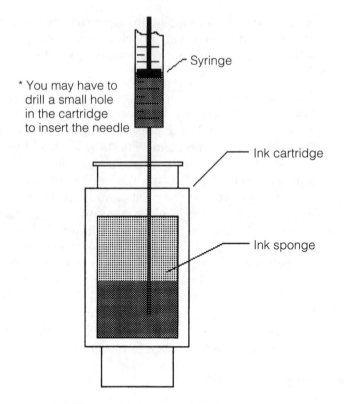

* You may have to
drill a small hole
in the cartridge
to insert the needle

Syringe

Ink cartridge

Ink sponge

■ **7-17** *Filling an ink jet cartridge.*

chasing ink cartridges: the sealed (or "shelf") life and the printer life. New ink cartridges are typically marked with a "use by" date. The difference between the current date and the use by date is the shelf life. If you intend to stock a quantity of ink cartridges, be sure to purchase fresh units. A typical shelf life is 18 months from the date of manufacture. Once an ink cartridge is opened and installed in a printer, the ink has a printer life (usually no more than three months from the date it is opened). Of course, regular use will probably exhaust the ink cartridge before the three months has expired. The major idea to remember here is that ink will become unusable in a matter of months.

Evaluating the print head

The first step in all diagnosis is examination. In the case of ink jet print head testing, you should print a sample page that will place the print head under a certain amount of stress. You can discern glaring problems while printing ordinary text, but to really push the print head's capabilities, the printer should generate a demanding graphic. You can use the companion utility PRINTERS to generate a print head test pattern, such as a test page or the large black rectangle shown in figure 7-5. By observing inconsistencies in the resulting image, you can make an accurate determination of the print problems. Of course, if you have other utilities that will generate similar results, feel free to use them as well.

Troubleshooting the ink jet heads

Contemporary ink jet print heads are a mixed blessing for technicians. Their simple, modular design and internal ink supply make them a snap to replace; the overall process requires no more than 10 seconds. The ink cartridge is also reasonably reliable because there are essentially no moving parts. However, the proliferation of microscopic nozzles in today's ink jet heads are almost impossible to keep clean over the long term (and clogged nozzles are readily noticeable). Even when a nozzle does not clog completely, an accumulation of foreign matter can deflect the droplets and cause splattering, ruining the fine detail in high-resolution graphics. When an ink jet print head problem does occur, the entire head will have to be replaced.

Symptoms

Symptom 1 Print quality is poor. Print appears smeared, faint, or smudged. Print quality problems are often related to the characteristics of the paper being used. Ideally, liquid ink should dry immediately as it sticks to the paper, resulting in a neat, crisp appearance that will not smudge when touched. Just how quickly the ink actually dries depends on the characteristics of the ink, as well as the paper. Paper that absorbs ink readily might drink up too much ink from the page surface. This can result in a faded or dull appearance. On the other hand, glossy paper or paper that does not ink well might allow ink to smudge or smear through rollers or when touched. Typical ink jet printer paper is impregnated with clay or chemicals that will absorb ink quickly on contact, but not so fast that it disappears into the paper. You can usually identify "ink jet" paper by its shiny, smooth appearance on one side, and its normal, dull appearance on the other side.

Inspect your ink supply and ink nozzles carefully. In most contemporary ink jet designs, the ink supply is manufactured into a disposable cartridge along with the ink nozzles. A low ink supply or partially clogged nozzle can cause some nozzles to spit or sputter, resulting in a smudged appearance. Follow procedures to clean, prime, and retest the print head. Replenish your ink supply or replace the entire ink cartridge if necessary. Replaceable ink jet cartridges, as shown in figure 7-18, can often be cleaned and primed easily. Use a clean cotton swab dipped lightly (if necessary) in ethyl alcohol to wipe the face of each nozzle. When nozzles are clean, use the stick to push gently on the ink bladder. This forces fresh ink through the head (be sure to wipe up any spillage before reinstalling the ink cartridge). The head should now be ready to retest.

Hewlett-Packard Co.

■ **7-18** *A disposable ink jet cartridge.*

Finally, check your print head spacing. If your print head is too close, ink impacting the page might spatter or run. A head that is too far away might produce print that appears to sag across a printed line. Most printers allow head spacing to be adjusted by several thousandths of an inch using a mechanical lever assembly, such as in figure 7-6. Your user's manual probably lists the optimum distance settings for various paper thicknesses.

Symptom 2 Print contains one or more missing lines. This also happens when printing self-tests or test pages. Refer to Chart 7-5. Begin by checking your ink supply. If the supply is marginally low, there might not be enough ink to supply every nozzle evenly. Noz-

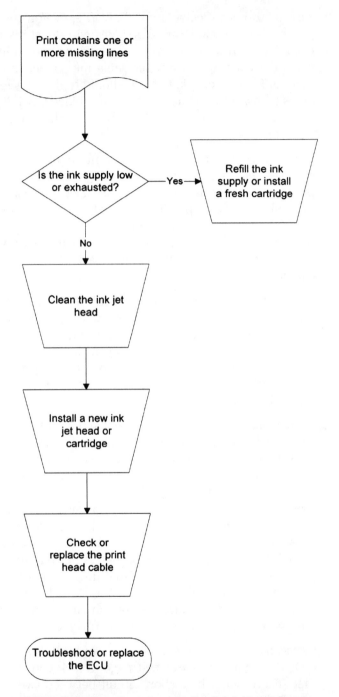

Chart 7-5 *Flowchart for Ink Jet Print Head Service Symptom 2.*

zles that do not receive ink will not fire, even if they are working properly. Replaceable ink cartridges with local ink reservoirs are usually easy to check. If the ink bladder appears low, simply replace the cartridge. Nonreplaceable ink jet heads might be a bit more difficult to check, but you should replenish the reservoir if it appears low. You might also want to replace any ink jet cartridges that have expired, or that have been in the printer longer than three months.

If the ink level is satisfactory, one or more ink nozzles might be obstructed with foreign matter or clogged with dried ink. Clean, prime, and retest the print head. Use a cotton swab dipped lightly in ethyl alcohol to wipe off each nozzle, then use its wooden end to gently press on its ink bladder. This forces fresh ink through the head. Make sure that fresh ink beads up on every nozzle. It should now be ready to retest. If cleaning does not help, try a new ink jet cartridge.

Missing lines can be caused by a fault in the print head cable. Intermittent or broken signal lines can disable any of the nozzles. Unplug the printer and use your multimeter to check continuity across each conductor in the print head cable. You might have to disconnect the print head cable at one end to prevent false continuity readings. Replace the print head cable if you find it to be excessively worn or defective.

If a new ink cartridge does not resolve the problem and everything else checks properly up to this point, the trouble is probably in your driver circuits, or in the main logic of the ECU. If you have an oscilloscope available, you can check the input and output signals of the drivers that correspond to the suspected nozzles. If a logic input signal is present but a high-power signal is not, the driver stage is bad and should be replaced. If the logic input signal is missing, the logic of the ECU is probably at fault and should be replaced. Of course, if an oscilloscope is not handy, you should replace the driver circuits first. If that does not correct the problem, replace the main logic portion of the ECU. In printers where the main logic and driver circuits are combined onto the same board, you should replace the entire ECU outright.

Symptom 3 Print contains one or more black lines. This also happens during a self-test or test page. Check your print head cable first. There might be a short circuit between two or more nozzle signals. Turn off and unplug the printer, then use your multimeter to check continuity between individual cable wires. You should disconnect the cable from at least one end to prevent false continuity readings. Ideally, all print signal wires should be isolated from one another (infinite resistance). Replace the print head cable or connectors if they appear shorted.

If problems persist, the fault might be in your ink jet print head. Try a new ink jet cartridge. If that fails to correct the problem, you can be confident that the trouble is located in the driver circuits or main logic. Try replacing the driver assembly first. If problems continue, replace the main logic board. In printers where the driver circuits and main logic are integrated onto the same board, you should just replace the entire ECU outright.

Symptom 4 Printer does not print under computer control. Operation appears correct in its self-test mode. Make sure that your printer is actually on-line with the host computer. An indicator will usually be lit on the control panel when your printer is selected (on-line). If your printer is off-line for any reason, it can not communicate with the computer, even if everything is working properly. A printer can be off-line for several reasons. Paper might be exhausted. This inhibits operation until paper is replenished and reselected manually. In order to access any of the functions available from a printer's control panel (e.g., font style, character pitch, line width, etc.), the printer must go off-line, then be reselected after its modes have been changed. If functions are selected incorrectly, or you forget to reselect the printer, it will not operate under computer control. Finally, make sure that your computer software is configured properly for your particular printer. You might have to select a new printer driver.

The communication interface cable might have become loose or detached at one end. If it is a new or untested cable, make sure that it is wired correctly for your particular equipment interface. Cables that are prone to flexing or bending might have developed a faulty connection. Disconnect the cable at both ends and use your multimeter to measure continuity through the cable. Repair or replace any defective communication cable.

Examine your DIP switch settings next. DIP switches are used to change optional settings, such as serial communication format, character sets, or automatic line feed. If this is a new installation, or you have just set up the printer for a different computer, there might be an error in these settings. DIP switches also tend to become unreliable after many switch cycles. If you suspect a faulty switch, rock it on and off several times, then retest the printer. If problems continue at this point, there is probably a fault in the

ECU's communication circuitry. You might troubleshoot or replace the ECU at your discretion as discussed in Chapter 9.

Symptom 5 Print is intermittent or absent. All other functions appear correct. Intermittent or missing print can be the result of a low ink supply or trouble with the ink flow, so inspect your nozzles and ink flow carefully. A low ink supply might not be able to provide adequate ink to all nozzles at all times, especially while printing dark graphics. You might see this as randomly occurring areas of missing dots. In severe conditions, print might disappear entirely. A clog or restriction in the ink supply can also limit ink that is available at the nozzles. If the head has been unused for a prolonged period of time, the nozzles might be clogged with dry ink. Replace your ink cartridge or replenish the ink supply as necessary.

Replaceable ink jet cartridges, as in figure 7-18, can usually be cleaned and primed easily. Use a clean swab dipped lightly in ethyl alcohol to wipe the face of each nozzle, then use the swab's wooden end to push gently on the ink bladder. This forces fresh ink through each nozzle. When you see ink bead up on every nozzle, the print head is ready for use. Be sure to wipe away any residual ink from the cartridge's face before reinstalling it in the printer.

A faulty print head cable can cause intermittent or missing print. Unplug the printer and use your multimeter to check continuity across each print head signal wire. You might have to disconnect the cable from at least one end to prevent any false readings. Wiggle the cable to stimulate any possible intermittent connections. Replace any defective print head cable.

When problems persist, there is likely to be an intermittent fault in the printer's driver circuits or main logic. Unfortunately, troubleshooting intermittent systems can be an extremely time-consuming and frustrating exercise, so it is often preferable to simply replace the suspect subassemblies. Start by replacing the ink jet driver circuit board first, then replace the main logic board. For printers that integrate the drivers and main logic onto the same board, simply replace the entire ECU.

Power supply service techniques

8

ALL ELECTRICAL AND ELECTRONIC COMPONENTS IN YOUR printer, as well as every other piece of electronic equipment, require electrical power in order to function (figure 8-1). Power is always supplied in the form of voltage and current, and there must be adequate amounts of both to ensure the proper operation of each component. Unfortunately, the commercial ac power available from the wall outlets of your home, shop, or office, is not directly compatible with any components in a printer. As a result, ac power must be manipulated and converted into values of voltage and current that are suitable for your specific equipment. This is the task of a power supply.

■ **8-1** *A Hewlett-Packard DeskJet 660C printer.* Hewlett-Packard Co.

The name *power supply* is rather a misleading one. A power supply (often designated *PS*) does not actually create power. Instead, it converts commercially generated and supplied ac power into one or more voltage levels that are better suited to particular tasks and components. Typical printer circuits require sources of 5, 12, 15, or 24 Vdc, usually several different voltages. In some printers (such as EP printers), the power supply can provide voltage levels much higher than the ac signal.

Power supplies are generally rugged and reliable devices, so much so that they are often overlooked or disregarded as possible problem sources. Yet, problems in the power supply can interfere with the printer's normal operation, cause erratic or intermittent printer behavior, or shut down the afflicted printer outright. Luckily, most supplies are reasonably simple to follow, and can be repaired or replaced with relative ease.

There are many different PS circuits. Each design is optimized to suit the needs of the specific circuits that they must supply. In spite of the array of PS arrangements, there are only two typical operating modes: linear and switching. Both modes refer to the way in which a PS controls its output(s). This chapter shows you the construction, operations, characteristics, and repair considerations for both power supply types.

Power supply hazards

All power supplies have one thing in common: they all handle raw ac power at some point. As detailed in Chapter 6, raw ac power can be extremely dangerous under the right circumstances. Technicians have been hurt (even killed) as a result of carelessness, so the threat of electrocution is very real while the printer is open and the power supply is exposed. The risk is greatest while taking power supply measurements on a running printer. Reducing the risk of electrocution is the main reason why you should keep the printer turned off and unplugged as much as possible during your repair. However, you can easily minimize these risks by taking some simple precautions as outlined in Chapter 6. If you have not yet read Chapter 6, take the time to review it now. If you feel that the risks inherent in power supply troubleshooting are just too great, do not attempt to troubleshoot the power supply. You can still correct power supply problems by replacing the entire power supply module outright while the printer is off and unplugged.

Linear power supplies

The term *linear* means line or straight. As you might see from the block diagram of figure 8-2, a linear PS operates in essentially a straight line from ac input to dc output. Components and power capacity vary radically between manufacturers and models, but all linear power supplies contain the same three basic subsections: a transformer, rectifier, and filter. The regulator block is also found in the majority of linear power supplies, but it is not mandatory for a minimum "working" supply. For the purposes of this book, all supplies will be regulated. The following sections outline the purpose of each functional area.

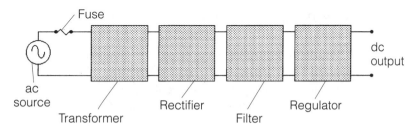

■ 8-2 *Block diagram of a linear power supply.*

Transformers

A transformer is used to alter the ac voltage and current characteristics of ac input power. This allows ac to be converted into more useful levels of voltage and current. This important process of transformation is accomplished through the principles of magnetic coupling, as illustrated in the schematic of figure 8-3. Transformers use two coils of solid wire wrapped along opposite sides of a common metal structure (called a *core*). Although figure 8-3 only illustrates two leads for each coil, many transformers offer a number of available leads (or taps) from both the primary and secondary coils. The core is built from laminated plates of "permeable" material (metal that can be magnetized). This serves not only as a physical base, but it is critical in concentrating magnetic fields around the transformer as well.

An ac voltage (or primary voltage), usually at 120 Vac, is applied across the primary winding of a transformer. The ac voltage causes current to fluctuate through the primary winding. In turn, this sets up a varying magnetic force field in the primary. The core concentrates this magnetic field and helps to transfer magnetic force to the secondary winding. Note that a solid core is not mandatory;

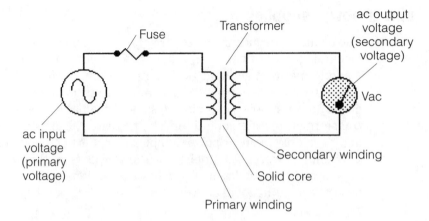

■ **8-3** *Principles of transformer operation.*

magnetic coupling between two coils can occur across an air gap, but solid cores make coupling much more efficient. This fluctuating magnetic field in the core cuts across the secondary winding, where it induces a secondary ac voltage between its terminals. Figure 8-3 shows an ac voltmeter measuring this voltage. This is the "transformer principle."

Voltage across the secondary winding is directly proportional to the ratio of primary-to-secondary windings. For example, if there are 1,000 turns of wire in the primary coil and 100 turns of wire in the secondary coil, the ratio (called *turns ratio*) would be 10:1. Because there are fewer secondary windings than primary windings, that transformer is known as a *step-down transformer*. If 120 Vac were applied across the primary, its secondary would ideally yield [120 Vac/(10/1)] 12 Vac. If the situation were reversed with 100 primary turns and 1,000 secondary turns, the transformer would be a 1:10 step-up transformer. An input of 12 Vac to this primary would result in an output of [12 Vac/(1/10)] 120 Vac across the secondary. An isolation transformer has a 1:1 turns ratio; the number of primary and secondary turns is equal, so secondary voltage will ideally equal primary voltage.

A transformer also steps current, but it is stepped in reverse to the voltage ratio. If voltage is stepped up, current is stepped down by that same ratio, and vice versa. In this way, power taken from a transformer's secondary will always (roughly) equal the power provided to its primary. As an example, suppose the transformer of figure 8-3 has 120 Vac at 0.1 A supplied to its primary. Primary power would then be [$P = I \times V$] or [120 Vac × 0.1 A] 12 W. With a 10:1 step-down transformer, its secondary voltage would be 12

Vac, but [0.1 × (10/1)] 1 A of current would be available. This results in a secondary power of [12 Vac × 1 A] 12 W. On paper, power output always equals power input. In reality, however, output power is always slightly less than input power due to losses in the core and coil resistance. Severe losses can cause excessive heating in the transformer. The ratio of output power to input power [P_o/P_i] is known as *efficiency*. Most solid core transformers can reach 80% to 95% efficiency, but never 100%.

You might wonder why transformers will not step dc voltages. After all, dc can produce a strong magnetic field in solenoids (impact print wires for example). While this is true, a magnetic field must fluctuate versus time in order to induce a potential (voltage) on another conductor. The dc would certainly magnetize the primary winding, but without constant fluctuation, no voltage would be induced across the secondary winding.

Rectifiers

Secondary voltage across the transformer secondary is still in an ac form; that is, the polarity swings between positive and negative voltages. An ac voltage must be converted into dc before it can be used by most electronic components. This conversion is known as *rectification*, where only one polarity of the input is allowed to reach the output, as shown in figure 8-4. Even though a rectifier's output varies greatly, the polarity of its signal will always remain within one polarity (thus the term *pulsating dc*). Diodes are ideal for use in rectifier circuits because they only allow current to flow in one direction. You will encounter three classical types of rectifier circuits: half wave, full wave, and diode bridge.

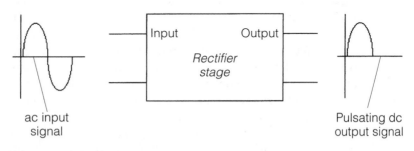

ac input
signal

Pulsating dc
output signal

■ **8-4** *Diagram of a generic rectifier.*

A half-wave rectifier circuit is shown in figure 8-5. It is the simplest and most straightforward type of rectifier circuit because it only requires one diode. As secondary ac voltage exceeds the turn-on

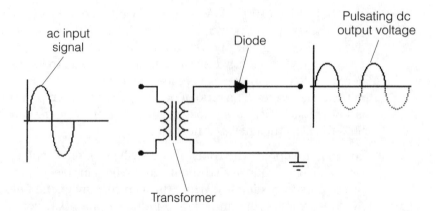

■ **8-5** *Diagram of a half-wave rectifier circuit.*

voltage of the diode (about 0.6 V for a silicon diode), it begins to conduct current. This generates an output that mimics the positive half of the ac input. If the diode were reversed, its output polarity would also be reversed. The disadvantage to this type of rectifier is that it is inefficient. A half-wave rectifier only deals with half of its ac input. The other half is basically ignored. The resulting gap between pulses results in a lower average output and a higher amount of ripple (ac noise) contained in the final dc signal. Half-wave rectifiers are rarely used in linear power supplies, but are used quite extensively in switching supply circuits.

Full-wave rectifiers, such as the one shown in figure 8-6, offer some substantial performance advantages over the half-wave design. By using two diodes in the configuration shown, both polarities of the ac secondary voltage input can be rectified into pulsating dc. Because a diode is at each terminal of the secondary, polarities at each diode will be opposite as shown. When the ac signal is positive, the upper diode conducts, but the lower diode is cut off. When the ac signal becomes negative, the lower diode conducts, but the upper diode is cut off. This means that one diode is always conducting, so there are no gaps in the final output signal. Ripple levels are lower and the average dc output voltage is higher. The disadvantage to a full-wave rectifier is its transformer requirement. A center-tapped secondary is needed to provide a ground reference for the supply. This often takes a slightly larger transformer, which is not a popular design choice among printer designers striving to reduce weight and bulk.

Diode bridge rectifiers use four rectifier diodes to provide full-wave rectification without the troubles of a center-tapped trans-

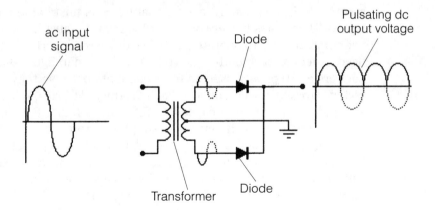

■ 8-6 *Diagram of a full-wave rectifier circuit.*

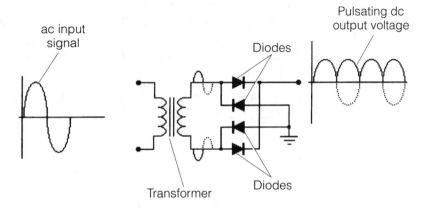

■ 8-7 *Diagram of a bridge rectifier circuit.*

former. Figure 8-7 illustrates a typical bridge rectifier stage. The ac voltage from the transformer secondary is connected to a series of diodes arranged in a "Wheatstone bridge" fashion. Diodes D1 and D2 provide the forward current paths, while D3 and D4 offer isolation between secondary voltage and a common reference point that serves as ground. When ac voltage is positive, diode D1 conducts because it is forward biased, and D4 provides isolation versus ground. As ac voltage becomes negative, D2 conducts while D3 supplies isolation versus ground. The complete bridge generates a full-wave pulsating dc output. Bridges are by far the most popular type of rectifier circuit.

Filters

By strict technical definition, pulsating dc is dc because its polarity remains consistent (even if its magnitude does change substantially). Unfortunately, even pulsating dc is unsuitable for any type of electronics power source. Voltage levels must be constant over time in order to operate electronic devices properly. A filter is used to achieve a smoothed dc voltage, as shown in figure 8-8. Capacitors are typically used as filter elements because they act as voltage storage devices, almost like light-duty batteries. When pulsating dc is applied to a capacitive filter, as in figure 8-8, the capacitor charges with current supplied from the rectifier. Ultimately, the capacitor charges to the peak value of pulsating dc. When a pulse falls off its peak (back toward zero), the capacitor will continue to supply current to a load. This action tends to hold up the output voltage over time (dc is filtered).

However, filtering is not a perfect process. As current is drained away from the capacitor by its load, voltage across the filter will also drop. Voltage continues to drop until a new pulse of dc recharges the filter for another cycle. This repetitive charge and discharge results in regular fluctuations of the filter output. These fluctuations are known as *ripple*, and ripple is an undesirable component of a smoothed dc output.

Figure 8-8 also shows a graph of voltage versus time for a typical filter circuit. The ideal dc output would simply be a constant, flat

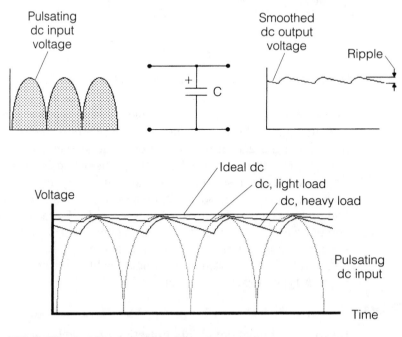

■ **8-8** *Effects of a capacitive filter stage.*

line at all points in time. In reality, there will always be some amount of filter ripple. Just how much ripple depends on the load. For a light load (a high resistance that draws relatively little current), discharge is less between pulses, so ripple is also lower. A large load (a low resistance that draws fairly substantial current) requires greater current, so discharge (and ripple) is greater between pulses. The relationship of dc pulses is shown for reference.

Additional filtering can be accomplished by adding more capacitance to the filter stage. This holds more charge, so load can be supplied with less discharge. As a general rule, more capacitance results in less ripple, and vice versa. Although this is true in theory, there are some practical limits to just how much capacitance can be used in a power filter. Size is always a big concern. Capacitors larger than 4,700 µF tend to be rather large and cumbersome. Above 10,000 µF, a filter can accept so much charging current on its initial charge (known as *inrush current*) that it might seem like a short circuit. Excessive inrush current can pop a fuse or even damage the rectifier stage.

In addition to electrocution hazards from exposed ac lines, you must understand the potential for a filter shock hazard. Power capacitors tend to accumulate a substantial amount of electrical charge and hold it for a long time. If you touch the leads of a charged capacitor, current will flow through your body. This is usually not dangerous when the printer is unplugged; however, it can be very uncomfortable or it could result in a slight burn. To remove any stored charge in your filter stage, charge must be bled away in a controlled fashion, as shown in figure 8-9. A large-value resistor (called a *bleeder resistor*) can be connected across the filter. This will slowly drain away any remaining charge. Note that some filter capacitors might already be built with a bleeder resistor. If a load remains connected across the filter, that too will discharge the filter after power is removed. Never attempt to discharge a capacitor using a screwdriver or wire. The sudden release of energy can actually weld a wire or screwdriver blade right to the capacitor's terminals, as well as damage the capacitor's internal structure.

Regulators

A transformer, rectifier, and filter form the absolute essentials of every linear power supply. These parts combined will successfully convert ac into dc that is capable of driving most electrical and electronic components. There are several troubles with these simple

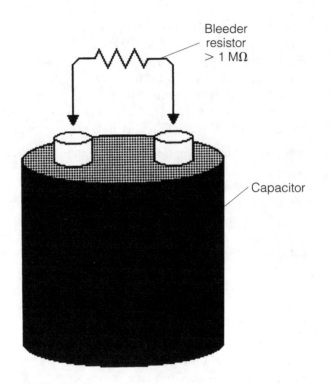

Bleeder
resistor
> 1 MΩ

Capacitor

■ **8-9** *Discharging a capacitor.*

"unregulated" supplies that make them undesirable. First, ripple is always present at a filter's output. Under some circumstances, this can cause erratic operation in even the most forgiving ICs. Second, output voltage varies with load. While load is fairly light, this effect might be negligible, but the effects of heavy loads can also cause unpredictable circuit performance. For reliable circuit performance, the filter's output must be stabilized to eliminate effects of ripple and loading. This is the task of a regulator.

Linear regulation is just as its name implies; current flows from the regulator's input to its outputs, as shown in figure 8-10. Voltage that is supplied to the regulator's input must be somewhat higher than the desired output voltage (usually by several volts). Internal circuitry within the regulator manipulates input voltage to produce a steady, consistent output level over a fairly wide range of loads and input voltages. If input voltage drops below some minimum value, the regulator falls out of regulation. In that case, the output signal tends to follow the input signal, including ripple.

In order to maintain a constant output voltage, the linear regulating circuit (or IC) will "throw away" extra energy provided by the

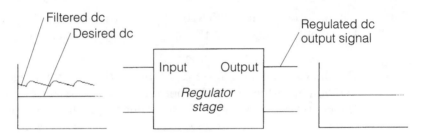

■ 8-10 *Diagram of a generic regulator.*

filter in the form of heat. This is why most regulators are often attached to large metal heat sinks. Heat is then carried away by the surrounding air. While linear regulation provides a simple and reliable method of operation, it is also very wasteful and inefficient. Typical linear regulators are only up to 50% efficient. This means that for every 10 W of power provided to the supply, only 5 W is provided to the load. Much of this waste occurs in the regulation process. Switching regulation is much more efficient, but that is covered later in this chapter.

You might encounter many various types of regulator circuits. Figure 8-11 illustrates a very simple series voltage regulator constructed with discrete parts. Input voltage is applied to the zener diode (Zd) through a current-limiting resistor (Rz). The zener diode works to "clamp" voltage to its zener level. In turn, this zener potential turns on the power transistor, which allows load current to flow. Output voltage equals zener voltage minus a small

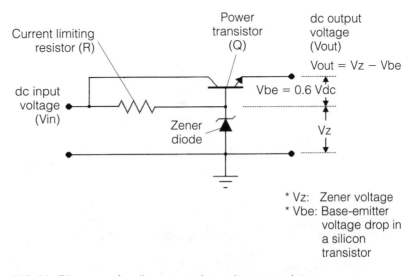

■ 8-11 *Diagram of a discrete series voltage regulator.*

voltage drop (usually 0.5 to 0.7 V) from the transistor's base-emitter junction. You can set the output voltage by changing the zener diode.

For the example of figure 8-11, suppose that input voltage is 10 V and you are using a 5.6-V zener diode. When power is applied to the circuit, zener voltage will be clamped at 5.6 V. Because input voltage is 10 V, the difference of [10 V–5.6 V] 4.4 V will appear across the current-limiting resistor Rz. Zener voltage saturates the transistor, so its output will be 5.6 V minus the transistor's base-emitter drop of 0.6 V, or 5.0 Vdc. As long as input voltage remains above the zener voltage, output voltage should remain steady regardless of load; output is effectively regulated. Load current can be substantial, so you will often find a power transistor used as the regulating transistor.

Regulator circuits can easily be fabricated as integrated circuits, as shown in figure 8-12. Additional performance features such as automatic current limiting and over-temperature shut-down circuitry can be included to improve the regulator's reliability. Input voltage must still exceed some minimum level to achieve a steady output, but IC regulators are much simpler to use. One additional consideration for IC regulators is the addition of high-frequency filter capacitors at both the input and output. These serve to filter out any high-frequency noise or signals that could interfere with the regulator's operation. HF filters are generally small-value, non-polarized capacitors (0.01 µF or 0.1 µF). A complete linear power supply is shown in figure 8-13.

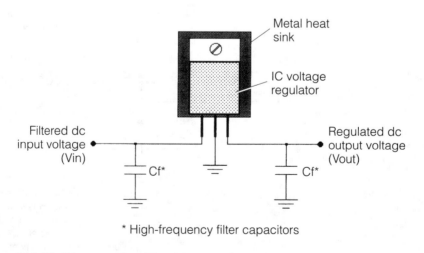

* High-frequency filter capacitors

■ **8-12** *Diagram of an IC voltage regulator.*

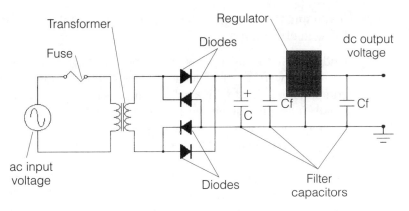

■ 8-13 *Schematic diagram of a basic linear power supply.*

Linear supplies in printers

Generally speaking, linear power supplies are simple, inexpensive, and reliable devices. In actual practice, however, few printers use linear power supply technology. You see, modern computer printers demand substantial amounts of power (especially EP printers), but the inherent inefficiency of linear power supplies would result in serious heat dissipation, which is undesirable. As a consequence, most commercial printers will employ switching power supplies as discussed later in this chapter. Linear supplies will usually be employed in low-power applications such as ac adapter modules for mobile printers. You might wonder why we have covered linear supplies in such detail when they are used so infrequently. The reason is that the principles behind switching power supplies are virtually identical to those of linear supplies; only the regulation technique is different. So by learning about linear supplies, you cover much of switching supplies as well. You will also see that portions of the switching supply are essentially linear supply circuits.

Troubleshooting a linear supply

An important factor to keep in mind when troubleshooting a linear supply is that they tend to run hot, sometimes very hot, so be sure to allow 10 or 15 minutes for the regulator heat sink(s) to cool before attempting to replace the component. If you will be taking measurements on the "live" circuit, be careful to avoid touching heat sink fins. Another caution is the filter capacitors, which tend to be rather large. Allowing a few minutes of idle time (with the printer off and unplugged) before checking a capacitor allows the filter to discharge and reduce the possibility of a jolt. Once again, if you must run powered tests on the supply, take all precautions to avoid electrocution.

Under most circumstances, linear power supplies are reasonably straightforward to troubleshoot. The relatively small number of components, along with a simple "input-to-output" circuit layout, makes following the circuit far less challenging than logic circuits. A sound procedure for linear supplies is often to use your voltmeter (or oscilloscope) to trace the presence of voltage from the output back through the supply. If an output has failed, start your measurements at that output and work backwards into the circuit until you find the appropriate signal again. The following symptoms and troubleshooting procedures present more details. Of course, you can also replace the power supply module outright.

Symptoms

Symptom 1 Power supply is completely dead. Printer does not operate and no power indicators are lit. Before you begin to disassemble the printer, check to make sure that you are receiving an appropriate amount of ac line voltage into the power supply. Use your multimeter to measure ac voltage at the wall outlet powering your printer. Normally, you should read between 105 and 130 Vac (210 to 240 Vac in Europe) for a power supply to operate properly. More or less ac line voltage can cause the power supply to malfunction. Inspect the printer where the ac line cord connects; you might notice a red line voltage switch that allows the power supply to be set for 120-V or 220-V operation. If your printer has a line voltage switch, see that it is set for the appropriate level. For example, if there is 120 Vac available, but the line voltage switch is set for 220 Vac, the power supply (and the whole printer) will not work. Exercise extreme caution whenever measuring ac line voltage levels. Review the hazards of electricity as discussed in Chapter 6.

When you determine that an appropriate amount of ac is available at the printer, the fault probably exists in the printer itself. Check the printer's power switch to be sure that it is turned on. Even though it sounds silly, this really is a common oversight. Next, check the printer's main line fuse, which is often located close to the ac line cord. Remember to unplug the printer before removing the fuse for examination. You should find the fusible link intact, but it is not always possible to see the entire link. Use your multimeter to measure continuity across the fuse. Normally, a working fuse should read as a short circuit (0 Ω). If you read infinite resistance, the fuse is defective and should be replaced. Use caution when replacing fuses. Use only fuses of the same amperage and voltage ratings. If a new fuse fails immediately when replaced, it suggests a serious failure (such as a short circuit) elsewhere in the power supply or the printer's internal circuitry. Do not continue to replace fuses if they continuously fail.

If everything checks properly up to this point, suspect a fault in the power supply module. You must disassemble your printer and work on the power supply. Take all precautions to protect yourself from electrocution hazards. First, check all connectors and wiring leading to or from the supply to rule out a broken wire or loose connector. Turn on printer power and use your multimeter to measure dc output voltage(s) from the supply. A typical printer power supply will deliver two voltages (usually +5 Vdc and +24 Vdc), and most printed circuit markings will give you some indication of what voltage should exist at each respective output. A low or nonexistent output indicates a problem. To make sure that the output(s) are not being shorted by their loads, disconnect the supply from its load and measure the output(s) again. If your readings climb up to a normal level, chances are that the power supply itself is functional, but there might be a short circuit elsewhere in the printer's electronics. If voltage readings remain low, you will have to troubleshoot the supply. For the purposes of this discussion, refer to the diagram of figure 8-13. If you do not have the time, equipment, or inclination to troubleshoot the printer's power supply, you should replace it outright.

A linear troubleshooting process

If the voltage output is completely zero, check for the presence of a dc (low voltage) protection fuse. Some power supplies, especially high-end supplies, fuse each output in addition to the ac input. You might find normally sized fuses in the power supply itself, but other fuses might also be located around the printer's ECU. Look carefully for any subminiature or "pico" fuses (resembling carbon film resistors). You will need to turn off and unplug the printer, then test each one.

Of all the components in your supply, the regulator carries the greatest stress. Use your multimeter to measure the dc input to the regulator. You should read several volts greater than the expected output. For example, a regulator with an output of +5 Vdc requires an input of +7 or +8 Vdc. When the regulator's input voltage is correct, but its output is not, the regulator is probably defective and should be replaced. A low or nonexistent regulator input suggests a faulty filter or rectifier.

A shorted power filter capacitor can pull down the output from a rectifier. Turn off and unplug the printer, remove at least one capacitor lead from its circuit, and test the capacitor as discussed in the test equipment section of Chapter 5. Replace any filter capacitor that appears open or shorted. Any power capacitor that appears hot or emits a strong pungent odor should be a clear indication of trouble. Be sure to replace filter capacitors with identical devices, and take care to observe any polarity markings.

Inspect the rectifier circuit carefully. A faulty rectifier diode can completely disable your supply. Turn off and unplug the printer, then test each rectifier diode as discussed in the test equipment section of Chapter 5. When a bridge rectifier fails, you will usually find the two forward diodes open circuited. Replace any diodes that appear open or short circuited. If your rectifier is built into a potted bridge module, the entire module must be replaced.

Finally, turn on printer power and check the ac voltages at the transformer's primary and secondary windings. Use extreme caution when measuring ac voltage levels. You should find about 120 Vac across the primary (about 220 Vac in Europe) and some lower amount of ac (usually about 30 Vac) across the secondary. Where there are multiple taps from the secondary winding, you should measure the output on each tap. An open circuit in either the primary or secondary winding can prevent any secondary output. Check for shorted transformer windings, and be suspicious of a transformer that becomes very hot after a short period of use, or one that emits an audible 50–60 Hz hum. Such a transformer might be developing a short circuit.

Consider the possibility of a printed circuit board failure, especially if the printer failed after being dropped or abused. Faulty soldering at the factory or on the repair bench can also cause a PC board problem. You might see printed circuits referred to as *printed wiring boards* (or PWBs). As figure 8-14 illustrates, there are three different kinds of problems that can plague a printed circuit: lead pull-through, trace break, and board crack.

Lead pull-through occurs when a component lead or wire is ripped away from its through-hole. Often, the soldering at the printed circuit pad might appear perfectly normal, but there will be a hole in the middle where the lead was. The lead might remain within its printed circuit hole, but it might not be fully connected. This kind of problem can easily result in bizarre, intermittent behavior, but it can be repaired simply by reheating the solder joint and applying fresh solder to reestablish the connection.

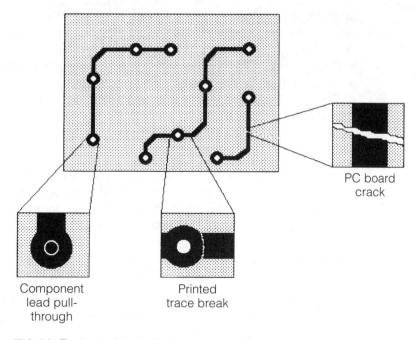

Component
lead pull-
through

Printed
trace break

PC board
crack

■ **8-14** *Typical printed circuit board problems.*

Trace break occurs commonly around large or awkward components that are too well soldered or mechanically attached to experience lead pull-through. Instead, physical force will break the solder pad away from its trace. Trace breaks are difficult to detect because they are usually so fine and clean that you might not see them during a visual inspection. You have to spot them by wiggling each lead individually. When a solder pad moves, but its trace does not, you will see the break location. Jumper between two adjacent solder pads to reliably repair this failure. Do not attempt to solder or jumper across the break itself. Chemical coatings applied to printed circuit boards prevent solder from sticking to trace areas.

Board cracks accompany such physical traumas as drops or other abuse. Impact forces can actually crack the circuit board. This, in turn, can split any traces that might be over it. Luckily, board cracks are relatively easy to spot because they leave a long, discolored line along the PC board. As with trace breaks, you must jumper between two adjacent solder pads to make a reliable repair rather than soldering across the break itself. If you detect an extensive board crack, however, you should replace the board rather than attempt to repair it.

Symptom 2 Supply operation is intermittent. Printer operation cuts in and out along with the supply. Begin by inspecting the ac line voltage powering your printer. If your line cord is loose at the wall plug or printer, it can play havoc with printer operation. Use extreme caution when dealing with ac line voltages to prevent injuries from electrical shock. Also check the integrity of any ac connections within the printer, along with any output connectors attached to the printer's internal circuitry. Tighten or replace any connectors that appear to be loose.

Check the power supply printed circuit board for any signs of failure, especially if the printer started malfunctioning after a drop or other physical abuse. Faulty soldering connections from the factory (or from your workbench) can also cause printed circuit failures. Review figure 8-14 for three typical printed circuit problems. If you do not have the time, equipment, or inclination to troubleshoot the power supply, you should replace it outright.

Lead pull-through is a fault that occurs anywhere a component lead or wire is soldered into a through-hole. Sudden, sharp force applied to the lead can overcome the connection's strength and rip the lead right away from its solder joint. This might or might not pull the lead from its hole entirely. Trace breaks can happen anywhere a round solder pad meets a printed trace. Sudden impacts that do not cause lead pull-through might cause a hairline fracture between the solder pad and its printed trace. These can be particularly difficult problems because it might not be possible to see trace breaks on visual inspection. You might have to wiggle each solder pad gently to reveal any trace breaks. Board cracks are fairly obvious problems, but large cracks can sever a great number of traces; your decision on whether or not to repair or replace the board really depends on the crack size and board complexity. As a general rule, breaks or cracks can be corrected by soldering jumpers between the solder pads across each defect.

Consider the possibility of thermal intermittents if your printer works fine when it is first turned on, but fails after some period of operation. Often, the printer must sit for a time with all power off before its operation will return. Test for thermally intermittent components by spraying suspect parts with a liquid refrigerant (available from just about any electronics parts store or mail-order house).

Begin by exposing the power supply. Apply power and operate the printer until it fails. Use your multimeter to measure each supply output before and after it fails. This will tell you what outputs are failing. When you have identified a faulty output, check its regulator for excessive heat. Never touch live components that might be hot or carrying high voltages; it is a certain opportunity for injury! Instead, smell around the regulator for any trace of smoke or unusually heated air. Spray the regulator with refrigerant, wait a moment, and recheck your output voltage. If normal voltage returns temporarily, you have isolated the problem. Replace the faulty regulator. Keep in mind that you might have to spray a component several times to cool it properly. One common mistake many novice troubleshooters make is their random use of refrigerant. If you spray several components at a time, you will not be able to determine where the fault is, so concentrate on only one component (such as the regulator) at a time.

Filter components and rectifier diodes are rarely subject to thermal problems. Transformer windings can open or short due to excessive heat, but only after a long period of breakdown. Use your multimeter to measure voltages through the remainder of the supply in order to track down any further problems.

Symptom 3 Printer is not operating properly. It might be functioning erratically or not at all. Power indicators might or might not be lit. Use your multimeter to measure the ac line voltage reaching the printer. Under normal circumstances, you should measure 105 to 130 Vac (some European countries use 210 to 240 Vac). On the average, 120 Vac should be available. If line voltage drops below 105 Vac, power supply outputs can begin to fall out of regulation. As a result, printer circuits might not receive enough voltage or current to ensure proper operation. This can cause erratic operation that can disable (or even damage) the printer. High input voltages (over 130 Vac) can force more current into the supply than desirable. Additional current generates heat, which can cause premature breakdowns in the power supply.

Turn off and unplug the printer, then check for any loose connectors or wiring that might be interrupting circuit operation. They might have been improperly installed at the factory, or you might have reinstalled them incorrectly during a previous repair. Restore printer power and use your multimeter to measure voltage at each supply output. If you locate a defective output, troubleshoot the supply from its output, back to its transformer as discussed in Symptom 1, or replace the power supply module entirely. If all supply outputs appear correct, there might be a fault in the printer's ECU, so troubleshoot your printer electronics according to the procedures outlined in Chapter 9, or replace the ECU outright.

If you detect a faulty supply output, use your multimeter to measure input voltage at that regulator. It should be several volts higher than the expected regulator output. When a regulator's input appears normal, but its output does not, try replacing the regulator. A low or nonexistent regulator input voltage might be caused by a fault in the filter or rectifier stages. Unplug the printer and check your filter capacitor(s) and rectifier diodes for open or short circuits, as discussed in the test equipment section of Chapter 5. Replace any filter capacitors or rectifier diodes that appear defective, or replace the power supply.

Switching power supplies

The great disadvantage to linear power supplies is their tremendous waste. At least half of all power provided to a linear supply is literally thrown away as heat; most of this waste occurs in the regulator. Ideally, if there was just enough energy supplied to the regulator to achieve a stable output voltage, regulator waste could be reduced almost entirely, and supply efficiency would be improved dramatically.

Understanding switching regulation

Instead of throwing away extra input energy, a switching power supply senses the output voltage provided to a load, then switches the ac primary (or secondary) voltage on or off as needed to maintain steady levels. A block diagram of a typical switching power supply is shown in figure 8-15. There are a variety of circuit configurations that are possible, but figure 8-15 illustrates one possible design. You can see the similarities and differences between a switching supply and the linear supply shown in figure 8-2.

An ac line voltage entering the supply is immediately converted to pulsating dc, then filtered to provide a "primary dc" voltage. Notice that ac is not transformed before rectification, so primary dc can reach levels approaching 170 V. Remember that ac is 120 V_{rms}. Because capacitors charge to the peak voltage (peak = rms × 1.414), dc levels can be higher than your ac voltmeter readings. This is as dangerous as ac line voltage, and should be treated with extreme caution.

On start-up, the switching transistor is turned on and off at a high frequency (usually 20 to 40 kHz), and a long duty cycle. The switching transistor breaks up this primary dc into chopped dc, which can now be used as the primary signal for a step-down

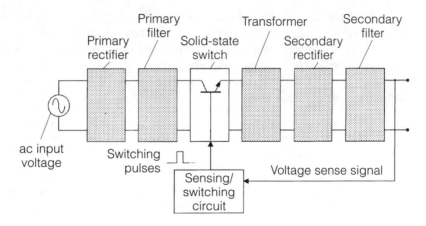

■ 8-15 *Block diagram of a switching power supply.*

transformer. The duty cycle of chopped dc will affect the ac voltage level generated on the transformer's secondary. A long duty cycle means a larger output voltage (for heavy loads), and a short duty cycle means lower output voltage (for light loads). Duty cycle itself refers to the amount of time that a signal is "on" compared to its overall cycle. Duty cycle is continuously adjusted by the sensing/switching circuit. You can use an oscilloscope to view switching and chopped dc signals.

The ac voltage produced on the transformer's secondary winding (typically a step-down transformer) is not a pure sine wave, but it alternates regularly enough to be treated as ac by the remainder of the supply. Secondary voltage is rerectified and refiltered to form a "secondary dc" voltage that is actually applied to the load. Output voltage is sensed by the sensing/switching circuit, which constantly adjusts the chopped dc duty cycle.

As load increases on the secondary circuit (more current is drawn by the load), output voltage tends to drop. This is perfectly normal; the same thing happens in every unregulated supply. However, a sensing circuit detects this voltage drop and increases the switching duty cycle. In turn, the duty cycle for chopped dc increases, which increases the voltage produced by the secondary winding. Output voltage climbs back up again to its desired value; output voltage is regulated.

The reverse will happen as load decreases on the secondary circuit (less current is drawn by the load). A smaller load will tend to make output voltage climb. Again, the same actions happen in an unregulated supply. The sensing/switching circuit detects this increase in voltage and reduces the switching duty cycle. As a result, the duty cycle for chopped dc decreases, and transformer secondary voltage decreases. Output voltage drops back to its desired value; output voltage remains regulated.

Switching supply trade-offs

Consider the advantages of a switching power supply. Current is only drawn in the primary circuit when its switching transistor is on, so very little power is wasted in the primary circuit. The secondary circuit will supply just enough power to keep load voltage constant (regulated), but very little power is wasted by the secondary rectifier, filter, or switching circuit. Switching power supplies can reach efficiencies higher than 85% (35% more efficient than most comparable linear supplies). More efficiency means less heat is generated by the supply, so components can be smaller and packaged more tightly.

Unfortunately, there are several disadvantages to switching supplies that you must be aware of. First, switching supplies tend to act as radio transmitters. Their switching frequencies can wreak havoc in radio and television reception, not to mention its own circuits inside the printer. This is because most switching supplies are somehow covered or shielded in a metal casing. It is critically important that you replace any shielding removed during your repair. Strong electromagnetic interference (EMI) can disturb the printer's electronics. Second, the output voltage will always contain some amount of high-frequency ripple. In many applications, this is not enough noise to present interference to the load. In fact, virtually all printers use switching power supplies. Finally, a switching supply often contains more components, and its "feedback loop" configuration is more difficult to troubleshoot than a linear supply. This is often outweighed by the smaller, lighter packaging of a switching supply.

A dot-matrix switching supply

Now that you understand the important concepts and operations behind switching power supplies, you can take a detailed look at two practical switching supplies. The power supply for a Tandy DMP 203 is illustrated in figure 8-16. You can see the demarcation line between the primary and secondary sides of the supply at the transformer T1. Raw ac power enters the supply when switch SW1 is closed. The inductor L1 serves as a filter, and diode bridge DB1 is the primary rectifier. If you look at the positive side of C2 with

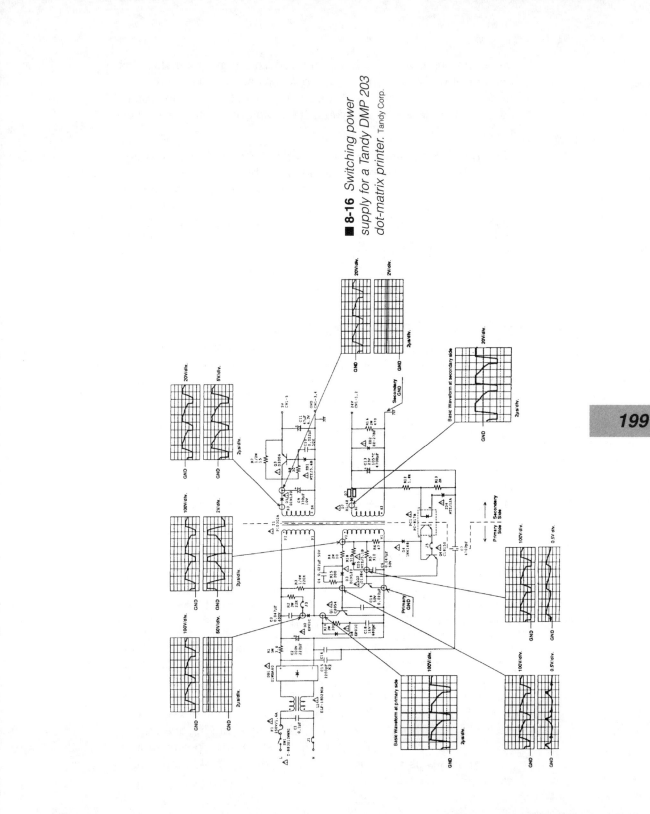

■ 8-16 *Switching power supply for a Tandy DMP 203 dot-matrix printer.* Tandy Corp.

an oscilloscope, you will find about 170 Vdc. Remember that this amount of dc is every bit as hazardous as ac, so take all precautions to protect yourself from injury. If you look at the junction of J3 and D2 in the schematic, you will find an even higher level of dc (about 275 V).

Now, the high dc signal must be "chopped" before it can be transformed by T1. Transistors Q1 and Q2 (with their associated biasing components) form an oscillator that breaks up dc into repetitive signals (such as the 12-V_{p-p} signal at transformer primary winding V2). There are some interesting points to note about this process. First, the chopped signals are not sinusoidal. While that would yield the best results, the repeating signals are close enough to support transformation across T1. Second, there are two primary and two secondary windings on T1; in effect, T1 acts as two separate transformers.

Now remember that the switching circuit requires a sensing signal that is fed back from the output. The feedback signal is supplied from the optoisolator PC1, which is driven from the 24-V output. The advantage of an optoisolator is that the primary and secondary circuits are entirely isolated, so a fault in one side will have less damaging effects on the other side. It is also important to note that the optoisolator and its adjacent components are vital to the proper operation of the supply. A fault in PC1, ZD4, Q4, and so on, will disable the entire supply.

Once the chopped signals are transformed by T1, the secondary signals are fed to two separate secondary power circuits, a 5-V circuit and a 24-V circuit. As you might notice, each of the secondary circuits is basically a simple linear power supply. For the +5-V output, diode D5 forms a half-wave rectifier, while C9 provides filtering (you'll see about +7 V across C9 with your oscilloscope). Transistor Q3 and zener diode ZD1 act as a series voltage regulator, which clamps the output to 5 V. Capacitor C11 merely provides additional filtering to the output signal.

The 24-V power supply is quite similar. Because the 24 V is intended to operate high-power devices, such as motors and print solenoids, the output does not need high-performance regulation. Diode D7 acts as a half-wave rectifier, and the pulsating dc is filtered by C13. Zener diode ZD2 provides a low level of regulation, basically minimizing electrical variations by clamping the output to 24 V. Resistor R16 works as a load resistor for the power supply, which doubles as a bleeder resistor for the beefy 4,700-μF filter capacitor.

An EP switching supply

The power supply circuit shown in figure 8-17 illustrates the same basic principles as figure 8-16, but this new design incorporates a series of sophisticated enhancements. Because this supply is intended for an EP printer, it must be capable of providing more power and stability than the circuit of figure 8-16. As a consequence, much of the added complexity is included to accommodate those two objectives. But ultimately, the supply in figure 8-17 will provide three outputs: a +5-Vdc output for the ECU logic, a +24-Vdc output to support the printer's motors and electromechanical clutches, and a separate +24-Vdc output designed to drive the fusing heater.

All of the detailed circuit descriptions are not covered here; that would take a small chapter itself. However there are some important similarities and differences that you should recognize. The ac enters the supply and is initially fused by F101. The arrangement of inductors and capacitors around the supplemental fuse F102 serve as surge and electrical noise suppressors. The small circuit arrangement below CN101 is the switching regulator circuitry that controls the fusing heater voltage. For the +5- and +24-V circuits, initial rectification is provided by the bridge rectifier D101. The bulk of active components around Q106, Q102, and IC101 form the sensing and switching network that breaks the high dc level into the chopped ac, which feeds transformer T1.

The single secondary output of T1 feeds ac to the +24- and +5-V regulator circuits. Note that both output circuits are fed from one half-wave rectifier (D201). As with figure 8-16, the secondary circuitry in the +24-V arm is responsible for developing and filtering the actual output voltage. Note the added complexity that is needed to support a more demanding motor system, as well as a laser and scanning system. To achieve optimum performance, the +24-V output is adjustable through VR1. Note also the added complexity of the +5-V circuit. This is necessary to support the extensive logic used in an EP printer. Voltage must be maintained within very tight margins in spite of the large current load; this is the responsibility of IC301 and associated circuitry. The +24-V output is fused by F201, while the +5-V output is fused by F301. With a total of four fuses in the power supply, you can see why it is so important to check the fuses before wasting time by troubleshooting a supply, which might not be defective. Another point to observe is that isolation between the primary and secondary circuits is provided by three individual optoisolators (PC1, PC2, and PC3) in-

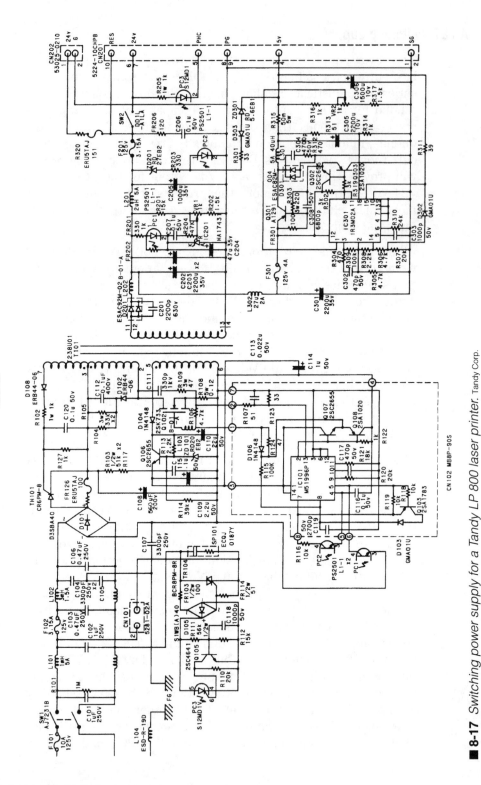

■ 8-17 *Switching power supply for a Tandy LP 800 laser printer.* Tandy Corp.

Power supply service techniques

stead of the single optoisolator in figure 8-16. This affords independent control over each output.

Troubleshooting a switching supply

When you consider the intricacies of a switching power supply, you can start to understand the difficulties associated in troubleshooting such a circuit. Tracing high-energy analog signals that feed back on themselves in this manner can be a time-consuming challenge even for experienced technicians. As a result, you should evaluate the economics of replacing the supply outright, rather than spending the time to troubleshoot it. Of course, enthusiasts with the time and inclination to perform such troubleshooting might find the challenge very rewarding. Still, subassembly replacement remains a viable option. The following section outlines a series of procedures for dealing with switching supply problems. For the purposes of this discussion, refer to the supply shown in figure 8-16.

Symptoms

Symptom 1 Power supply is completely dead. Printer does not operate, and no power indicators are lit. Refer to Chart 8-1. Before you begin to disassemble the printer, check to make sure that you are receiving an appropriate amount of ac line voltage into the power supply. Use your multimeter to measure ac voltage at the wall outlet powering your printer. Normally, you should read between 105 and 130 Vac (210 to 240 Vac in Europe) for a power supply to operate properly. More or less ac line voltage can cause the power supply to malfunction. Inspect the printer where the ac line cord connects; you might notice a red line voltage switch that allows the power supply to be set for 120 V or 220 V operation. If your printer has a line voltage switch, see that it is set for the appropriate level. For example, if there is 120 Vac available, but the line voltage switch is set for 220 Vac, the power supply (and the whole printer) will not work. Exercise extreme caution whenever measuring ac line voltage levels. Review the hazards of electricity, as discussed in Chapter 6.

When you determine that an appropriate amount of ac is available at the printer, the fault probably exists in the printer itself. Check the printer's power switch to be sure that it is turned on. Even though it sounds silly, this really is a common oversight. Next, check the printer's main line fuse (i.e., fuse F1 in figure 8-16), which is often located close to the ac line cord. Remember to un-

203

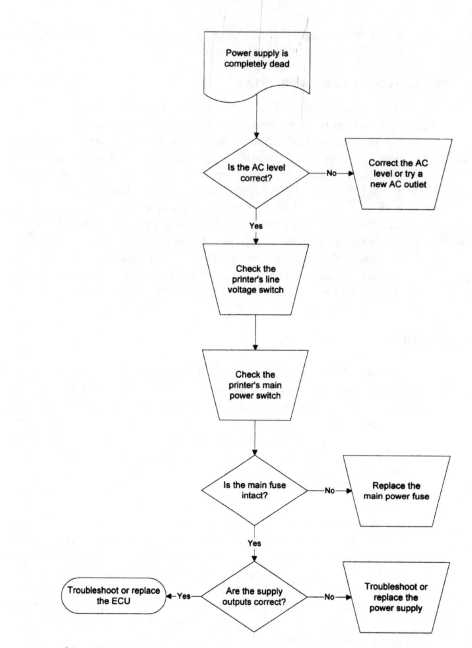

Chart 8-1 *Flowchart for Switching Power Supplies Symptom 1.*

204

plug the printer before removing the fuse for examination. You should find the fusible link intact, but it is not always possible to see the entire link. Use your multimeter to measure continuity across the fuse. Normally, a working fuse should read as a short circuit (0 Ω). If you read infinite resistance, the fuse is defective and should be replaced. Use caution when replacing fuses. Use only fuses of the same amperage and voltage ratings. If a new fuse fails immediately when replaced, it suggests a serious failure (such as a short circuit) elsewhere in the power supply or the printer's internal circuitry. Do not continue to replace fuses if they continuously fail.

If everything checks properly up to this point, suspect a fault in the power supply module. You must disassemble your printer and work on the power supply. Take all precautions to protect yourself from electrocution hazards. First, check all connectors and wiring leading to or from the supply to rule out a broken wire or loose connector. This is especially important for switching supplies because proper switching depends on having the proper load attached. If the load circuit (i.e., the ECU) is disconnected, you might not get proper outputs (if any) from the supply, or the outputs might fluctuate wildly. Turn on printer power and use your multimeter to measure dc output voltage(s) from the supply. The supply in figure 8-16 will deliver two voltages (+5 Vdc and +24 Vdc), and most printed circuit markings will give you some indication of what voltage should exist at each respective output. A low or nonexistent output indicates a problem. If you do not have the time, equipment, or inclination to troubleshoot the printer's power supply, you should replace it outright.

A switching troubleshooting process

When switching supply outputs continue to measure incorrectly with all connectors and wiring intact, your problem is probably inside the supply. With a linear supply, you generally begin testing at the output, then work back toward the ac input. But for a switching supply, you should begin testing at the ac input, then work toward the defective output. Refer to figure 8-16.

Measure ac entering the supply (i.e., across the inductor L1 entering the bridge rectifier DB1). If ac is absent, the current filter L1 might be defective and should be checked and replaced if necessary. Be sure to take all precautions to protect yourself from injury when measuring ac voltages. Next, check the dc level across the output of diode bridge DB1. Because the bridge is rectifying raw (untransformed) ac, you will read at least 170 Vdc. If this level is low or absent, the diode bridge might be defective, and should be checked and replaced.

Now, check the chopped ac levels across the transformer's primary windings. You can see typical oscilloscope patterns shown in figure 8-16. If chopped ac is missing, there is a defect in the sensing/switching circuit that you will need to isolate by turning off all power, then checking the active switching components (i.e., Q1, Q2, Q4, PC1, and their associated diodes). A fault in any one of these components will interrupt the oscillator circuit and disable the entire supply. Also keep in mind that high-frequency oscillator operation can fatigue capacitors in the switching circuit as well, and I have seen capacitors exhibit thermal intermittent operation (where the supply works when first turned on, but gradual heat buildup during operation causes the marginal capacitor(s) to fail and shut down the supply). As a rule, the probability of component failure is: transistors, diodes, capacitors, resistors, and transformers (with transistors the most likely devices to fail).

If you find the chopped ac signal available on the primary windings, check for secondary signals. If the secondary ac signal is missing, the transformer has failed, or there is a problem with the PC board contacts. For the circuit of figure 8-16, it is unlikely that both secondary windings (and thus both outputs) will fail simultaneously, so when only one output is lost, you can usually begin testing at the transformer. However, common secondary designs, like figure 8-17, will disable all outputs when the secondary fails.

From here on, you are fundamentally testing individual linear supply circuits. For the +5-V circuit, rectification is supplied by diode D5, filtering is handled by capacitor C9, and regulation is provided by the combination of Q3 and ZD1 (biased by R7, R8, and C10). The capacitor C11 merely supplies some supplemental filtering. If secondary ac is present, and the output is low or absent, the fault lies in one or more of those components, but is most likely in the regulator components which suffer the greatest stress.

You would deal with the +24-V output in a similar fashion, starting with the diode D7, the capacitor C13, and the regulator ZD2. Resistor R16 forms a load/bleeder to stabilize the output, and is an unlikely candidate for failure. If your measurements remain inconclusive or confusing, your best course is often just to replace the power supply module entirely.

Symptom 2 Supply operation is intermittent. Printer operation cuts in and out with the supply. Refer to Chart 8-2. Begin by inspecting the ac line voltage into your printer. Be sure that the ac line cord is secured properly at the wall outlet and printer. You need not worry about any fuse(s) in the supply. If the printer comes on at all, at least the main ac line fuse (i.e., F1 in figure 8-16) has to be intact. Turn off and unplug the printer, then expose your power supply. Carefully inspect every connector or interconnecting wire leading into or out of the supply. A loose or improperly installed connector can interfere with the operation of the supply. Pay particular attention to any output connections. A switching power supply must often be connected to its load circuit in order to operate. Without a load, the supply might cut out or oscillate.

In many cases, intermittent operation is the result of a PC board problem, such as the ones shown in figure 8-14. PC board problems are often the result of physical abuse or impact, but they can also be caused by accidental damage during a repair. Lead pull-through occurs when a wire or component lead is pulled away from its solder joint, usually through its hole in the PC board. This type of defect can easily be repaired by reinserting the pulled lead and properly resoldering the defective joint. Trace breaks are hairline fractures between a solder pad and its printed trace. Such breaks can usually render a circuit inoperative, and they are almost impossible to spot without a careful visual inspection. Board cracks can sever any number of printed traces, but they are often very easy to spot. The best method for repairing trace breaks and board cracks is to solder jumper wires across the damage between two adjacent solder pads; however, in the face of extensive damage, it might be more reliable to replace the damaged module outright.

Some forms of intermittent failures are time- or temperature-related. If your printer works just fine when first turned on, but it fails only after a period of use, then it spontaneously returns to operation later (or after it has been off for a while), you might be faced with a thermally intermittent component. A component might work when cool, but fail later, after reaching or exceeding its working temperature. Thermal intermittents have been known to plague poorly designed switching power supplies. After a printer quits, check for any unusually hot components. Never touch an operating circuit with your fingers; injury is almost certain. Instead, smell around the circuit for any trace of burning semiconductor or unusually heated air. If you detect an overheated component, spray it lightly with a liquid refrigerant. Spray

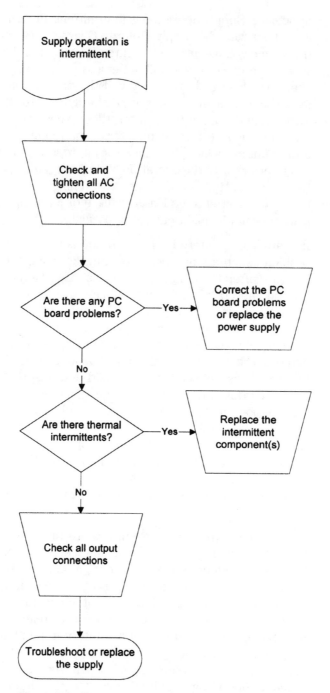

Chart 8-2 *Flowchart for Switching Power Supplies Symptom 2.*

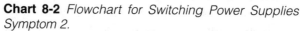

in short bursts for the best cooling. If normal operation returns, then you have isolated the defective component. Replace any components that behaved intermittently. If operation does not return, test any other unusually warm components.

If problems persist to this point, you are certainly free to begin troubleshooting the supply (from input to output, as described in the previous symptom). From a practical standpoint, however, troubleshooting an intermittent circuit (especially a feedback-type circuit) can be a particularly difficult exercise. Of course, you are welcome to take the challenge if you have the time and inclination to do so. But when the printer simply has to be fixed as soon as possible, you are often best advised to just replace the power supply module outright.

Symptom 3 Printer is not operating properly. It might be functioning erratically or not at all. Power indicators might or might not be lit. Use your multimeter and check the ac line voltage first. Take all precautions to protect yourself from injury. Normally, printers require a line voltage between 105 to 130 Vac to ensure proper operation. If line voltage is too low, the supply output(s) might not be able to maintain regulation. As voltage falls, circuitry in the ECU might begin to behave erratically or not at all; it could even damage some delicate printer circuits. You should also check to see if a 120/240 Vac power selector switch is available. If it is, it should be switched to the appropriate setting.

Check all wiring and connectors leading to and from the power supply to be sure that everything is tight and installed correctly. Pay particular attention to connector orientation. Loose, missing, or incorrectly inserted connectors can easily disable your printer, or at least cause unpredictable operation. Remember that switching power supplies usually require a load circuit (i.e., the ECU) to be connected. Otherwise, its output(s) might oscillate or shut down totally.

Switching power supplies produce a relatively large amount of electromagnetic interference (EMI) as a by-product of their switching operation. When a logic circuit (such as an ECU) is exposed to strong EMI, random errors might occur that will result in unpredictable or intermittent operation. To combat the effects of EMI, many switching power supplies are contained in some sort of metal enclosure or shroud that is wired to chassis (earth) ground. This is done to contain any EMI generated by the supply. After you are done taking measurements or replacing the power supply, make sure that all original shielding is in place and securely wired to ground. If it is not, EMI might interfere with the operation of other printer circuits to cause erratic or random behavior.

Use your multimeter or oscilloscope to measure each supply output. If all outputs measure correctly, then your trouble is most likely to be in the ECU. Refer to the troubleshooting procedures for electronic circuits contained in Chapter 9, or replace the ECU outright. Supply outputs that are low or nonexistent suggest a problem in the power supply itself. You might troubleshoot the supply, or replace it outright. Refer to figure 8-16.

Measure ac entering the supply (i.e., across the inductor L1 entering the bridge rectifier DB1). If ac is absent, the current filter L1 might be defective and should be checked and replaced if necessary. Be sure to take all precautions to protect yourself from injury when measuring ac voltages. Next, check the dc level across the output of diode bridge DB1. Because the bridge is rectifying raw (untransformed) ac, you will read at least 170 Vdc. If this level is low or absent, the diode bridge might be defective, and should be checked and replaced.

Now, check the chopped ac levels across the transformer's primary windings. You can see typical oscilloscope patterns shown in figure 8-16. If chopped ac is missing, there is a defect in the sensing/switching circuit that you will need to isolate by turning off all power, then checking the active switching components (i.e., Q1, Q2, Q4, PC1, and their associated diodes). A fault in any one of these components will interrupt the oscillator circuit and disable the entire supply. Also keep in mind that high-frequency oscillator operation can fatigue capacitors in the switching circuit as well, and I have seen capacitors exhibit thermal intermittent operation (where the supply works when first turned on, but gradual heat buildup during operation causes the marginal capacitor(s) to fail and shut down the supply). As a rule, the probability of component failure is: transistors, diodes, capacitors, resistors, and transformers (with transistors the most likely devices to fail).

If you find the chopped ac signal available on the primary windings, check for secondary signals. If the secondary ac signal is missing, the transformer has failed, or there is a problem with the PC board contacts. For the circuit of figure 8-16, it is unlikely that both secondary windings (and thus both outputs) will fail simultaneously, so when only one output is lost, you can usually begin testing at the transformer. However, common secondary designs, like figure 8-17, disable all outputs when the secondary fails.

From here on, you are fundamentally testing individual linear supply circuits. For the +5-V circuit, rectification is supplied by diode D5, filtering is handled by capacitor C9, and regulation is provided by the combination of Q3 and ZD1 (biased by R7, R8, and C10). The capacitor C11 merely supplies some supplemental filtering. If secondary ac is present, and the output is low or absent, the fault lies in one or more of those components, but is most likely in the regulator components, which suffer the greatest stress.

You would deal with the +24-V output in a similar fashion, starting with the diode D7, the capacitor C13, and the regulator ZD2. Resistor R16 forms a load/bleeder to stabilize the output, and is an unlikely candidate for failure. If your measurements remain inconclusive or confusing, your best course is often just to replace the power supply module entirely.

High-voltage power supplies

The high-voltage power supply is a vital element of EP printers. High voltage is necessary to operate the primary and transfer coronas, as well as the development unit (where toner is transferred to the drum). Older EP printers using discrete corona wires needed thousands of volts for proper operation, so the "SX-type" engines would receive –6,000 V for the primary corona, –600 V for the development unit, and +6,000 V for the transfer corona from the high-voltage power supply (HVPS). Newer EP engines fitted with charge rollers require a bit less voltage, and the corresponding HVPS will supply –1,000 V for the primary charge roller, –400 V for the development unit, and +1,000 V for the transfer roller. It is important to note that an HVPS uses a +24-V output from the dc power supply rather than a connection to raw ac, so if the dc power supply quits, so does the high-voltage unit. You can see a simple "black box" representation of the HVPS in figure 8-18.

Troubleshooting the HVPS

The HVPS presents a technician with several important problems that just can't be ignored. First, it is impossible to measure the output(s) of a high-voltage supply with conventional test instruments such as multimeters or oscilloscopes. Most conventional instruments do not accept more than 1,000 V, so higher voltages can easily damage the instruments. Most test leads only insulate to 600 V, so you stand an excellent chance of being electrocuted right through the test leads anyway. Ultimately, you will need specialized test equipment and test leads to check an HVPS. The other

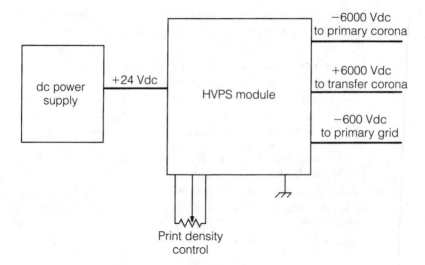

■ **8-18** *Block diagram of a high-voltage power supply.*

problem is an availability of parts. High-voltage circuitry demands very specialized components—devices rated for thousands of volts. These are often difficult and expensive parts (when you can find them). As a consequence, this book treats the HVPS as a black-box subassembly that has to be replaced rather than repaired. The EP printer's symptoms will suggest when to suspect an HVPS problem.

Warning: Under no circumstances should you attempt to measure the outputs of high-voltage power supplies using conventional test instruments; damage and injury will almost certainly result. Also do not open the HVPS, and stay clear of it while running powered tests on an EP printer. If you suspect an HVPS fault, replace the entire unit outright.

Replacing the HVPS

When you determine that the HVPS must be replaced, there are some important issues to keep in mind. First, a high-voltage source will retain a significant charge, even after the printer is turned off. To create the very safest conditions for an HVPS replacement, make sure that the printer is turned off and unplugged. Allow at least 15 minutes for the entire printer to cool and for power supplies to discharge.

To replace the HVPS, you need to unbolt several screws (and perhaps one or more grounding straps). This detaches the supply from the printer's chassis. You can then remove any wiring harnesses and place the old HVPS aside. When you install the new HVPS and attach all of the wiring harnesses (be careful to install all connectors properly and completely), you must be certain to install all ground straps and mounting screws securely. Failure to secure the high-voltage power supply to the chassis could result in a shock hazard, or unstable HVPS operation.

Symptoms

Identifying a failed HVPS is not always such a simple task, especially when you cannot measure its outputs. But once you realize that its outputs drive the primary charging area, the development unit, and the transfer area, you can easily understand that the HVPS plays a key role in the image formation process. A low or absent –6,000 (or –1,000) V will disable the primary charging mechanism (corona or roller). The net result would be no conditioning charge on the drum, and all pages would be black. A low or absent +6,000 (or +1,000) V will disable the transfer charging mechanism (corona or roller). No transfer means that no toner would be drawn to the page, so the pages would be white. Of course, the loss of supplemental voltages (i.e., –600 or –400 V) would adversely affect the primary grid (for corona assemblies) and the development unit, which would also whiten any latent images.

Remember that you should suspect the HVPS only when indicated by symptoms in the print, so refer to Chapter 11 for image formation problems. As you review those symptoms, you will see when it is advisable to replace the HVPS.

Electronic service techniques

REGARDLESS OF THE PARTICULAR TECHNOLOGY AT WORK in your printer (figure 9-1), each and every operation is controlled by a set of electronic circuits that is commonly called the *Electronic Control Unit* (or ECU). The specific architecture and components used in an ECU vary radically between printer models and manufacturers. A straightforward dot-matrix impact printer might only need simple control circuits, while a sophisticated EP printer requires a much more complicated ECU. This chapter shows you the internal workings of an ECU, and presents a series of troubleshooting procedures.

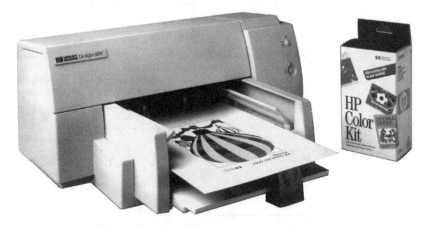

■ **9-1** *An HP DeskJet 600C printer.* Hewlett-Packard Co.

Regardless of their complexity, all ECUs must be capable of performing the same range of tasks. They must communicate with the outside world (i.e., the user as well as the computer), and translate communicated data and control codes, often in the ASCII format, into characters or patterns of dots that can be used to drive a print head. An ECU also directs such physical tasks as carriage transport and paper advance. It interprets sensor information regarding paper supply, carriage position, and head temperature. Finally, each of these tasks (and more) must be coordinated to work together. A typical ECU can be broken down into six functional areas: communication (or the interface), memory, control panel, drivers, main logic, and sensors. Figure 9-2 is an actual block diagram for a Hewlett-Packard QuietJet series printer. The circuit board layout for this block diagram is implemented in figure 9-3. Notice that these areas can often be fabricated onto a single PC

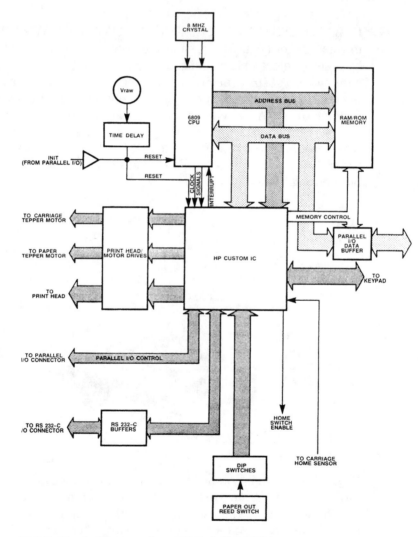

■ **9-2** *Block diagram of an HP QuietJet ECU.* Hewlett-Packard Co.

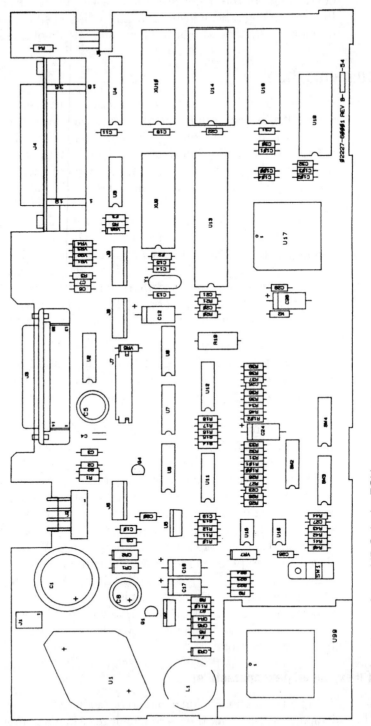

■ 9-3 *Board layout for an HP QuietJet ECU.* Hewlett-Packard Co.

217

Electronic service techniques

board. Each of the major areas of an ECU are represented here. Before you begin troubleshooting an ECU, you should thoroughly understand the operations and key components of these sections.

Communication

By itself, the printer serves little practical purpose. To be of any use at all, the printer must communicate with the outside world to receive the characters or graphics data that it must print. Binary data representing this information is sent from the host computer to your printer over one of several possible communication links. The computer also receives commands and status information back from the printer. These return signals are used to regulate the flow of data. Although there are many variations of communication links, data is transferred using either a serial or parallel technique. Data is sent over a parallel link as whole characters; that is, all the binary digits (or bits) that compose a character are sent at the same time over multiple signal wires. A serial link transfers data one bit at a time over a single wire. This part of this chapter explains the important concepts of printer communication.

A computer typically sends three types of information to a printer: text characters, control codes, and graphics data. Not all printers will accept character or graphic data or interpret them in the same way. Character data simply represents text, letters (in any language), numbers, punctuation, or other text symbols. Graphics data are the individual dots (or sequences of dots) that compose a much larger graphic image. Control codes are used to send commands to the printer. They can set general operating modes such as font style, enhancements, or pitch, but they can also direct immediate operations, such as form feed or line feed. Using control codes eliminates the need to operate a control panel manually while a document prints. Other control codes can switch the printer into and out of various graphics modes. When a printer is placed in a graphics mode, subsequent information sent to the printer will be processed as individual dots before being sent along to the print head or writing mechanism. An opposing control code, or reset control, will return the printer to its character (or text) mode. Printing under Windows or Windows 95 is done exclusively in the graphics mode.

ASCII explained (text characters)

Before any communication can take place, both the printer and the computer must speak the same language; when a computer sends out the character "H," its printer must be able to recognize that character as an "H." Otherwise, it will just print unintelligible garbage. Because each character and control instruction is represented by its own unique numerical code, both printers and computers must use a common set of codes that describe some minimum number of characters. In the early days of computers, each manufacturer had their own character code set. You can probably imagine how difficult it was to combine equipment made by different manufacturers. As the electronics industry matured and printers became more commonplace, the demand for equipment compatibility forced manufacturers to accept a standard character code set.

The American Standard Code for Information Interchange (known as ASCII) has come to represent a single, standard code set for computer/printer communication. Table 9-1 represents a conventional ASCII table showing characters, their code numbers, and a binary representation of those numbers. The standard ASCII code covers letters (upper- and lowercase), numbers, simple symbols for punctuation and math functions, and a few basic control codes. For example, if you want to print a capitol "D," your computer must send the number "68" to the printer. The printer would then translate "68" into a dot pattern that reflects the selected font, character pitch, and enhancements, to form the letter "D." To print the word "Hello," a computer would have to send a series of numbers: "72," "101," "108," "108," and "111." Pure (original) ASCII uses codes 0 to 127.

Due to the way character codes are actually sent, however, most computers can also use ASCII codes ranging from 128 to 255, but keep in mind that any code over 127 is not pure ASCII. Instead, codes from 127 to 255 are sometimes called an *alternate character set*. Such an alternate character set can contain single block graphic characters, Greek symbols, or other language characters. In some cases, codes 128 to 255 just duplicate codes 0 to 127. If your computer sends a code from 128 to 255, you might be printing characters that are different than those on your computer screen.

Control codes (the great ESC)

Not only must a computer specify what to print, it must also specify how to print. Carriage returns, line feeds, font styles, and enhancements are just some of the controls that a computer must exercise to automate the printing process. Just imagine the confu-

■ Table 9-1 Standard ASCII chart (0 to 127).

Character	Decimal	Hex	Character	Decimal	Hex
NUL	0	00h	SOH	1	01h
STX	2	02h	ETX	3	03h
EOT	4	04h	ENQ	5	05h
ACK	6	06h	BEL	7	07h
BS	8	08h	HT	9	09h
LF	10	0Ah	VT	11	0Bh
FF	12	0Ch	CR	13	0Dh
SO	14	0Eh	SI	15	0Fh
DLE	16	10h	DC1	17	11h
DC2	18	12h	DC3	19	13h
DC4	20	14h	NAK	21	15h
SYN	22	16h	ETB	23	17h
CAN	24	18h	EM	25	19h
SUB	26	1Ah	ESC	27	1Bh
FS	28	1Ch	GS	29	1Dh
RS	30	1Eh	US	31	1Fh
SP	32	20h	!	33	21h
"	34	22h	#	35	23h
$	36	24h	%	37	25h
&	38	26h	'	39	27h
(	40	28h	)	41	29h
*	42	2Ah	+	43	2Bh
,	44	2Ch	-	45	2Dh
.	46	2Eh	/	47	2Fh
0	48	30h	1	49	31h
2	50	32h	3	51	33h
4	52	34h	5	53	35h
6	54	36h	7	55	37h
8	56	38h	9	57	39h
:	58	3Ah	;	59	3Bh
<	60	3Ch	=	61	3Dh
>	62	3Eh	?	63	3Fh
@	64	40h	A	65	41h
B	66	42h	C	67	43h
D	68	44h	E	69	45h
F	70	46h	G	71	47h
H	72	48h	I	73	49h
J	74	4Ah	K	75	4Bh
L	76	4Ch	M	77	4Dh
N	78	4Eh	O	79	4Fh
P	80	50h	Q	81	51h
R	82	52h	S	83	53h
T	84	54h	U	85	55h
V	86	56h	W	87	57h

Character	Decimal	Hex	Character	Decimal	Hex
X	88	58h	Y	89	59h
Z	90	5Ah	[	91	5Bh
\	92	5Ch	]	93	5Dh
^	94	5Eh	_	95	5Fh
`	96	60h	a	97	61h
b	98	62h	c	99	63h
d	100	64h	e	101	65h
f	102	66h	g	103	67h
h	104	68h	i	105	69h
j	106	6Ah	k	107	6Bh
l	108	6Ch	m	109	6Dh
n	110	6Eh	o	111	6Fh
p	112	70h	q	113	71h
r	114	72h	s	115	73h
t	116	74h	u	117	75h
v	118	76h	w	119	77h
x	120	78h	y	121	79h
z	122	7Ah	{	123	7Bh
		124	7Ch	}	125	7Dh
~	126	7Eh	DEL	127	7Fh

221

sion if you stand by your printer to change these modes manually while a document was printing! Unfortunately, control codes are often the cause of some incompatibilities between computers and printers. When ASCII was first developed, printers were extremely primitive by today's standards. Multiple fonts and type sizes, graphics, and letter-quality print had not even been considered. Few controls were needed to operate these early printers, so only those few critical controls were incorporated into ASCII. You might recognize such controls as Form Feed (FF), Line Feed (LF), or Carriage Return (CR) from Table 9-1.

With the inclusion of advanced electronic circuitry, a greater amount of "intelligence" became available in printers. This, in turn, has made so many of their current features possible. ASCII codes are still standard, but there simply are not enough unused codes to handle the wide variety of commands that are needed. Manufacturers faced the choice of replacing ASCII (and obsoleting an established and growing customer base), or developing a new scheme to deal with advanced control functions. Ultimately, manufacturers responded to this by devising a series of multicode control sequences. These were known as *escape sequences* because the ASCII code "27" (Escape) is used as a prefix. Printer capabilities can vary greatly between models and manufacturers. As a result, escape codes are not standard. If computer software is not written or configured properly for its particular printer, control codes sent by the computer might cause erratic or unwanted printer operation.

Escape sequences are typically two or three ASCII codes long, and each begins with ASCII code "27." It is the escape character that tells the printer to accept subsequent characters as part of a control code. For example, to set a printer to "Compressed Print," a computer might have to send an ASCII code "27," followed by an ASCII code "15" (SI). Software in the printer's main logic would interpret this code sequence and alter the appropriate modes of operation accordingly. A typical sequence to set a new "Character Pitch" might be "27" (Escape), followed by a "103" (letter "g"). Multicode sequences will certainly become more common as printers get even more sophisticated. The PRINTERS utility allows you to test printer features by entering manual escape code sequences.

Number systems

Not only must a computer and printer exchange codes that they both understand, but every code must be sent using a number system that is compatible with electronic (digital) circuitry. You already know the decimal (or base 10) number system. The symbols 0 through 9 are used in combinations that can express any quantity. The symbols themselves are irrelevant; ten other symbols could just as easily have been used, but 0 through 9 are the ones we have accepted down through the centuries. What is important is the quantity of characters in a number system. In decimal, one character can express 10 unique levels or magnitudes (0–9). When the magnitude to be expressed exceeds the capacity of a single character, the number carries over into a higher representative place, which is equal to the base of the system raised to the power of that place. For example, the number 276 has a 2 in the "hundreds" place, a 7 in the "tens" place, and a 6 in the "ones" place. You have worked with this system since grammar school.

If electronic circuits could recognize ten different levels for a single digit, then digital electronics would be directly compatible with our human decimal system, and ASCII codes would be exchanged directly in their decimal form. However, digital electronics can only recognize two signal levels. These conditions are On or Off (True or False). This is known as the *binary* (or base 2) number system. Because only two conditions can be expressed, only two symbols are needed to represent them. The symbols "0" (Off) and "1" (On) have come to represent the two possible conditions for a binary digit (commonly called a *bit*). ASCII codes must be sent as sets of binary digits.

As with the decimal number system, when a quantity to be expressed exceeds the capacity of a character (in this case it is a "1"), the number carries over into a higher place that is equal to the base of the system (base 2) raised to the power of the place. You have probably seen binary signals expressed as $2n$, where n is the bit's place position. As figure 9-4 illustrates, the decimal number "20" equals the binary number "10100." A lowercase "u" with an ASCII code of "117" would be expressed as "1110101" in a digital system. Seven bits can express numbers from 0 to 127. Eight bits can express numbers up to 255.

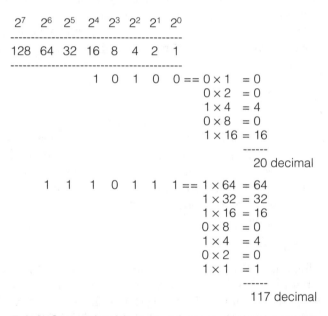

$$
\begin{array}{cccccccc}
2^7 & 2^6 & 2^5 & 2^4 & 2^3 & 2^2 & 2^1 & 2^0
\end{array}
$$

----

128 64 32 16 8 4 2 1

```
              1   0   1   0   0 == 0 × 1  = 0
                                   0 × 2  = 0
                                   1 × 4  = 4
                                   0 × 8  = 0
                                   1 × 16 = 16
                                            ------
                                            20 decimal

      1   1   1   0   1   1   1 == 1 × 64 = 64
                                   1 × 32 = 32
                                   1 × 16 = 16
                                   0 × 8  = 0
                                   1 × 4  = 4
                                   0 × 2  = 0
                                   1 × 1  = 1
                                            ------
                                            117 decimal
```

■ **9-4** *Converting binary numbers to decimal numbers.*

Binary digits

In order to have any meaning at all in electronic circuits, there must be a clearly defined relationship between a binary digit and a voltage level. Because a binary "1" is considered to be an On condition, it usually indicates the presence of a voltage. A binary "0" is considered Off, so it denotes the absence of a voltage. In reality, the actual amount of voltage that describes a 1 or 0 depends on the logic family in use. Common digital circuits using conventional TTL (Transistor-Transistor Logic) ICs classify a 0 as 0 to +0.8 Vdc, and a 1 as +2.4 to +Vcc (the voltage powering the IC).

Practical printer communication

There are literally hundreds of ways that you might be able to implement a communication interface. A great many versions have been tried and abandoned since the early days of commercial printers. The evolution of technology favors the best methods and techniques, so those that work well and grow with advances in technology can sometimes develop into standards that other manufacturers adopt in the future. Standards are basically a detailed set of rules and performance characteristics that clearly define the construction, connection, and operation of a circuit or system, in this case a communication interface. By adopting an established standard, manufacturers can be sure that printer brand "Y" will operate just fine with computer brand "X," and vice versa. This book covers two types of interfaces: parallel (Centronics) and serial (RS232).

The parallel interface

The Centronics Corporation was one of the original printer manufacturers, so they were able to establish an early lead in the marketplace. Their printers used a parallel interface of their own design. Its speed and simplicity soon made it a "de-facto" standard for other manufacturers to duplicate. It is commonly called the *Centronics interface*, although most PC technicians simply refer to it as the parallel port or LPT port.

Parallel architecture

A parallel port is easiest to understand because of its straightforward operation. You can see a simplified diagram of a parallel interface shown in figure 9-5. Notice that eight bits of an ASCII code are transferred simultaneously (D0 through D7). Data lines alone, however, are not enough to transfer information successfully. Both the computer and printer must be synchronized so that the printer will accept data when it is offered, or ask the computer to wait until it is ready. Synchronization of a parallel link is accomplished using several control wires, in addition to data lines. Some control lines signal the printer, and others will signal the computer. This mutual coordination is known as *handshaking*.

Parallel operation is really quite fast. The printer will accept information as fast as the computer can send it, often operating at

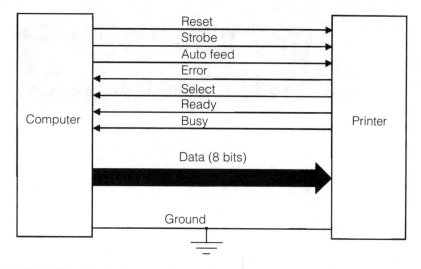

■ **9-5** *Simplified diagram of a parallel interface.*

speeds exceeding 1,000 CPS. At eight bits per character, that amounts to more than 8,000 bits per second. As discussed later in this chapter, the printer's internal data bus is also eight bits, so an eight-bit parallel word can be accepted into a data buffer with very little additional circuitry; parallel communication circuits are usually very simple. The main disadvantage to parallel links is its limited cable distance. With so many high-speed data signals running together in the same cable, its effective length is just a few feet. Beyond that, electrical noise and losses can cause distortion and loss of parallel data.

A Centronics-type interface connects a 25-pin subminiature D-type connector at your computer to a 36-pin Centronics-type connector at your printer. The size and shape of this 36-pin connector is standard (Amphenol Corporation #57-30360), so anytime you find a Centronics-type connector, as shown in figure 9-6, you will know the printer supports parallel communication. You might wonder why a 25-pin connector is used at the PC, while a 36-pin connector is used at the printer; what happens to the extra 11 pins? Well, the extra pins on the 36-pin connector are not used, or they are simply tied to ground.

A Centronics-type interface uses standard TTL levels to transfer information, so a binary "0" will be between 0 and +0.8 Vdc, and a binary "1" will be between +2.4 Vdc and $+V_{cc}$. These logic levels can easily be measured with a multimeter, but a logic probe or oscilloscope would probably be needed to observe actual data oc-

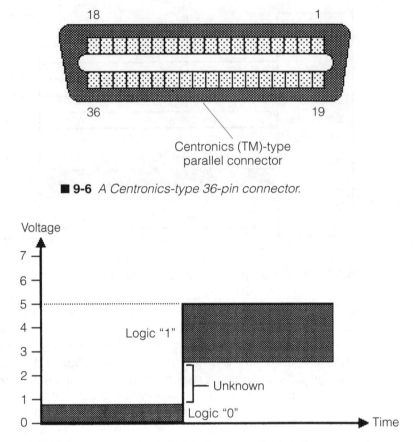

■ **9-6** *A Centronics-type 36-pin connector.*

■ **9-7** *Voltage versus logic levels for a parallel interface.*

curring on each signal line. Figure 9-7 illustrates the relationship between voltage and logic levels.

Interface signaling

The interface wiring for a typical Centronics interface is illustrated in figure 9-8. As you look at each of the pins along the left (connector) side of the diagram, you will see that all 36 pins of the Centronics interface are accounted for. Another thing you might notice is that the parallel interface is remarkably simple; there is little more there than pull-up resistors and signal conditioning circuitry. In fact, each of the pins on the right (internal) side of the diagram are connected to a single gate array IC (an ASIC), which handles all of the parallel communication processing. (The gate array is not shown in the schematic segment of figure 9-8.) Aside from the gate array and signal conditioning components, there is little else that can fail in the parallel interface.

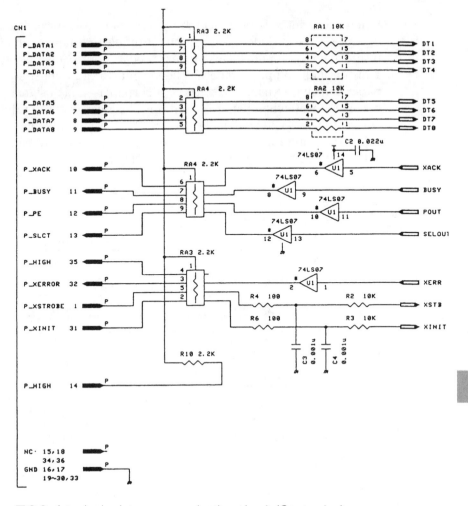

■ 9-8 *A typical printer communication circuit (Centronics).* Tandy Corporation

When the interface is first turned on, the P_SLCT (select) signal becomes a logic 1 to indicate that the printer has been selected. P_SLCT will remain logic 1 as long as paper is available and the printer is left on-line (it can be taken off-line through the control panel). The computer sends data bits representing a character or control code simultaneously over data lines P_DATA1 through P_DATA8. Once data bits are sent, the computer generates a brief logic 0 on its P_XSTROBE (strobe) line. A P_XSTROBE pulse tells the printer that data is valid, and that it should be accepted and stored in the data buffer memory. Once data is accepted by the printer, it generates a logic 1 P_BUSY (busy) signal. The host computer stops sending information until the P_BUSY signal returns to logic 0.

When the printer is no longer busy, it generates a brief logic 0 P_XACK (acknowledge) signal, which requests a new character from the host computer. This process repeats approximately every 1 ms until all computer data has been sent. Printing must stop if paper becomes exhausted, so the P_PE (paper empty) signal will become logic 0 when paper is used up. This deselects the printer automatically, so you will have to press the "on-line" button to re-select the printer after paper is restored. Figure 9-9 shows a typical timing chart for a normal printing cycle.

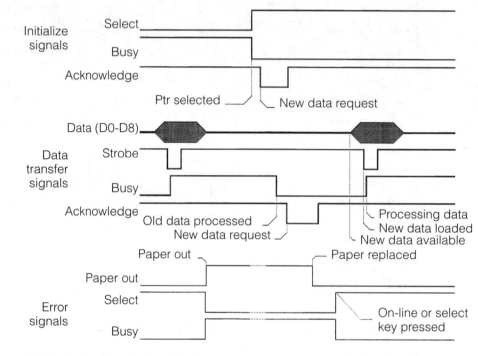

■ **9-9** *Centronics timing signals.*

IEEE 1284

By the end of the 1980s, it was becoming clear that "conventional" bidirectional parallel ports were simply not adequate to handle the new generations of faster peripherals that were appearing for the parallel port (i.e., CD-ROMs, tape drives, and laser printers). The 150KB/s parallel transfer rates that were once considered speedy were now severely limiting the new peripherals. In 1991, a group of major PC manufacturers (including IBM, Lexmark, and Texas Instruments) formed the Network Printing Alliance (NPA) in an attempt to develop a new parallel port architecture. In 1994, the IEEE (in conjunction with the NPA) released the Standard Signaling Method for a Bidirectional Parallel Peripheral Interface, also known as IEEE standard 1284. You will encounter this standard when installing new motherboards, multi-I/O boards, and parallel port devices.

The IEEE 1284 does not define a single parallel approach, but instead outlines five different operational modes for the parallel port: compatibility mode, nibble mode, byte mode, ECP mode, and EPP mode. All five modes offer some amount of bidirectional capability (known under IEEE 1284 as *back channel communication*). When the 1284-compliant parallel port is initialized, it checks to see which operating mode is appropriate.

Compatibility mode

IEEE 1284 is fully backward-compatible with conventional parallel port technologies where data is sent along eight data lines, the status lines are checked for errors and to see that the device is not busy, then a Strobe signal is generated to "push" the data into the device. As with ordinary parallel ports, the output of a single byte requires at least four I/O instructions. Data bandwidth is limited to 150KB/s.

Nibble mode

The nibble mode (four bits at a time) is a simple means of receiving data back from a peripheral device in fewer I/O instructions, though it is very inefficient. When used by itself, the nibble mode is limited to about 50KB/s. In most practical implementations of IEEE 1284, the nibble mode will rarely be used for more than gathering brief diagnostic or status information about the peripheral.

Byte mode

The byte mode allows the PC to disable the hardware drivers normally used to operate parallel data lines, which allows the data lines to be used as an input port to the PC. When in the byte mode, a peripheral can send a full byte to the PC in only one I/O cycle. Thus, it is possible to acquire data from a peripheral much faster than would be possible in the nibble mode.

ECP mode

The Enhanced Capabilities Port (ECP) allows bidirectional data transfer within a single I/O cycle. When a transfer is requested, the port's hardware will automatically perform all of the port synchronization and handshaking operations formally handled by software-driven I/O cycles in the compatibility mode. When properly implemented, an ECP port can run from 800KB/s to 2MB/s depending on the device at the port and the cable between them.

EPP mode

The Enhanced Parallel Port (EPP) is the apex of IEEE standard 1284. Like the ECP mode, EPP operation facilitates bidirectional data transfer in a single I/O cycle with the port hardware itself handling all synchronization and handshaking. EPP operation also can run from 800KB/s to 2MB/s. However, EPP operation takes another step forward by treating the parallel port as an extension of the system bus; this allows multiple EPP devices to exist on the same port.

The cable

Conventional parallel ports are limited to cable lengths of about ten feet. Beyond that, crosstalk in the parallel cable can result in data errors. Ideally, high-quality, well-shielded cable assemblies can extend that range even more, but the cheap, mass-produced cable assemblies that you often find in stores are rarely suited to support communication over more than six feet. To support the high-speed communication promised by IEEE 1284, a new cable specification also had to be devised. This is hardly a trivial concern, especially considering that IEEE 1284 seeks to extend parallel port operation to as much as 30 feet (about 10 meters).

IEEE 1284 considerations

Unfortunately, while the potential and promise of IEEE 1284 offers a lot of appeal, there are some serious considerations involved in configuring an enhanced port arrangement. Specifically, you will require an IEEE 1284-compliant parallel port, cable, and peripheral (i.e., printer, tape drive, hard drive, and so on) to take full advantage of enhanced capabilities.

Installing an IEEE 1284 parallel port is certainly not a problem; most current multi-I/O boards and late-model motherboards are now providing IEEE 1284 ports. The trouble is that using a $5 printer cable with your old Panasonic KX-P1124 dot-matrix printer will just not provide any advantages. To start benefiting from an IEEE 1284 port, you will need at least an IEEE 1284 cable and a device with significant memory capacity (such as a laser printer). At that point, you might start to see some speed improvements, but the additional speed will still fall far short of the projected figures. Ultimately, you will need to install IEEE 1284-compliant peripherals that will provide ID information to the port and allow optimum performance.

The serial interface

A serial approach emerged when the Electronics Industry Association (EIA) developed a comprehensive serial standard that they dubbed "Recommended Standard 232," or simply RS232 as we know it today. Since its original inception, the RS232 standard has been updated and revised a number of times. For the purpose of this discussion, however, it will simply be called RS232. You should know that RS232 was not designed just for use in printers, but as a universal interface scheme for any serial device, such as a modem, video monitor, or keyboard. As a comprehensive standard, it offers many handshaking and signal lines, only a few of which are really needed for printer communication. When RS232 was first introduced, many manufacturers chose to use only the signal lines that they thought were necessary for serial operation. This quickly fragmented the use of RS232, and created incompatibilities between computers and printers. Today, the use of RS232 signal and handshaking lines is much more readily accepted.

Serial architecture

A serial link might appear simpler because of the reduced wiring, as shown in figure 9-10, but its actual operation is somewhat more involved. You can see from figure 9-10 that two wires are used to transfer information. One of these wires carries data from computer to printer, while the other carries data from printer to computer. Because data can travel in both directions, this is known as a *bidirectional* data link. Only one wire is available to send (or receive), so a character must be sent only one bit at a time. Serial data must also be synchronized between the computer and printer. To accomplish this over a single wire, synchronization bits (i.e., start and stop bits) are added at the beginning and end of each character. An extra bit (known as a *parity bit*) might also be included to allow basic error checking. This is discussed in greater detail later in the chapter.

Serial handshaking can be provided either through hardware or software. Software handshaking takes advantage of the bidirectional nature of serial communication by allowing the printer to

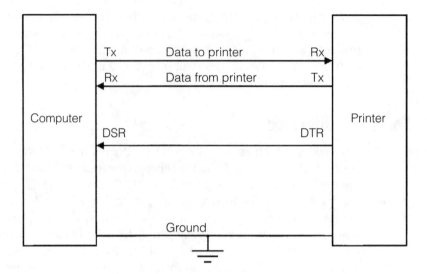

■ 9-10 *Simplified diagram of a serial interface.*

transmit control codes back to the computer. Two codes used commonly for software handshaking are "XON" and "XOFF." Older serial handshaking might use the codes "ETX" and "ACK." Hardware handshaking does not support data transfer from printer to computer. Instead, an additional handshake line signals the computer that the printer is busy. The DSR/DTR line is used for hardware handshaking in figure 9-10, but some interfaces carry more than one handshaking line. As a general rule, you can expect to see a variety of handshaking schemes between printer generations, so pay particular attention to the wiring in your serial printer's cable.

In spite of their added operating complexity, serial communication is extremely popular because of its bidirectional nature, flexibility, and its ability to work well over long distances. Typical applications of RS232 use subminiature D-type connectors, like the one shown in figure 9-11, at both the printer and computer ends. However, because a serial data link can be implemented with as few as three wires, some computers use 15-pin or 9-pin subminiature D-type connectors.

The data frame & signaling

Data inside a computer (and printer) is processed and stored as complete characters, but in order to transmit a complete serial word, it must be disassembled for transmission, then reassembled after being received. This process of parallel-to-serial and serial-to-parallel conversion is handled by a circuit called a UART (or

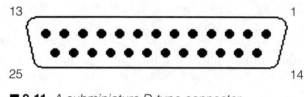

■ **9-11** *A subminiature D-type connector.*

Universal Asynchronous Receiver/Transmitter). Because data is asynchronous, communication can take place at any time, so a receiving UART must be able to tell when a word starts, when it stops, and whether or not it is correct. To handle this tracking process, a UART automatically adds start, stop, and parity bits to the data word (referred to as the *data frame*). Figure 9-12 shows a bit configuration found in a typical serial word.

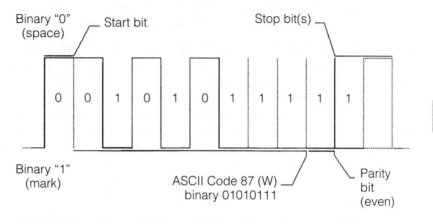

■ **9-12** *Bit configuration of a typical serial word.*

When the transmit or receive line is idle, it rests at a binary 1 level at –10 V or so (called a *mark*). A UART inserts a binary 0 at +10 V (or a *space*) as a start bit to signal the receiving UART that a word is transmitting. The first space is always considered to be the start bit. The next seven or eight bits contain the actual ASCII code. Most serial interface circuits can switch between seven or eight data bits using a dip switch selection. Keep in mind that both the computer and printer interfaces must be configured exactly the same way for successful communication to take place.

After all data bits have been sent, the UART might insert an error-checking bit (called a *parity bit*) that will aid the receiving UART in determining if the word it received is correct. Even parity will add an extra bit to produce an even number of binary 1s in the data word. Odd parity will add an extra bit to produce an odd number of 1s in the data word. For example, if even parity is selected and the number of 1s in the data word is odd, a binary 1 will be inserted as the parity bit. If the word already contains an even number of 1s, a binary 0 will be inserted as the parity bit. If odd parity is selected and the number of binary 1s is even, a binary 1 will be inserted as the parity bit. If the number of 1s is already odd, the parity bit will be a binary 0. The receiving UART counts 1s in the data word, then calculates its own parity bit. If this result matches the received parity bit, the data word is assumed to be correct. If the result does not match the received parity bit, the UART flags an error code to the computer. A serial interface can be set for even, odd, or no parity through dip switch settings.

Finally, the data word must be ended. A UART will add 1 or 2 stop bits to the end of the data stream. Stop bits are always binary 1s. After the selected number of stop bits is sent, the data line remains in its logic 1 state, and a new word can be sent at any time. A serial interface can be set for 1 or 2 stop bits. Another important consideration is the rate at which asynchronous bits are sent. This is known as the *baud rate*. Serial ports offer data rates from 120 bps to more than 38,400 bps, but serial ports for printer applications rarely run at more than 9,800 bps. Data rates can often be adjusted through dip switch settings.

Notice that the signal levels of serial data and control lines are also different from those of a parallel interface. Serial signals are bipolar; that is, one logic level is represented by a positive voltage, while the opposing level is represented by a negative voltage. This kind of bipolar operation allows serial interfaces to carry data over greater distances than parallel interfaces. Although figure 9-12 shows bipolar (loosely referred to as *analog*) signals ranging from +10 to –10 V, a serial interface can utilize voltage levels from ± 5 to ± 15 V. You can measure these serial levels with an oscilloscope.

Unfortunately, bipolar voltages are not compatible with the digital logic devices at work in the printer, so data and handshaking signals must be translated between bipolar and TTL levels as required. This transition is accomplished by a set of devices known as *line transceivers*. For example, bipolar bits received from a computer are converted to TTL levels using a line receiver, while TTL bits must be converted to bipolar levels using a line driver.

Serial circuitry

A typical serial communication circuit is shown in figure 9-13. While the array of lines and interconnections might appear daunting, the circuit's operation is remarkably straightforward. Data enters the printer through the Rx line (pin CN3-18) where it is translated from a bipolar signal into a TTL signal by the line receiver IC6. From IC6, data is passed directly to the gate array (IC10), which contains all of the UART and data handling functions needed by the printer. Similarly, data to be passed back to the printer is converted to serial form by the gate array, then fed to the line driver (IC8), which converts TTL signals into bipolar signals for transmission. Bipolar serial data leaves the printer through Tx line (pin CN3-9). Communication parameters can be set through DIP switch DSW1. DTR and CTS handshaking signals are processed through IC7 and IC5. The proliferation of resistors and capacitors in the circuit serve largely as signal conditioning and filter components.

Isolating the communication circuits

A communication interface involves much more than printer circuits. The successful transfer of data requires proper operation of a computer and interconnecting cable as well. Trouble in any one of these three areas can interrupt the flow of data. Before you disassemble your printer, you should isolate the problem to the printer itself. The quickest and most certain way to do this is to test a known-good printer on your existing computer using the same parallel or serial interface. If a new printer works properly, then you have ruled out the computer, cable, or software program. If a known-good printer also fails to operate, you might have a problem in your computer, its software configuration, or the cable.

Once you have isolated the problem to your printer, run a printer self-test or use the PRINTERS utility. This will test a printer's motors, carriage, print head, power supply, and most of its logic and driving circuitry. If the test pattern looks good, you can be pretty certain that the printer's interface circuit is defective. If the test pattern is faulty, then your printer is suffering from a defect elsewhere.

Troubleshooting communication circuits

Symptom 1 The printer's self-test looks correct, but the "parallel port" printer does not print at all under computer control. Refer to Chart 9-1. A "printer not ready" error might occur at the computer. Check the printer's on-line status; if the printer is off-line, it

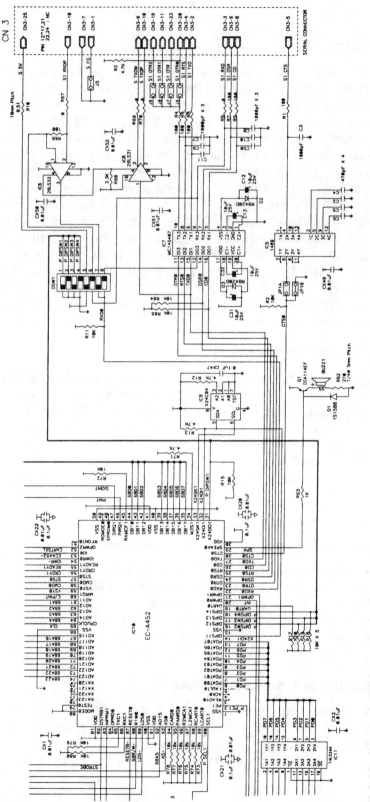

236

■ 9-13 *A typical serial communication circuit.* Tandy Corporation

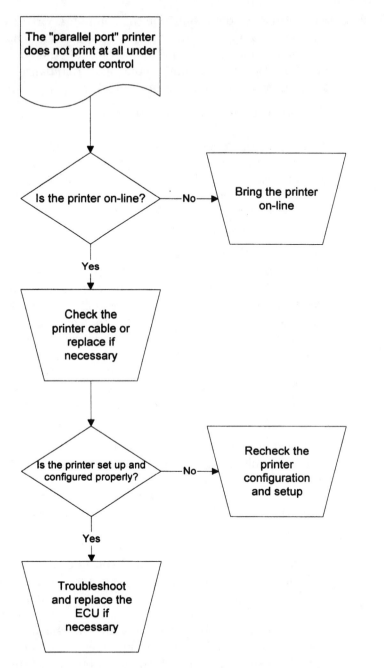

Chart 9-1 *Flowchart for Printer Communication Symptom 1.*

will not receive data sent from the computer. Virtually all printers have an LED on the control panel, which is illuminated when the printer is on-line. If the printer is off-line (the "on-line" LED is dark), try pressing the on-line button or reinitialize the printer. If the printer refuses to go on-line, check the paper supply.

Examine your interface cable. If it is loose at either end, data and handshaking signals might not reach the printer. If you have run another printer successfully using your current cable, then the cable is almost certainly good. If you are in doubt, disconnect the cable, and use your multimeter to measure continuity across each conductor (or just try another known-good cable). Wiggle the cable to stimulate any possible intermittent connections. Replace any defective interface cable.

Next, check the printer setup parameters in the PC's printing application, especially under the Windows or Windows 95 "printer" dialog(s). Make sure that the correct printer driver is loaded, and see that your printer is selected as the "default" printer. Also check that the correct "printer port" is selected. For example, if Windows is attempting to print to a Panasonic DMI printer on LPT2, but you have an HP DeskJet printer on LPT1, the printer simply will not work. Check any unique DIP switch settings in the printer to be certain that they are appropriate. If another parallel printer works, check its DIP switch settings against those of the defective printer.

If everything is configured properly up to this point, disassemble your printer and expose its communication circuitry. Refer to the parallel circuit of figure 9-8. With the printer connected and running, use your logic probe or oscilloscope to examine each handshaking line. There are four key status signals that communicate the current operating conditions of your printer: Busy (pin 11), Select (pin 13), Paper Out (pin 12), and Error (pin 32). Table 9-2 illustrates the interaction of these status lines versus the printer's "on-line" or "off-line" condition. If you find these four signals at their appropriate levels when the printer is on-line, the printer should be ready to accept data. If on-line conditions are incorrect, a problem exists in your interface or main logic circuitry. Check the supporting logic that provides your handshaking signals. In figure 9-8, the 74LS07 buffer IC provides signal conditioning for many handshaking signals. If the inputs to any of those gates are different than the outputs, the buffer IC might be faulty. If handshaking signals are correct back to the gate array IC (not shown in figure 9-8), replace the ASIC or gate array that directs handshaking signals.

■ Table 9-2 Handshaking conditions in a parallel interface.

Condition	BUSY	SELECT	PE	ERROR
On-line	0	1	0	1
Off-line	1	0	0	0
Paper out	1	0	1	0

If your examination proves inconclusive, or if you do not have the time or inclination to perform this type of component-level examination, replace the main logic portion of your ECU. For impact and ink jet printers, this is typically a single board. For EP printers, you should replace the main controller board or communications submodule.

Symptom 2 The printer's self-test looks correct, but the "serial port" printer does not print at all under computer control. A "printer not ready" error might occur at the computer. Check the printer's on-line status; if the printer is off-line, it will not receive data sent from the computer. Virtually all printers have an LED on the control panel that is illuminated when the printer is on-line. If the printer is off-line (the "on-line" LED is dark), try pressing the on-line button or reinitialize the printer. If the printer refuses to go on-line, check the paper supply.

Examine your interface cable. If it is loose at either end, data and handshaking signals might not reach the printer. If you have run another printer successfully using your current cable, then the cable is almost certainly good. If you are in doubt, disconnect the cable, and use your multimeter to measure continuity across each conductor (or just try another known-good cable). Be careful. Many serial cables flip (or reverse) the Tx and Rx lines between computer and printer, so you must take that into account when making measurements. Wiggle the cable to stimulate any possible intermittent connections. Replace any defective interface cable.

Next, check the printer setup parameters in the PC's printing application, especially under the Windows or Windows 95 "printer" dialog(s). Make sure that the correct printer driver is loaded, and see that your printer is selected as the "default" printer. Also check that the correct "printer port" is selected. For example, if Windows is attempting to print to a Panasonic DMI printer on COM2, but you have an HP DeskJet printer on COM1, the printer simply will not work.

Serial communication requires a fairly large number of parameters to specify the structure and speed of each serial character. Word length, stop bit(s), parity, baud rate, and handshaking method are some of the more common options that can be selected when setting up a serial communication link. However, each option must be set exactly the same way at both the computer and printer. If not, a printer cannot interpret just where data starts and ends. The resulting confusion will cause an erratic jumble of unintelligible print (if it prints at all). Communication parameters are usually set by a series of jumpers or DIP switches within the printer. Check these settings against those listed in your user's manual. If you have run another printer, check its configuration and compare settings.

If problems persist, there is probably a fault in the serial communication circuit. Refer to the circuit fragment of figure 9-13 for typical troubleshooting. There are only a handful of components involved in serial communication: a line receiver, line driver, gate array, and several handshaking gates. If the printer appears online, use your oscilloscope and check the DTR (Data Terminal Ready) line. If the printer is ready to receive data, DTR should be at a positive voltage (+5 to +15 V depending on the serial interface design). If the DTR line is at a negative voltage, the computer will not send data because the printer is not ready. If DTR is locked low, check the line transceiver circuit (i.e., IC7). If the DTR0 (IC7 pin 11) is high but DTR0 at the output (IC7 pin 10) is low, IC7 is probably defective and should be replaced. If DTR0 (IC7 pin 11) is locked low, suspect a fault in the gate array (IC10).

If DTR (CN3-20) is high as expected and communication still fails to take place, suspect a problem in your line transceivers. Use your oscilloscope to check the receive (R_x) signal (CN3-3) and follow that signal to the line receiver (IC7 pin 9). Check the line output (IC7 pin 12). If received data is present at the input, but missing at the output, the line receiver (IC7) is defective and should be replaced. If data is present at the line receiver output, the DIP switches at DSW1 might be set improperly, or there is a problem in the gate array (IC10) that should be replaced. You can also check the transmit (T_x) signal (CN3-2) and follow that signal to the line driver (IC7 pin 5). If the T_x signal is missing, check the line driver input (IC7 pin 15). If the outgoing data to the line driver is present, the line driver (IC7) is probably defective and should be replaced. If the outgoing data is missing, there is probably a fault in the gate array (IC10) that should be replaced.

If your examination proves inconclusive, or if you do not have the time or inclination to perform this type of component-level examination, replace the main logic portion of your ECU. For impact and ink jet printers, this is typically a single board. For EP printers, you should replace the main controller board or communications submodule.

Memory

A printer accepts data and control codes from its host computer, processes and interprets that information, then operates its print head and transport mechanisms to transcribe that information into a permanent form. Solid-state memory plays an important role in this operation. Figure 9-2 illustrates a typical arrangement of memory (marked RAM/ROM memory) within an ECU.

A printer operates on a fixed set of instructions, its own internal program that tells main logic how to communicate, how to operate its print head and transports, what function each dip switch represents, how to interpret control panel operations, and so on. This program resides permanently inside the printer, so it is stored in a permanent memory device. Other data, such as font styles and enhancements, can also be stored in permanent memory. On the other hand, some information changes constantly; characters and control codes received from the computer are only stored until they are processed. Main logic also requires locations to store control panel variables and results from calculations. Temporary memory devices are used to hold rapidly changing information during printer operation.

Memories are arranged as a matrix of storage cells, each specified by their own unique location (or address). The number of available addresses in a memory device depends on the number of address lines available. For example, a memory IC with eight address lines can access 2^8 or 256 locations. If your IC has 12 address lines, it can access 2^{12} or 4,096 locations. Each address can hold one, two, four, eight, or more bits, depending on the organization of the particular device. A one megabit (Mbit) memory could be arranged to hold one bit in 1,048,576 (1M) individual locations, four bits in 262,144 (256K) locations, eight bits in 131,072 (128K) locations, or some other arrangement. While PCs typically use large, low-bit memory devices (e.g., $1M \times 1$ bit), conventional printers frequently use "wider" devices (e.g., $32K \times 8$ bit).

Permanent memory

As the name suggests, information in permanent memory is retained at all times, even while power is removed from its circuit. You might hear permanent memory referred to as *nonvolatile* or *read-only* memory. There are three basic classes of nonvolatile memory that you should be familiar with: ROM, PROM, and EPROM. Your printer might employ any of these memory types, although the simpler ROM type devices are encountered most frequently.

ROM

A ROM (Read-Only Memory) is the oldest and most straightforward class of permanent memory. Its information is specified by the purchaser, but the actual IC must be fabricated already programmed by the IC's manufacturer. ROMs are rugged devices. Because their program is actually a physical part of the device, it can withstand a lot of electrical and physical abuse, yet still maintain its contents. However, once a ROM is programmed, its contents can never be altered. If a program change is needed, an entirely new device must be manufactured with any desired changes, then installed in the circuit.

PROM

The PROM (Programmable Read-Only Memory) can be programmed by a printer manufacturer instead of relying on a ROM manufacturer to supply the programmed devices. A PROM can be programmed (or *burned*) once, but it can never be altered. Factory-fresh PROMs are built as a matrix of fusible links. An intact link produces a binary 0, while a burned link produces a binary 1. One link is available for each bit in the device. A special piece of equipment called a *PROM Programmer* is fed the desired data for each PROM address. It then steps through each PROM address and burns out any links where a binary 1 is desired. When the PROM is fully programmed, it contains the desired data or program.

EPROM

An EPROM (Erasable PROM) can be erased and reprogrammed many times. Binary information is stored as electrical charges placed across MOS (Metal Oxide Semiconductor) transistors. One transistor is provided for each bit in the device. An absence of charge is a binary 0, while the presence of charge is a binary 1. Programming is very similar to that of a PROM. An EPROM programmer is loaded with the desired information for each location. It then steps through each location and locks charges into the appropriate bit locations. To erase an EPROM, you must remove charges from every bit location. That is accomplished by exposing the memory device (the die itself) to a source of short-wavelength ultraviolet light for a prescribed period of time. Light is introduced through the transparent quartz window on top of the IC package.

An example of permanent memory

The memory layout for a Tandy DMP203 dot-matrix printer is illustrated in figure 9-14. IC U4 is the permanent memory (a PROM). The 17 address lines support 131,072 (128K) locations, and the eight data lines indicate eight bits per address. In total, the ROM is classified as a 128KB ROM. Of course, address and data lines alone are not enough to run the IC. Control lines must also be added. The ROM uses two control signals: a `Chip Enable (`CE) signal, and an `Output Enable (`OE) signal. When the ROM is needed to supply instructions and data, the `Chip Select signal is driven low by the managing gate array (U3). Then data at the selected address will be available. Finally, the `Output Enable signal is driven low by the microprocessor (U7) to put data on the data bus (D0 to D7).

Temporary memory

Digital information contained in temporary memory can be altered or updated frequently, but it will only be retained as long as power is applied to the device. If power fails, all memory contents will be lost. This kind of memory device is referred to as *volatile, read-write*, or *Random Access Memory* (RAM). The term *random access* refers to the fact that the device can be accessed for reading or writing operations as needed. The two basic types of RAM that you should know are static and dynamic.

SRAM

A static RAM (SRAM) uses conventional logic flip-flops (called *cells*) to store information. One cell is provided for each bit. A read/write control line is added to select between a read or write operation. During a write operation, any data bits existing on the data bus are loaded into the cells at the address specified on the address bus. If a read operation is selected, data contained at the selected address is made available to the data bus. Once data is loaded into a static RAM, it will remain until it is changed, or until power is removed. SRAMs are used heavily in printers.

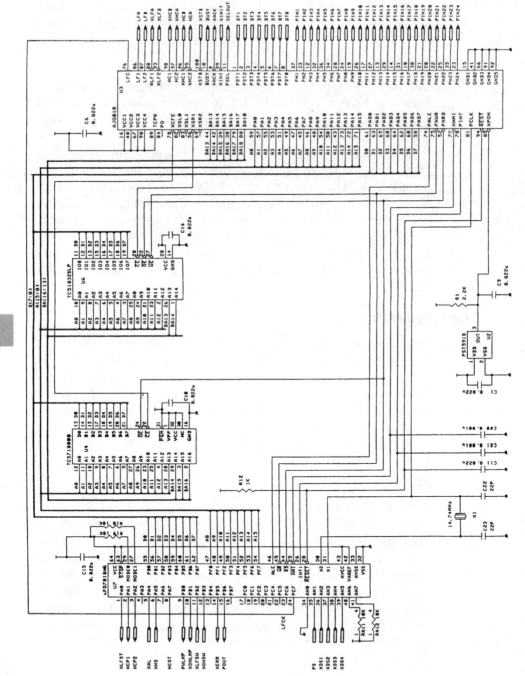

■ 9-14 *Main logic schematic fragment with memory devices.* Tandy Corporation

DRAM

Dynamic RAM (DRAM) devices use small MOS cells to store data in the form of electrical charges. While reading and writing operations remain virtually identical to those of SRAMs, DRAMs must be "refreshed" every few milliseconds, or data will be lost. Refresh is provided by a combination of external circuitry and circuits within the DRAM chip itself. Although the need for refresh increases the complexity of a memory circuit, MOS technology offers very low power consumption and a large amount of storage space as compared to most SRAM devices. DRAM is used primarily in PCs, but you might encounter DRAM in EP printers that demand substantial amounts of memory.

An example of temporary memory

You can see an example of temporary memory in figure 9-14. IC U6 is an SRAM. With 15 address lines, the RAM can access 32,768 (32K) addresses. Eight data lines provide eight data bits to each address, so the SRAM is 32Kx8 (32KB) device. Like the ROM, an SRAM uses a `Chip Enable and `Output Enable signal to manage the flow of data. In addition, the SRAM needs to differentiate between reading and writing, so a `Write Enable (`WE) line is included. When the signal is low, data on the data bus will be written to the selected address. When the signal is high, data at the selected address will be placed on the data bus.

Troubleshooting memory

Memory is usually one of the most reliable sections of an ECU, but when a failure does occur, the results can manifest themselves as lost characters, occasional operating hang-ups for no apparent reason, or up-front initialization failure. The difficulty in testing memory is that it is virtually impossible to tell for sure just what location or bit the problem is coming from. In order to properly test a RAM device, a known pattern of data would have to be written to each location, then read back and compared to what was written. If there is a match, that location is assumed to be good. If there is no match, that location (and the entire IC) is defective. Most printers perform a memory check on initialization. Unfortunately, there is no way of performing this sort of test in the printer during operation.

ROM devices are even more difficult to test. Each location would have to be read and compared against a listing of data at each address. If there is a match, the location is valid. If there is no match, the location (and the entire ROM) is defective. Even if you could check each ROM address, you have no listing of its program for comparison. The bottom line here is that unless your printer specifically checks memory on initialization and reports such errors through error messages, LED flashes, or beep patterns, it is extremely difficult to test memory directly. Still, there are some memory symptoms that you should be familiar with.

Symptom 1 A RAM error is indicated with a text message or blinking LED. While an error message might be relatively easy to understand, blinking LEDs can be a bit more difficult to follow. Typically, the Power LED will blink on and off three times quickly, go dark for about 500 ms, then repeat. Of course, your own printer might use a different error sequence. For relatively simple printers, such as the DMP 203 shown in figure 9-14, a RAM error indicates that the SRAM (U6) has failed. However, the microprocessor also frequently uses a small amount of internal RAM, so if the problem persists after replacing the SRAM, try replacing the microprocessor. If problems still continue, or you prefer to avoid component-level work, replace the ECU board entirely.

The problem is a little more complicated with EP printers, as you can see in the schematic fragment of figure 9-15. The Tandy LP800 uses a series of 8-bit DRAM modules. However, if you study the memory layout closely, you will note that the DRAM modules are paired, allowing a 16-bit memory data bus to the main processing gate array (IC25). This design is further complicated by the addition of refresh controllers (IC36 and IC37) constantly working to keep the printer's data intact. The refresh signals can typically be identified by RAS and CAS prefixes. You can insert a mix of 0.5MB and 1MB DRAM modules, but you must set the DIP switches in DSW2 in the proper configuration to recognize the corresponding modules. When a memory fault occurs here, there is a lot more ground for you to cover.

If the DRAM modules are hard-soldered to the main controller board, your best course is often just to replace the main controller entirely; otherwise, you would systematically have to replace each DRAM unit until the defective one was discovered. When the DRAM units are available in replaceable modules, it is possible to try a new module in each location systematically until the defect disappears. If DRAM problems persist, suspect a fault in the refresh controllers (IC36 and IC37) or the managing gate array (IC25). If the problem persists, replace the main controller board.

Symptom 2 A ROM checksum (or similar) error is reported. Permanent memory holds the instructions that run a printer, as well

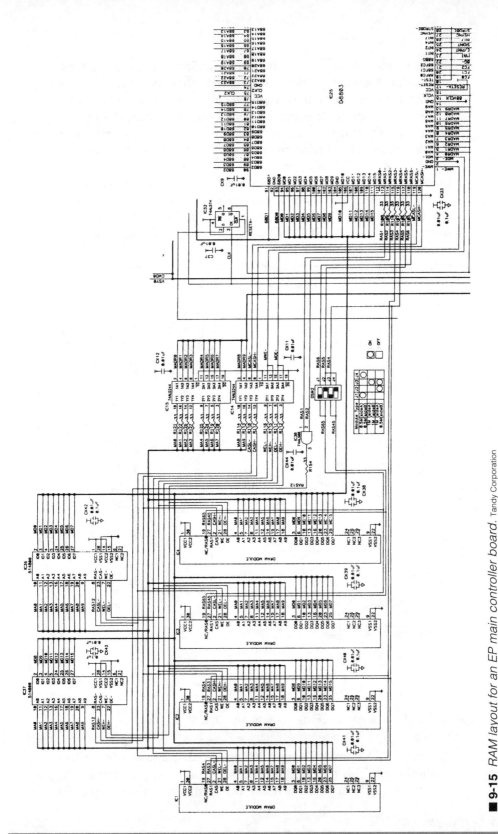

■ 9-15 *RAM layout for an EP main controller board.* Tandy Corporation

as any resident fonts. When the printer is first initialized, a check-sum value for each ROM is calculated, then compared to the checksum stored in each ROM. If the calculated value matches the stored value, the ROM is considered good; otherwise, the ROM is assumed bad, and an error message is generated. For relatively simple printers, the firmware and font data can be stored on a single IC (such as U4 in figure 9-14). For more complex printers, however, a number of permanent memory devices are needed, as shown in figure 9-16. The EP printer's firmware is stored on two ROMs (IC26 and IC28). Note that these ROMs are not independent; one is the odd ROM (data bits 0 through 7), while the opposing ROM is even (data bits 8 through 15). This supplies a 16-bit data path. There are also two sets of font ROMs: outline font ROMs (IC31 and IC30) and filled font ROMs (IC27 and IC29).

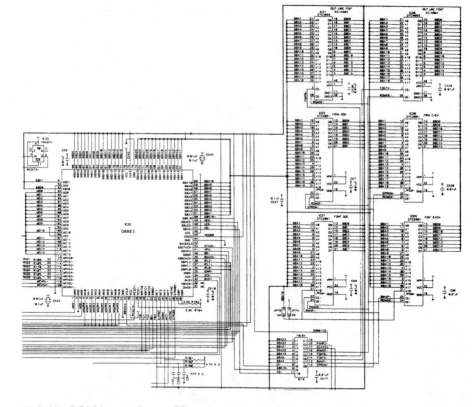

■ **9-16** *ROM layout for an EP main controller board.* Tandy Corporation

Typically, ROMs are DIP-type ICs inserted into sockets on the motherboard. This means that ROMs are often easy to replace, if you can find them. Replacement ROMs are not always easy to come by. The first ROMs to suspect are the firmware ROMs because they hold the instructions for your printer. A firmware failure will allow the printer to malfunction during operation. Font ROM problems are generally less catastrophic. At worst, certain font styles and sizes might appear distorted. If you cannot locate replacement ROMs (or new ROMs have no effect), replace the ECU or main controller board.

Control panel

The control panel serves several purposes. It allows you to operate certain immediate functions, such as form feed, line feed, reset, or on/off line. Certain key combinations let you alter options and running modes. In some printers, for example, pressing the reset and form feed keys together allow you to select a different font style or character pitch. As another example, pressing form feed and line feed together might cause the printer to execute a self-test cycle. Your user's manual will specify the exact key strokes and their effect. Finally, indicators are included to display various printer status conditions, such as "on-line" or "paper out." More advanced control panels offer an LCD that displays full text prompts and information. Figure 9-17 illustrates a schematic for a simple control panel. Most control panels connect directly to ASIC or microprocessor circuits as figure 9-2 shows.

As you might expect, there is little that can go wrong with the "control panel" of figure 9-17. LEDs and sealed switches are hardly sensitive components. However, switch operation is very important to the printer; marginal switch operation can make operating the printer a frustrating chore. The "control panel" connector (CN7) is attached to the ECU main board. For simple printers, the control panel is attached directly to the main microprocessor or microcontroller. You can see the control panel signals on the left side of the microprocessor (U7) in figure 9-14.

While the control panel of figure 9-17 appears pretty simple, there are much more complicated designs as shown in figure 9-18. EP printers employ microcontrollers that are custom-designed to support an LCD module (not shown in figure 9-18). In addition to a microcontroller (IC1), there are many more switches (SW1 to SW8) to deal with, and several more LEDs (D1 to D4). It is interesting to realize that the switches and LEDs are tied directly to the microcontroller, so IC1 is not only a processing element, but it also includes the circuitry to drive real-world devices. This high level of integration reduces the need for multiple ICs. Communication between the

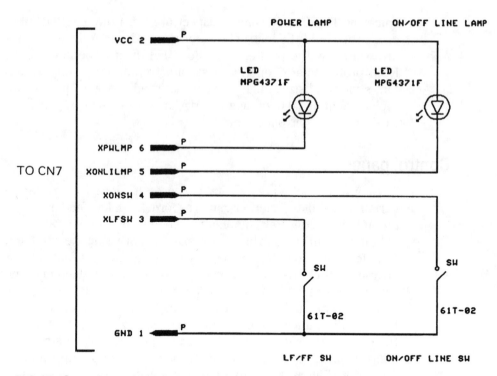

9-17 *Control panel schematic for a simple dot-matrix printer.* Tandy Corporation

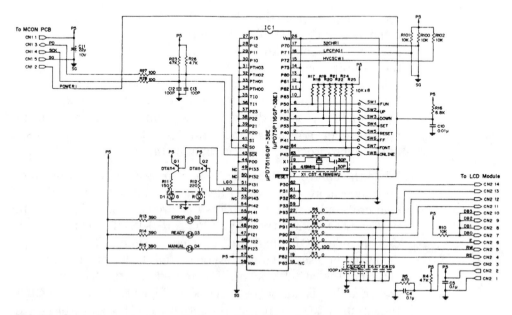

9-18 *Control panel schematic for an EP laser printer.* Tandy Corporation

microcontroller and the remainder of the ECU (typically the mechanical controller board in an EP printer) is accomplished through a single serial line (marked PD at CN1 pin 3). Thus, when a message or error code must be displayed, the condition is transmitted serially to the control panel microcontroller, which interprets the message, drives LEDs, and displays the actual text on the LCD module. Conversely, the microcontroller will transmit messages to the mechanical controller as various switches and switch combinations are pressed. You can see this data with an oscilloscope or logic probe, although the data stream is hardly meaningful to the casual observer.

Understanding sealed switches

Most printers (conventional and EP alike) use sealed membrane-type switches covered by a solid plastic strip containing the graphics for each key. A cross-sectional diagram of this arrangement is shown in figure 9-19. Membrane switches use a flexible metal diaphragm mounted in close proximity to a conductive base electrode at the switch bottom. Ordinarily, the diaphragm and base do not touch, so the switch is open. When you touch the proper location on a desired graphic, a solid plunger deforms the metal diaphragm and causes it to contact the base electrode. This closes the switch. After you release the graphic, the metal diaphragm below returns to its original position and opens the switch again. The diaphragm's design might "snap" a bit when pressed to provide you with a tactile sensation of positive contact.

Unfortunately, membrane switches are subject to breakdown with age, use, and environmental conditions. Although membrane

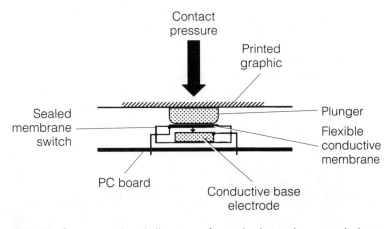

■ 9-19 *Cross-sectional diagram of a typical membrane switch assembly.*

switches are sealed to prevent disassembly, most are not hermetically sealed to keep out moisture and dust. Over time, oxidization can occur that prevents positive contact between the membrane and electrode. Regular use can also wear away at both contact surfaces and eventually cause bad or intermittent contacts; the switch might not always respond when you press it. Finally, regular use can cause the diaphragm to stretch or dislodge from its mounting. This can lead to a short circuit if the diaphragm fails to snap open when released. Ultimately, switch problems occur quite frequently in older control panels.

Troubleshooting a control panel

Symptom 1 The control panel does not function at all. No keys or indicators respond. Otherwise, the printer appears to operate normally under computer control. Open your printer enclosure and expose the control panel circuit. Make sure that any connector(s) or wiring from the panel are installed properly and securely. If you have just finished reassembling the printer, perhaps you forgot to reconnect the control panel, or reconnected it improperly. Interconnecting wiring might have been crimped or broken during a previous repair.

It is rare that a simple control panel will fail outright (such as the circuit in figure 9-17). When a total failure occurs in this case, the trouble is almost always in the ECU rather than the control panel; the microprocessor or microcontroller that manages the keyboard has probably failed. Try replacing the ECU. However, when problems occur with a sophisticated control panel (such as the circuit in figure 9-18), it is more likely that the control panel's local microcontroller has failed. It is the local microcontroller that handles the LEDs, switches, and LCD panel (if included). Try replacing the local microcontroller. If this is not possible, replace the entire control panel module. In most cases, a new control panel should solve your problem, but in the rare event that problems persist, the fault will be in the ECU (specifically the mechanical controller board for EP printers). Your best course then is simply to replace the mechanical controller board.

Symptom 2 One or more keys is intermittent or defective. Excessive force or multiple attempts might be needed to operate the key(s), but the printer appears to operate normally otherwise. You will find this symptom most often in older, heavily used printers. In almost every instance, this symptom is simply the result of faulty keys. Typical printer keys are built using pressure-sensitive contact switches. With age and use, the switch contacts become unreliable, even stubborn. Ideally, the preferred course would be to replace the questionable switches. Unfortunately, replacement switches are often not available. If you cannot replace the defective keys, you will have little choice but to replace the entire control panel module.

Symptom 3 One or more LED indicators fail to function. Printer appears to operate normally otherwise. LEDs are notoriously reliable devices (lifetimes of 30,000 hours or more is not uncommon), so it is unlikely that the LED itself has failed. Rather, you should first suspect the circuit that manages the LED. Consider the circuit of figure 9-18, and examine the LEDs D2, D3, and D4. Notice that they are all powered from a common source (usually +5 V). Start your examination by checking this voltage with your multimeter. If the supply voltage is low, check your dc power supply.

Now, when supply voltage is present, each LED will be on when the cathode side of the LED is grounded (performed by circuits inside IC1). Use a multimeter or logic probe and check the cathode side of each diode. If the LED is off, you should see about +4.5 V (or a logic 1) on the LED's cathode. If the LED is on, you should read about +0.8 V or so. If the cathode voltage is low but the LED is dark, the LED is defective and should be replaced. If the LED should be on but the cathode voltage is still high, the corresponding control circuit (in this case IC1) might be defective and should be replaced.

If you are unable to troubleshoot LED signals, you can replace the suspected subassembly. For simple control panel systems, problems are typically in the ECU, so try replacing the ECU entirely. For EP-type control panels, you can try replacing the control panel module. In the rare event that problems continue, try replacing the mechanical controller board.

Symptom 4 The LCD display is dark or displaying gibberish. This symptom arises in high-end printers that employ LCDs in addition to LED indicators. Start by checking the LCD module's connector to see that it is installed properly and completely. If the control panel is working otherwise, chances are that the LCD module is defective, so try replacing the LCD unit. If the control panel has failed (i.e., switches not responding), the local microcontroller has probably failed. Replace the microcontroller. If you cannot replace the microcontroller, replace the control panel subassembly entirely.

Driver circuits

Printers depend on electromechanical devices (e.g., motors and solenoids) for proper operation. Unfortunately, logic circuits simply cannot provide the power needed to drive an electromechanical device directly. Instead, driver circuits must be inserted between the ECU logic and the particular device to be driven. For example, a driver circuit can convert a 5-V, 10-mA TTL pulse, into a 24-V, 1.5-A pulse needed to fire a DMI print wire. Conventional printers use three sets of drivers: print head drivers, carriage motor drivers, and paper advance motor drivers. Electrophotographic printers often use a single set of motor drivers to operate the drum, along with drivers to actuate pickup and registration roller clutches. Before you review some actual driver circuits, you should understand the operation of transistor switches.

Transistor switches

Transistors can be used as linear amplifiers or solid-state switches. When configured as a linear amplifier, a transistor's output signal would appear as a larger duplicate of its input signal. Digital signals, however, are only on or off, so there is no real advantage to providing linear amplification. Instead, transistor circuits are configured as "switches" that operate either totally on or totally off depending on the condition of its input signal. Consider the main logic circuit in your printer. It executes preprogrammed steps based on input from the computer, control panel, and any sensors available in the circuit. Suppose at some point in its program, main logic needs to activate a pick-up roller clutch or operate a motor. TTL devices alone can only source or sink a few milliamps of current, hardly enough power to drive these real-world devices. A "driver" is added that will accept logic signals, and amplify them enough to power such things as a motor or relay. The transistor is ideal for this function because it needs very little base current, and it can carry relatively large currents in its collector. Figure 9-20 is just one typical example of this type of interface.

When the driving gate's output is off (logic 0) at point "A," no current flows into the transistor's base, so it is turned off (or *cutoff*), and it acts like an open switch (no collector current flows and the relay remains off). If the driving gate turns on (logic 1) at point "A," its TTL output supplies enough current into the transistor's base to fully activate (or *saturate*) the device. The transistor then acts like a closed switch (current flows, which turns on the relay). This is the same technique used to operate many electromechani-

254

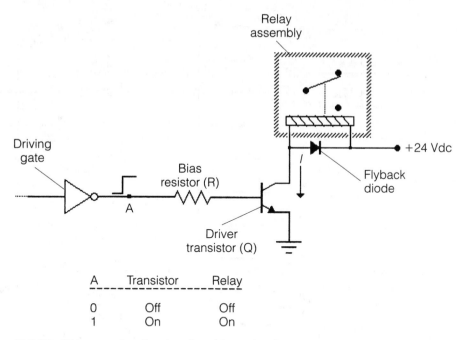

A	Transistor	Relay
0	Off	Off
1	On	On

■ **9-20** *Diagram of a simple relay driver circuit.*

cal devices in your printer. Although discrete transistors are often used as drivers, some printers use integrated drivers—driver transistors fabricated onto IC structures.

Print head drivers

Moving-carriage print heads form images by firing discrete points as the head passes across the page surface. Each "discrete point" must be fired independently by the ECU, and driver circuits are required to interface the logic to the print wires, ink nozzles, or whatever mechanism the particular print head is using. Figure 9-21 shows you the driver schematic for a 24-pin DMI print head. As you see, the driver circuit is really quite straightforward; a separate transistor switch is used for each print wire. Each of the "PIN" signals (PIN1 through PIN24) is generated by a gate array in the ECU's main logic (the IC marked U3 in figure 9-14). When a PIN signal becomes logic 1, the corresponding transistor saturates, causing current to flow in the print wire solenoid. This action fires the print wire. Note that the output connector (marked CN6) is where the print head cable attaches to the ECU.

When a print wire driver fails, it can fail in an open or shorted condition. The open failure is most common, and prevents the corre-

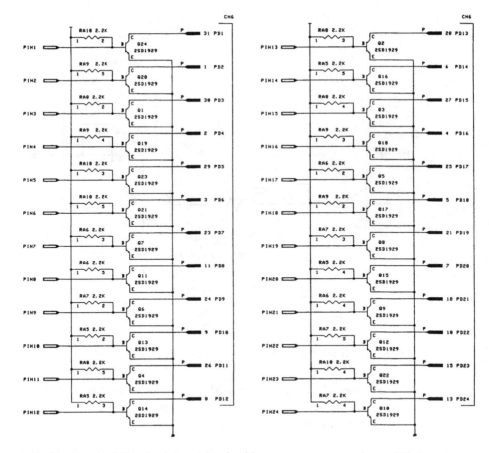

■ 9-21 *A typical 24-pin dot-matrix pin driver array.* Tandy Corporation

sponding print wire from firing. The result is a horizontal white
line in the print. On the rare occasions when a print wire driver
fails in the shorted condition, the print wire remains extended.
This often appears as a horizontal black line in the print, but it
might also tear the ribbon. You can test a driver transistor with any
oscilloscope. Set your oscilloscope to measure TTL logic levels,
and check the base (B) of the suspect transistor. If the TTL signal
is missing, the gate array IC in the printer's main logic is probably
defective and should be replaced. If the TTL logic signal is present, set your oscilloscope to measure 20–30-V signals, and check
the collector (C) of the transistor. You should see a driver signal,
such as the one in figure 9-22. If this signal is absent, the transistor has probably failed and should be replaced. If this signal is present (but the print wire is not working), the print head cable is
defective, or the print wire is jammed or broken, so try cleaning or
replacing the print head assembly.

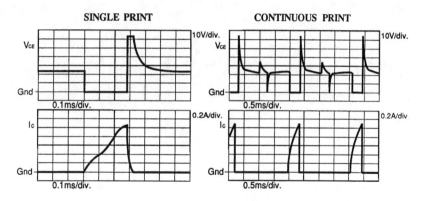

SINGLE PRINT

CONTINUOUS PRINT

■ **9-22** *Typical pin driver oscilloscope signals.* Tandy Corporation

Carriage motor drivers

The carriage transport system is responsible for carrying the print head back and forth across the page. At the heart of this system is the carriage return (CR) motor, which is typically a four-phase stepping motor. The circuit in figure 9-23 is used to drive a CR motor. There are two major portions to this circuit: the phase drivers and the voltage drivers. You will probably recognize the four-phase driver transistors (Q38 to Q41) immediately. These transistors provide the high-energy pulses that actually step the CR motor. While this might be all that is needed for a CR driver, there are some additional factors to consider; most important is that the carriage must be stepped at different rates depending on what state the printer is in. As a result, the CR motor must be set to different supply voltages. The upper portion of figure 9-23 is used to set the operating voltage (and thus the stepping speed) of the carriage. The step signals (HCA, HCB, XHCA, and XHCB) are produced by the managing gate array (U3 in figure 9-14), while the step rate signals (HCP1, HCP2, and HCST) are generated by the microprocessor (U7 in figure 9-14). All of the output signals (marked CN4) are connected directly to the CR motor assembly.

Carriage problems can typically be traced to either a fault in the phase drivers, or a problem in the step rate control circuit. The phase drivers are straightforward to troubleshoot (the same approach used for the pin drivers of figure 9-21), and you should find collector signals similar to those shown in figure 9-24. When trouble strikes the step rate circuit, you will often find the trouble in high-energy components, such as D3 or D4, or Q42 through Q45.

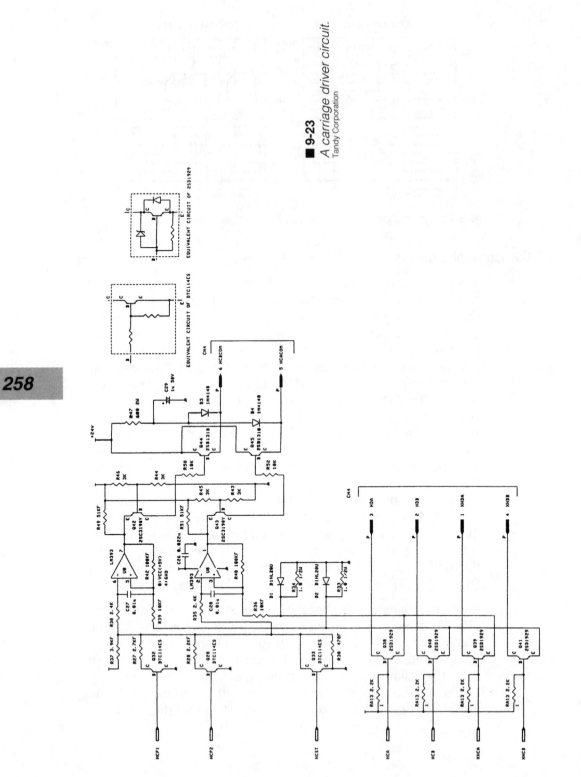

■ **9-23**
A carriage driver circuit.
Tandy Corporation

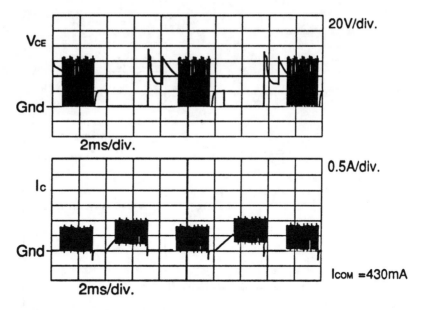

20V/div.

V_{CE}

Gnd

2ms/div.

0.5A/div.

I_c

Gnd

I_{COM} =430mA

2ms/div.

■ **9-24** *Typical carriage driver oscilloscope signals.* Tandy Corporation

Line feed motor drivers

The paper transport system is responsible for carrying each page through the printer. Regardless of whether the transport is a tractor-feed or friction-feed mechanism, both are driven by a single line feed (LF) motor. The LF motor itself is typically a four-phase stepping motor, and the LF driver circuit is shown in figure 9-25. Notice the similarities between this circuit and the CR motor circuit of figure 9-23. Basically, the LF driver circuit is composed of four-phase driver transistors that actually provide the high-energy pulses that run the LF motor. There is also a voltage cutoff circuit (composed of Q30, Q35, and their related components) that is used to supply power to the LF motor windings. The phase control signals (LFA, LFB, XLFA, and XLFB) are produced by the managing gate array (U3 in figure 9-14), while the power enable signal (XLFST) is supplied directly from the microprocessor (U7 in figure 9-14). All of the output signals (marked CN5) are connected directly to the LF motor.

Line feed problems can typically be traced to either a fault in the phase drivers, or a problem in the line feed enable circuit. The phase drivers are straightforward to troubleshoot (the same approach used for the pin drivers of figure 9-21), and you should find collector signals similar to those shown in figure 9-26. When trou-

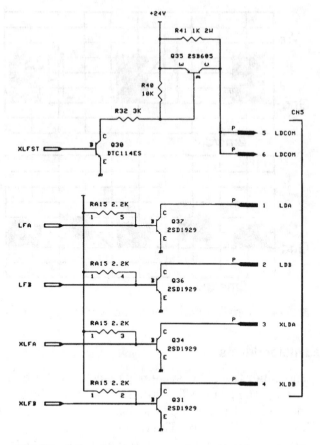

■ 9-25 *A line feed (paper advance) driver circuit.*
Tandy Corporation

ble strikes the line feed enable circuit, you will often find the trouble in high-energy components, such as Q30 or Q35.

EP driver circuits

Electrophotographic printers use much of the same electronic wizardry that moving-carriage printers do. The main motor is the heart of an EP mechanical system, and it provides all of the force needed to pick up a page and carry it through the printer. Normally, the main motor runs throughout the printing cycle, but pick-up and registration units only need to work for brief periods; thus, some means is needed to engage and disengage the pick-up and registration assemblies as needed. To address this problem, solenoids are used as "clutches" to transfer force to the pick-up and registration assemblies as needed. The mechanical controller board provides the drivers needed to operate the main motor and

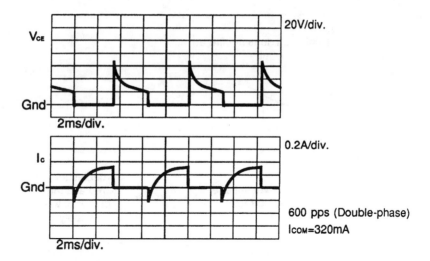

■ **9-26** *Typical line feed driver oscilloscope signals.* Tandy Corporation

solenoid clutches. When main motor or clutch operation problems arise, you should suspect a fault in the mechanical controller board.

Troubleshooting a driver circuit

Symptom 1 Print head does not fire at all. Refer to Chart 9-2. Other printer operations appear correct. The printer also fails to print during a self-test. Review the print head driver circuit illustrated in figure 9-21. There are typically four reasons why the moving-carriage print heads fail to operate: the print wires have jammed or internal wiring has failed, the print head cable has failed, the print head drivers have failed, or the managing gate array (ECU logic) has failed. When a complete failure occurs (the print head fails to work at all), chances are that your interruption has occurred in a common signal wire, such as a ground or voltage line.

To best determine the nature of this problem, reinitialize the printer and use your multimeter or oscilloscope to measure the driver supply voltage. For a pin driver circuit, such as figure 9-21, you should expect to find +24 V on pins 14, 16, 17, 19, 20, and 32 of CN6. If the +24 V is low or missing, you might have a fault in the printer's power supply. Also check the +5-V supply output to be sure that appropriate voltage is available to the printer's logic. Address the symptoms of Chapter 8 to troubleshoot the power supply in more detail.

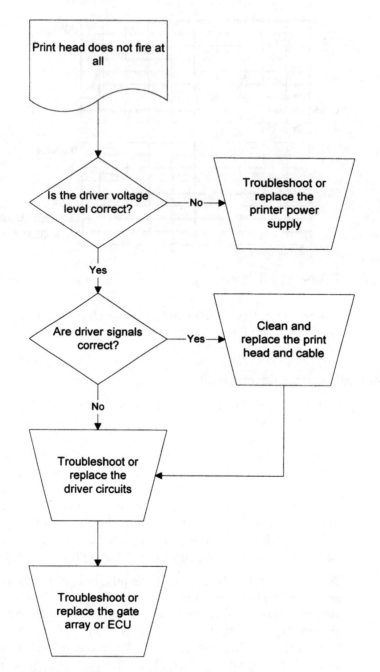

Chart 9-2 *Flowchart for Driver Circuits Symptom 1.*

Electronic service techniques

If the supply voltage is correct, run the printer (use the PRINT-ERS utility to generate a test page) and use your oscilloscope to measure the pin signals at the collector (C) of each driver transistor. In normal operation, you should find pin signals similar to the waveform of figure 9-22. If you find each pin signal as expected (but the print head still does not print), either your print head or print head cable has failed. Check the installation and continuity across your print head cable, and replace the cable if necessary. Then try replacing the print head.

If the pin signals are missing, check the logic signals at the base (B) of each driver transistor. If the base signals are missing, the gate array responsible for generating those signals has failed. Try replacing the gate array, or replace the entire ECU. If the base signals are present (but the collector signals are absent), suspect a ground fault. Notice in figure 9-21 how the emitter (E) of each driver transistor is tied to a common ground. Inspect the circuit carefully and note if any circuit interruptions occurred in the ground line; this can disable the entire driver array. If you are unable to locate a printed wiring fault, replace the ECU.

Symptom 2 Print contains one or more white (missing) or black lines. All other printer operations appear correct. Symptoms persist during a self-test. Always start by powering down the printer and cleaning the print head. It is not unusual for accumulations of ink and paper residue to jam a print wire. If problems persist, use your oscilloscope to measure the driver pulses at the collector (C) of the suspect driver transistor(s) such as in figure 9-21. You should observe pulses similar to those shown in figure 9-22. If the pulses are present (but the print wire does not fire), there might be a fault in the print head cable or the print head wire itself. Inspect the print head cable and replace it if necessary. Otherwise, replace the print head assembly.

If the collector signal(s) are absent, use your oscilloscope to measure signals at the transistor's base (B) pin. When the base signal is present (but the collector signal is absent), the driver transistor has failed and should be replaced. Be sure to use an exact replacement part. If you cannot obtain a replacement part, try replacing the ECU. If the base signal is absent, suspect a fault in the gate array, which provides the base signal. Replace the gate array or replace the ECU outright.

Symptom 3 A paper or carriage advance does not function properly (if at all). Refer to Chart 9-3. Always start your investigation by inspecting the motor connections. Turn off and unplug the printer,

263

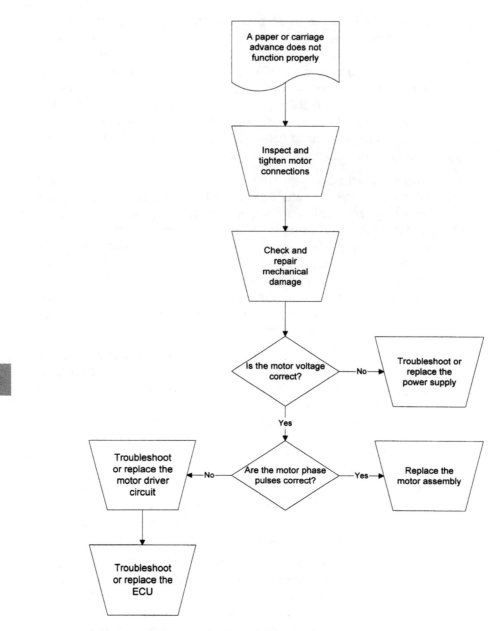

Chart 9-3 *Flowchart for Driver Circuits Symptom 3.*

then check the cables to your suspect motor. Try reinstalling the motor connector at the ECU. Also check for any obstructions or damage to the LF or CR mechanics. You might need to address mechanical problems as described in Chapter 10. Next, check the motor voltage by inspecting the +24-V output from the power supply. If this voltage is low or absent, troubleshoot or replace the power supply as described in Chapter 8. When the printer is running, there should be about +24 V at pins 5 and 6 of the CR or LF motor. If this motor winding voltage is missing, check the voltage control circuit feeding pins 5 and 6, or replace the ECU.

If the motor voltage is present, use your oscilloscope and measure the motor driver pulses. For the CR motor driver circuit of figure 9-23 and the LF motor driver circuit of figure 9-25, you can see that both stepper motors are driven by an array of four driver transistors. Check the motor pulse at the collector (C) or each driver. For a CR motor, you should see pulses similar to figure 9-24, while an LF motor pulse should appear close to the illustration of figure 9-26. When the motor winding voltage is correct and all the motor pulses are present (but the motor does not turn), the motor is probably defective and should be replaced. If one or more motor pulses are absent, check the signals at the base (B) of each driver transistor. When a base signal is present but a collector pulse is not, the corresponding transistor is defective and should be replaced. If the base signal is also missing, suspect the gate array used to operate the motor drivers. Try replacing the gate array, or replace the ECU outright.

Symptom 4 One or more solenoid clutches have failed. This manifests itself as either a pick-up or registration problem depending on which solenoid has failed. When dealing with a solenoid, it is either on or off. While the printer is idle, place your multimeter across the solenoid coil; you should read a low voltage. Now run a print cycle. If the voltage across the solenoid jumps up, but no movement takes place, the solenoid is defective or jammed. Check for obstructions that might be jamming the clutch mechanics, or replace the solenoid or solenoid board assembly outright.

If solenoid voltage remains low, however, there is likely a problem in the solenoid driver circuit. With very few exceptions, solenoid driver architectures follow a design similar to the one shown in figure 9-27. Connector CN22 receives logic signals from a gate array on the mechanical controller board. Connectors CM21 and CN20 attach to the physical pick-up and registration solenoids. The driver design for both solenoids is identical. When a logic signal

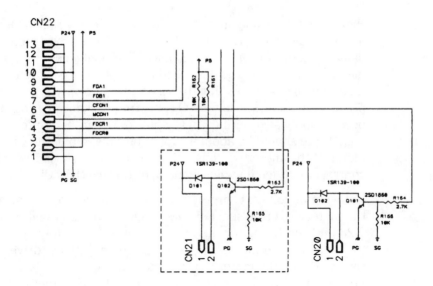

■ **9-27** *Solenoid clutch driver circuits.* Tandy Corporation

from the gate array (mechanical controller) becomes logic 1 (i.e., the CFON1 signal at CN22 pin 6), transistor Q101 saturates and turns on. This completes the solenoid circuit. When the CFON1 signal returns to logic 0, Q101 turns off, and diode D102 acts as a flyback diode for the solenoid.

Start your investigation by checking for the presence of +24 V. If the solenoid voltage is low or absent, troubleshoot or replace the dc power supply as explained in Chapter 8. If the operating voltage is correct, measure the base (B) signal at the switching transistor (i.e., Q101). If the logic signal becomes logic 1 but the solenoid voltage remains unchanged, Q101 is probably defective and should be replaced. Also check D102 and replace it if necessary. If there is no logic signal to operate Q101, the managing gate array on the mechanical controller board is probably defective, but it might be easier for you to replace the mechanical controller board.

Main logic

Main logic circuits are the heart and soul of your ECU. It typically includes a main microprocessor, one or more ASICs (gate arrays), at least one clock oscillator, and a variety of interconnecting "glue logic" components needed to tie these parts together. Figure 9-3 illustrates the relative locations of these components. The main logic is responsible for directing all aspects of printer operation.

Microprocessor operations

If a printer could be compared to a symphony orchestra, your main microprocessor would be the conductor. Technically speaking, a microprocessor is a programmable logic device that can perform mathematical and logical manipulation of data, then produce desired output signals. All microprocessors are guided by a fixed series of instructions (called a *program*), which is stored in the printer's permanent memory (the *ROM*). Although the microprocessor found in your printer is often less complex than those found in many computers, you can expect to find many of the same signals as shown in figure 9-28. Notice that a set of related signal wires (known as a *bus*) is often represented as a single wide line. This is done to simplify diagrams and schematics. Arrows are used to depict the possible flow of data. You will find three major buses: the address bus, the data bus, and the control bus.

Address lines

Address lines specify the precise location of instructions or data anywhere in the printer. Address locations might refer to memory locations, operating addresses of registers, ASICs/gate arrays, or other circuit-specific places. Because a microprocessor only generates address signals, it is always the controlling element in a main logic circuit. The Motorola 68000 microprocessor (IC16), shown in the Tandy LP800 main controller circuit fragment of figure 9-28, provides 24 address lines (A0 through A23), which can specify 2^{24} or 16,777,216 (16M) unique locations. Simpler printers (such as the logic circuit of figure 9-14) supply only 16 address lines (A8 to A15, data lines D0 to D7 are multiplexed to serve as the lower eight address lines), which provides 2^{16} or 65,536 (64K) unique locations.

Data lines

A microprocessor can read data from or write data to any location specified by the address line. Data is sent over the data bus. For example, when the printer is first turned on, its microprocessor will automatically generate an address (usually 0000 hex) and attempt to read whatever character is available from that address. The microprocessor automatically assumes this to be its first instruction. If memory is defective, or the character at that address is somehow incorrect, the microprocessor will become hopelessly confused. This leads to erratic or unpredictable operation.

During a write operation, the microprocessor will generate an address, then place a valid character on that address. Any device

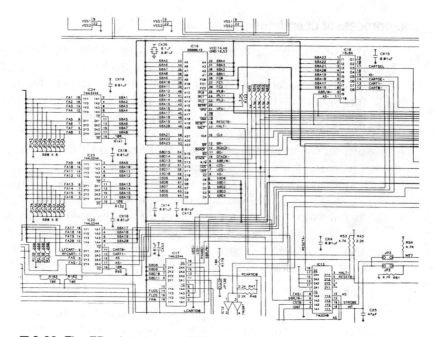

■ 9-28 *The EP printer's main CPU.* Tandy Corporation

that is active at that address (such as a RAM IC) will accept this character. Although a microprocessor is capable of writing to any address, not all addresses are able to accept data (such as ROM locations). Suppose a paper advance of 25 steps is required at a paper advance stepping motor. The microprocessor will address its mechanical gate array or ASIC, then place a binary "25" on the data bus. This number would then be translated by the gate array into a series of motor pulses, and delivered to a motor through driver circuits.

Control lines

Finally, a microprocessor is equipped with a number of control lines. A read/write (R/W) signal informs the system whether the microprocessor is performing a read or a write operation at its selected address. An interrupt request (`IPL0 to `IPL2) is sent to the microprocessor whenever the printer must deal with an immediate problem or condition. There might be several levels or interrupts depending on the particular IC in use. Other control lines that you might encounter are the halt (`HALT) and reset (`RESET) signals. There might be many other control lines depending on the age and sophistication of the particular microprocessor. Where address and data signals are often grouped together into bold lines, control signals are typically kept separate (or discrete).

The system clock

A microprocessor is a sequential device; instructions are executed one step at a time. As a result, timing becomes a very critical aspect of the microprocessor's operation. Timing signals are provided by a precision oscillator built into the microprocessor itself. This oscillator is known as the *master clock* or *system clock*. To achieve a precise and stable clock, a piezoelectric crystal is added externally (as shown by the crystal X1 in Fig 9-14), and the oscillator will run at some fraction of the crystal's resonant frequency. Crystal resonant frequencies are marked right on the part. You will be able to read these clock pulses with your oscilloscope. However, not all microprocessors use local crystals. The 68000 CPU in figure 9-28 receives timing pulses at the clock (CLK) line. Clock pulses are generated by a nearby crystal-controlled IC.

The role of ASICs or gate arrays

A single microprocessor or microcontroller cannot possibly handle the wide variety of operations required of a typical printer. The complex logical processing and operations needed to support modern printers demands the equivalent of hundreds of discrete ICs, diodes, and transistors. Unfortunately, the heat and power demands of so many components would make printers unwieldy. The solution to this problem is to integrate functions onto specialized, VLSI components, known as application-specific ICs (ASICs) or gate arrays. For our purposes, the terms *ASIC*, *gate array*, and *microcontroller* are all used interchangeably. You can see a classic example of the gate array in figure 9-29. IC1 is a 100-pin device that performs the bulk of functions needed by an EP mechanical controller board. Of course, there are some other ICs on the board, but IC1 is the single major processing component. Note the three piezoelectric crystals (X1, X2, and X3), locking in the IC's various timing signals. Ultimately, most of the logic signals being generated to manage the mechanical system are originating in a gate array.

Troubleshooting main logic

Most printer troubles do not involve a catastrophic failure of the main microprocessor or a gate array. When a main logic component does become defective, it can effectively disable a mechanical system, or shut down the printer entirely. This section will present symptoms that represent severe or complete defects in main logic. When you first turn a printer on, main logic performs an initialization procedure that checks the ROM, checks and clears all RAM locations, brings the print head to its "home" position, and

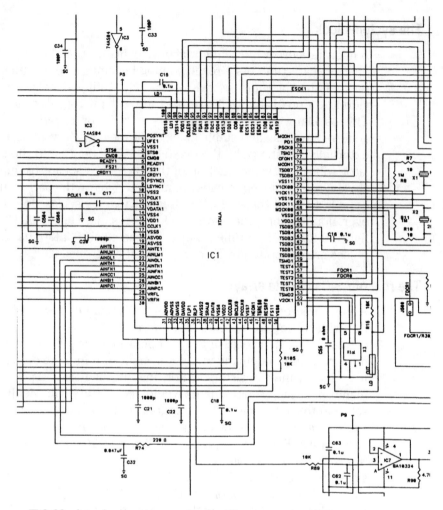

■ **9-29** *A typical system controller IC.* Tandy Corporation

establishes a communication link with the host computer. This procedure can be accomplished in just a few seconds with many conventional printers. An EP printer takes a bit longer because its photosensitive drum must establish a uniform conditioning charge, and the fusing roller must reach a stable operating temperature before printing can take place.

Symptom 1 The printer does not initialize from a "cold start" turn-on. Refer to Chart 9-4. There is no visible activity in the printer after power is turned on, but power indicators are lit. Self-test does not work. If a printer encounters an error condition during its initialization, there will be some visible or audible indication of the fault. Audible tones, flashing light sequences, or an alphanumeric

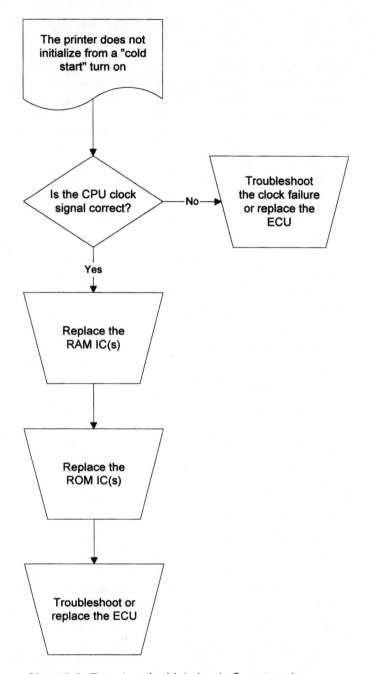

Chart 9-4 *Flowchart for Main Logic Symptom 1.*

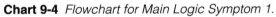

error code are just some typical failure indicators. Your user's manual will list any error codes and their meanings.

If the error is "expected" (that is, an error that is checked and handled by the printer's software), the printer will simply wait until the error is corrected. Paper out errors are a commonly "expected" error. However, "unexpected" errors can cause the printer to freeze or behave erratically for no apparent reason. ROM or microprocessor defects are considered unexpected failures; main logic has no way of dealing with such problems, so there is no way of knowing just how your printer will respond (if it works at all).

Suspect your power supply first. Use your multimeter to measure the logic supply voltage levels in your printer (usually +5 Vdc). If this voltage is low or absent, logic devices will not function properly (if at all). A low or missing voltage output suggests a defective power supply. Troubleshoot or replace the supply using the procedures of Chapter 8.

The next area to check is your system clock. Use your logic probe or oscilloscope to measure the oscillator signals on both sides of the piezoelectric crystal (make sure that your oscilloscope or logic probe can measure frequencies as high as the crystal's rating). If you are using an oscilloscope, there should be a roughly square wave at the frequency marked on the crystal. If the clock signal is low or missing, replace the crystal and stabilization capacitors. If this does not restore your clock source, replace the main microprocessor and retest the printer. If you do replace the microprocessor, be sure to install an appropriate IC socket first if possible. For ECU designs that generate the clock signal from a separate clock/timer IC, try replacing the clock IC. Otherwise, replace the ECU outright.

An initialization process can stall if the microprocessor detects a faulty RAM location. Unfortunately, there is no way to check a RAM IC completely without using specialized test equipment. Some printers might display an error code indicating a RAM error. Replace the RAM chip(s) and retest the printer. If normal operation returns, you have isolated the defective component(s). Be sure to install appropriate IC holders if possible when replacing RAM chips. Printer initialization can freeze if the printer's program ROM is defective. A bad data or instruction address can easily send the microprocessor into confusion. ROM chips are usually extremely reliable, so their failures are rare. You will have to obtain program ROMs from the manufacturer, or through a reputable distributor.

If symptoms continue, try replacing the main gate array. A failure here can prevent motor operation, disable sensor signals, and cut off communication. The main microprocessor must interact closely with gate arrays, so any fault can "hang up" the main microprocessor's operation. Of course, you can also just replace the ECU or main controller board outright.

Symptom 2 Printer operation freezes or becomes highly erratic during operation. Typically, you must activate the printer from a "cold start" to restore operation. The self-test might work until the printer freezes. Check the logic supply voltage with your multimeter. You should find about +5 Vdc. If this voltage is low or intermittently low, logic devices will behave erratically. Troubleshoot the power supply using the procedures of Chapter 8.

A microprocessor requires constant access to its program ROM in order to operate properly. Each instruction and data location must be correct, or the main microprocessor will become hopelessly misdirected. If you find that the printer only operates to some consistent point where it freezes or acts strangely, the ROM might be defective. Replace the firmware ROM and retest the printer thoroughly. If normal operation returns, you have probably isolated the problem. Install an appropriate IC socket if possible before inserting a new ROM IC.

Inspect all of your main logic ICs for any devices that might appear excessively hot, especially if the printer has only been on for a short time. Never touch potentially hot components in a live circuit. Instead, smell around the circuit for unusually heated air, or hold the palm of your hand over the circuit. If a part seems unusually hot, spray it with liquid refrigerant. If normal operation returns temporarily (you might have to reactivate the printer or press a reset or on-line button), replace the thermally intermittent component. If your search proves inconclusive, or problems persist, replace the ECU outright.

Sensors

To operate properly, a printer requires information from the "real world." For example, the printer must know if there is enough paper, toner, or ink. The printer also must be able to track head position, fusing temperature, or the page's progress through the printer. A sensor produces an electrical signal based on real physical conditions; main logic interprets these signals and makes real-time decisions on how to proceed. As you gain more experience with computer printers, you will learn that there are a wide variety of "sensors" in use. For the purposes of this book, however, there are three types of sensors: resistive, mechanical, and optical.

Resistive sensors

Temperature plays a key role in printer operations. Impact and thermal printers generate heat as part of their printing process. Thermal printers generate heat intentionally while impact printers produce heat as an unwanted side-effect, but both require head temperature to be carefully monitored. As discussed in Chapter 4, EP printers use a fusion roller assembly. The assembly applies heat and pressure to fix a toner image on paper. Fusion temperature must be carefully maintained to achieve an optimum melt. In each of these cases, a thermistor is used to sense temperature. Thermistors (or thermal resistors) are resistors whose values change in proportion to their temperature. Depending on their formulation, thermistors can be constructed to increase or decrease with temperature.

A typical fusing sensor circuit is shown in figure 9-30. The two-wire thermistor (not shown) is connected through a cable to the EP mechanical controller board at CN11. Pin TH-1 provides power to the thermistor, and the other end of the sensor attaches to pin TH-2. The resistance of the thermistor, along with R61 and R65 forms a stable network whose voltage output (filtered by C53) is measured by the gate array IC1. It is IC1 that interprets the voltage signal and regulates power to the fusing heater. If temperature falls, IC1 detects the change in temperature, and increases the power duty cycle to the fusing heater. Conversely, if temperature climbs, IC1 detects the temperature change and decreases the power duty cycle to the fusing heater.

Mechanical sensors

When position or presence must be detected, a mechanical switch can be used, as shown in figure 9-31. A set of mechanical contacts might be normally open or closed, then actuated by the presence of paper, the print head carriage, closed housings, and so on. Figure 9-31 illustrates a simple carriage "home" sensor made with a normally open switch. While the carriage is in motion, the switch is open, so its output equals the value of +V. As the carriage reaches its home position, it closes the switch, which grounds the sensor signal. An ASIC or microprocessor detects this condition and responds accordingly.

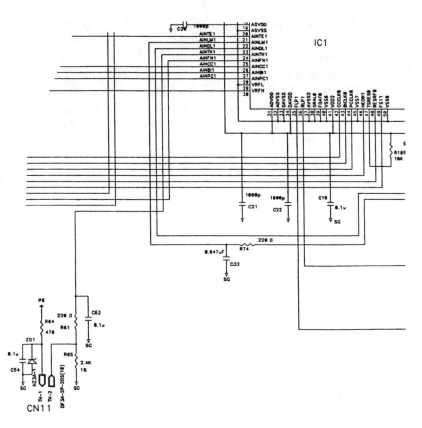

■ 9-30 *An EP thermistor signal conditioning circuit.* Tandy Corporation

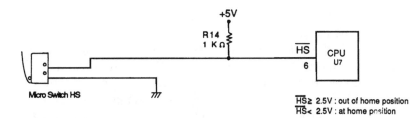

■ 9-31 *Diagram of a mechanical sensor.* Tandy Corporation

Optical sensors

Mechanical sensors are simple and inexpensive devices, but they lack reliability over long-term use. Electrical contacts wear out through use and environmental corrosion. They are also subject to electrical *ring*, an output that might vary on and off for several milliseconds before reaching a stable condition. Optical sensors are immune to these problems. A basic optical sensor (called an *optoisolator*) is shown in the paper-empty sensor of figure 9-32. It is made up of two parts: a transmitter and a receiver. Both are separated by a physical gap. The transmitter is usually an infrared (IR) LED kept on at all times. The receiver is typically a photosensitive transistor that is most sensitive to light wavelengths generated by the LED. When paper is present, light generated from the LED is reflected off the page to the phototransistor. The phototransistor is then saturated, and the PS signal at U7 is logic 1. When paper is missing, LED light is not reflected to the phototransistor, which remains off, so the PS signal becomes logic 0 and an error is generated.

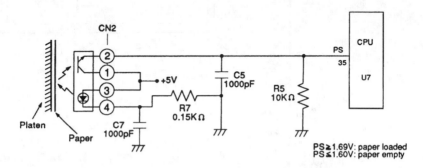

276

■ **9-32** *Diagram of an optical paper sensor.* Tandy Corporation

An optical encoder is a slightly more sophisticated application of optoisolators, as shown in figure 9-33. Although many DMI printers forgo optical encoders, more demanding printers (e.g., ink jet printers) often use an optical encoder to track head position. When a carriage advance motor moves, an optical encoder returns pulses to main logic. These pulses help the main logic determine if the motor has stepped as far as it was told. An incremental encoder uses two individual optoisolators separated by a code disk, which is tied mechanically to the carriage motor. The code disk itself is little more than a clear piece of plastic with opaque markings spaced evenly around it. Both optoisolators are aligned so that they cannot be turned on simultaneously. Less expensive encoders might use only one optoisolator. It will still report the same number of pulses, but it cannot determine direction.

As the carriage moves, it turns the code disk. The opaque lines turn each sensor on and off. Each on/off transition corresponds to a step of the carriage motor, so if the carriage motor turns 500 steps, a train of 500 pulses should be sent back to main logic where they are interpreted by the slave microprocessor or ASIC. If there are too

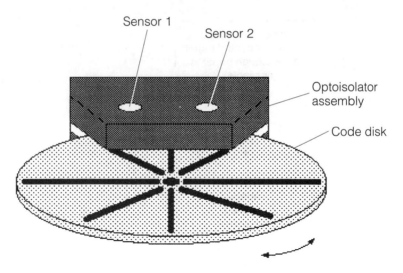

Sensor 1

Sensor 2

Optoisolator assembly

Code disk

■ **9-33** *Simplified diagram of an optical position encoder.*

few or too many pulses, the printer knows that its carriage is not in the desired position. Additional pulses can then be sent, if necessary, to correct the carriage position. Encoders using two offset optoisolators can be used to determine the carriage's direction as well as its length of travel. Because both sensors are offset, there is a slight time lag between them. Main logic can identify which pulse leads or lags. For example, if the carriage moves left, the pulses from signal 1 will lead the pulses from signal 2, and vice versa.

Troubleshooting sensors

Before performing sensor checks, be sure to examine any connectors or interconnecting wiring that ties the sensor into its conditioning circuit. If the sensor checks properly and its conditioning circuit appears functional, there might be a problem with the ASIC or slave microprocessor that interprets and processes sensor signals.

Symptom 1 Paper out alarm shows up on the control panel even though paper is available, or it does not trigger when paper is exhausted. Refer to Chart 9-5. If your paper sensor is a mechanical switch, place your multimeter across its leads and try actuating it by hand. You should see the voltage reading shift between a logic 1 and logic 0 as you trigger the switch. If you measure some voltage across the switch but it does not respond (or responds only intermittently) when actuated, replace the defective switch. If it responds as expected, check its contact with paper to be sure that it is actuated when paper is present. You might have to adjust the

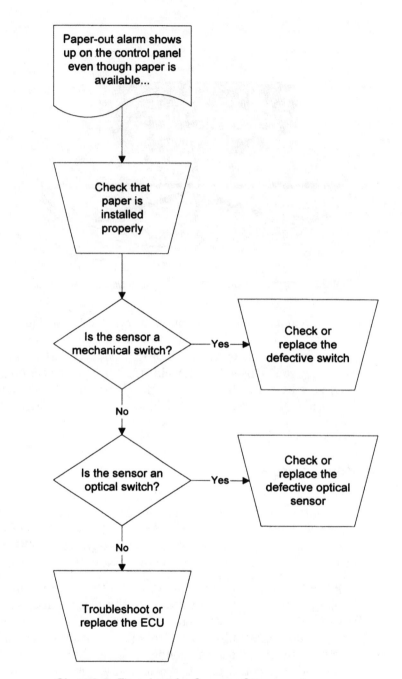

Chart 9-5 *Flowchart for Sensors Symptom 1.*

switch position or thread paper through again to achieve better contact.

Check an optical paper sensor by placing your multimeter across the photosensitive output and try to actuate the sensor by hand. This might involve placing a piece of paper or cardboard in the gap between transmitter and receiver. You should see the phototransistor output shift between logic 1 and logic 0 as you trigger the optoisolator. If it does not respond, check for the presence of dust or debris that might block the light path. If excitation voltage is present, but the phototransistor does not respond, it is probably defective. Replace the optoisolator. Measure voltage across the LED transmitter. You should find a typical voltage drop of 1.3 to 3.6 Vdc across the excited LED. If this voltage is high or equal to the logic supply voltage (usually +5 Vdc), the LED might be burned out, so replace the optoisolator. When a sensor responds correctly, the trouble is probably in your microprocessor or gate array (whatever is interpreting the sensor signal). Try replacing these components, or replace the ECU outright.

Symptom 2 Carriage does not find its home position. Refer to Chart 9-6. This might result in a frozen initialization or erratic print spacing. As with paper sensors, the sensing element can be mechanical or optical. If the home sensor is a mechanical switch, place your multimeter across its contacts and try actuating the switch by hand. You should see a voltage reading switch between a logic 1 and logic 0 as the switch is actuated. If voltage is present but the switch does not respond (or responds only intermittently), replace the switch. If it does respond, check its contact with the carriage to be sure that it actuates when the carriage is in its home position. You might have to adjust the switch position to achieve a better contact.

An optical home sensor can be checked in much the same manner. Place your multimeter across the phototransistor output and try to actuate the sensor by hand by blocking the optical gap with a piece of paper or cardboard. You should see the output voltage shift between a logic 0 and logic 1 as the sensor is actuated. If the phototransistor does not respond, check for dust or debris that might be blocking the light path. If excitation voltage is present but the phototransistor does not respond, replace the optoisolator. Measure voltage across the LED transmitter. You should find a typical voltage drop of 1.3 to 3.6 Vdc across an excited LED. If this voltage is high or equal to the logic supply voltage (usually +5 Vdc), the LED might be burned out, so replace the optoisolator.

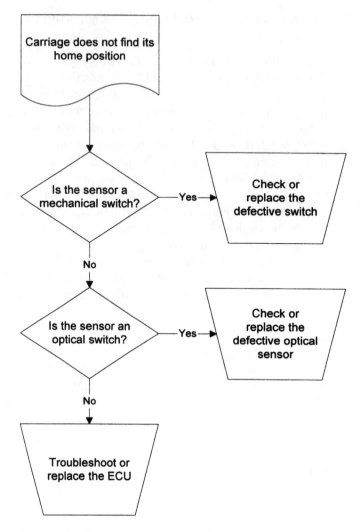

Chart 9-6 *Flowchart for Sensors Symptom 2.*

When your sensor responds as expected, the trouble is probably in your microprocessor or gate array. Try replacing these components, or replace the ECU outright.

Symptom 3 Carriage moves erratically or inconsistently. There are errors in print spacing. This symptom assumes that you have checked the carriage motor and mechanics. Use your oscilloscope to measure the pulses from each optoisolator output in the position encoder. If you have a dual-channel oscilloscope, place the signal from sensor 1 into channel 1, and the signal from sensor 2 into channel 2. As the printer runs, you will see these pulse signals changing in width and phase while the optical wheel spins. If one or both of these pulse signals are missing, use your multimeter to measure the LED voltage drop across each optoisolator. You should normally see a voltage drop between 1.3 and 3.6 Vdc for an excited LED. If this voltage is high or equal to the logic supply voltage (usually +5 Vdc), the LED might be burned out. Replace the suspect optoisolator, or the entire sensor module if it is a single unit.

Check the code disk to see that it is intact and clear, and rotating as the carriage moves. Also check for any accumulation of dust or debris that might be blocking the light path. If both pulses are available and operating as expected, there might be a problem in your microprocessor or gate array. Try replacing these components, or replace the ECU outright.

Note: This procedure suggests that two optoisolators are used in a position encoder. Keep in mind that the second sensor is needed only to sense carriage direction. Some printers might only use a single optoisolator in its encoder, but the concepts of optoisolator testing remain the same.

Symptom 4 Temperature control is ineffective; temperature never climbs, or climbs out of control. This might affect print quality or initialization for EP printers.

Warning: Always allow at least 15 minutes for the fusing system to cool before working in the fusing system. Unplug the printer and disconnect the thermistor at its connector. Use your multimeter to measure its resistance. A short or open circuit reading might indicate a faulty thermistor, so replace any suspect part. If you get some resistance reading, warm the thermistor with your fingers and see that the reading changes (even a little bit). A reading that does not change at all suggests a faulty thermistor. Never touch a hot thermistor with your fingers!

Chances are that the thermistor is intact. If this is the case, the gate array on your mechanical controller board has probably failed. You can try replacing the mechanical controller, or replace the mechanical controller board outright.

Mechanical service techniques

10

IT TAKES MUCH MORE THAN JUST A GOOD LOGIC CIRCUITRY or a pretty enclosure to make a first-class printer. A series of independent mechanical systems are needed to perform the variety of physical tasks that every printer must do (figure 10-1). The most obvious mechanism is the paper transport system responsible for moving paper through the printer. Every printer made must make provisions to handle paper. Another common mechanism is the carriage transport system for moving serial print heads back and forth across the page. Electrophotographic printers and printers using line-print heads do not need a carriage transport. Impact dot-matrix and thermal transfer printers use a media (also known as ink) ribbon that must be handled by a ribbon transport system. Still other printers use "secondary" mechanical systems, such as a paper cutter or sheet stacking mechanism.

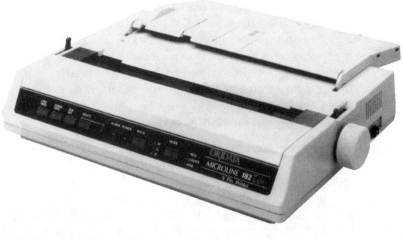

■ **10-1** *An Okidata Microline 182 printer.* Okidata

To fully appreciate the importance of a printer's mechanics, take a look at the exploded diagram of figure 10-2. The ECU and power supply module are in the left side of the illustration (marked 2-2 and 2-3 respectively). The remainder of the drawing shows a detailed layout of electromechanical and mechanical parts. You have already read about important parts in Chapter 2. This chapter expands that background material into a discussion on paper transports, carriage transports, and ribbon transports. It also contains an overview of mechanical preventive maintenance.

Friction & tractor paper transports

In order to print anything at all, paper must be moved in front of a print head. Control of the paper must be precise and consistent without inflicting any damage on the paper surface. This is a demanding application, but two types of inexpensive and reliable transport mechanisms have evolved to serve in today's moving-head printers: friction-feed mechanisms and tractor-feed mechanisms. Both approaches possess their own advantages and disadvantages.

Friction-feed

As its name suggests, a friction-feed paper transport uses the force of friction to "push" paper through the printer, as shown in figure 10-3. A single sheet of paper is threaded into the printer along a metal feed guide. The guide ensures that paper is maneuvered properly between the platen and pressure roller(s), then up in front of the print head assembly. A set of small, free-rolling bail rollers press gently against the paper to help keep paper flat while it passes around the platen. To allow the free passage of paper during threading, a lever is often included to separate pressure rollers from the platen (not shown in figure 10-3). After paper is positioned as desired, the lever can be released to reapply pressure. From then on, paper can only be moved by hand-turning the platen, or during the printer's actual operation.

Friction-feed paper transports are designed to use single sheets of standard paper. While early implementations demanded that the user manually insert and position each sheet, newer friction-feed designs (such as Hewlett-Packard's DeskJet series) have modified the mechanics to pick up single sheets from a free-standing stack, similar to the cassette approach used in EP printers. Single-sheet operation eliminates the need to use specialized (i.e., fan-fold perforated) papers, so letterhead and forms can also be used easily.

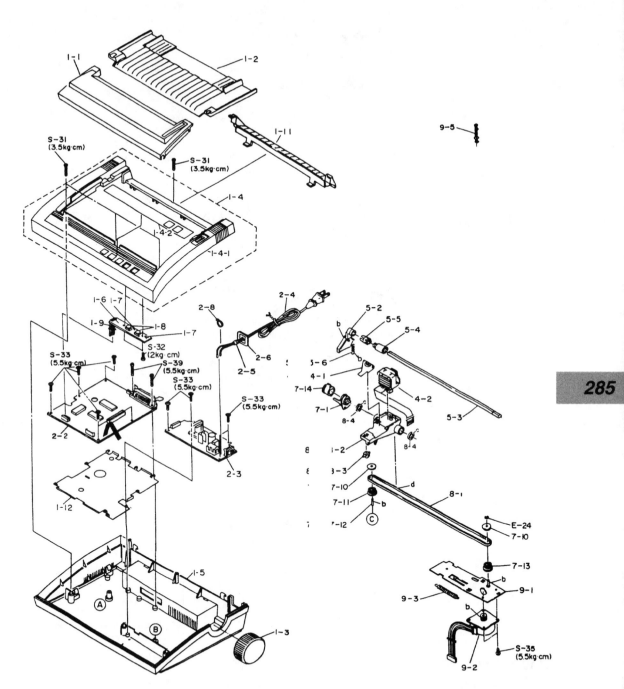

■ 10-2 *Exploded diagram of a dot-matrix printer.* Tandy Corporation

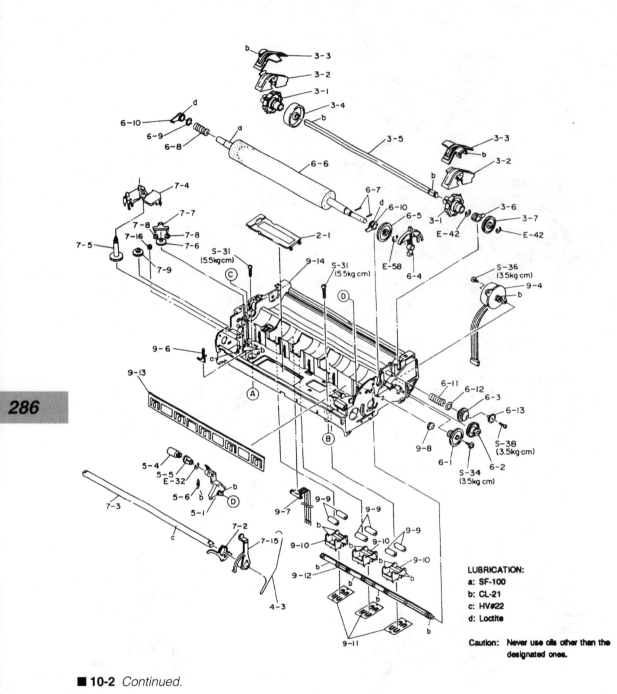

LUBRICATION:
a: SF-100
b: CL-21
c: HV#22
d: Loctite

Caution: Never use oils other than the designated ones.

■ **10-2** *Continued.*

Mechanical service techniques

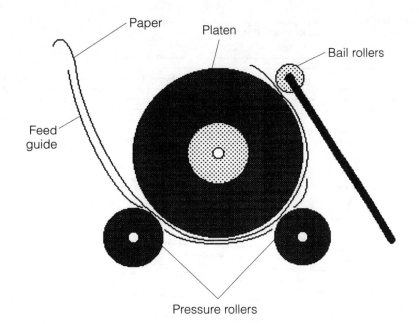

Paper Platen

Bail rollers

Feed
guide

Pressure rollers

■ **10-3** *Diagram of a friction-feed paper transport system.*

This combination of factors has made friction-feed systems quite popular.

Unfortunately, friction-feed transport systems still have their limitations in spite of mechanical advances. Probably the most pronounced and important problem is their tendency to "walk" the page as shown in figure 10-4. For a friction-feed mechanism to work, it must handle each page evenly. If contact pressure is not even, or some obstruction should occur in the paper path, the page will not feed evenly. Although figure 10-4 is rather an extreme case, it shows you how bad the problem can be. Specially coated papers, heavy-bond papers, and light-bond papers can all contribute to friction-feed problems.

Tractor-feed

Tractor-feed does not rely on friction to transport paper. Instead, a set of sprocket wheels are mechanically linked to the platen drive train. Pegs on each sprocket wheel mesh perfectly with specially made paper. This type of paper (also called *tractor* or *fanfold* paper) has holes perforated along both sides. Paper is threaded into the printer along a metal feed guide. There is very little resistance from its contact rollers, so paper can easily be fed through and secured into its sprocket wheels. Most sprocket

```
PRINTERS: The Printer Test and Alignment Utility  V.1.00
IMPACT PAPER WALK Test Pattern
```

■ **10-4** *An example of page "walk."*

wheels can slide left or right to accommodate a selection of paper widths or tractor feed label products. Bail rollers are included to help keep paper flat against the platen. Once paper is threaded, as shown in figure 10-5, it can only be advanced by hand-turning the platen knob, or in actual printer operation.

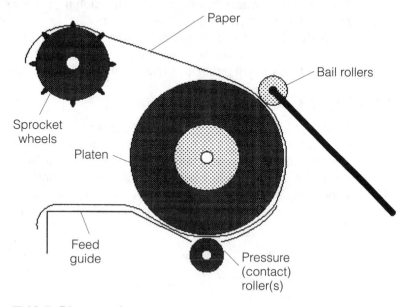

Paper

Bail rollers

Sprocket wheels

Platen

Feed guide

Pressure (contact) roller(s)

■ **10-5** *Diagram of a tractor-feed paper transport system.*

You can see the various mechanical components of a tractor feed system in figure 10-2. The frame (marked 9-14) forms the foundation for the entire mechanical system. A platen (6-6) is driven by the line feed motor (9-4) through a series of gears (marked 6-1, 6-2, 6-3, 6-5, and 3-7 respectively). You can also locate the tractor-feed sprocket wheels (marked 3-1) interconnected by a rod (3-5). This interconnecting rod is very important because the tractor assembly is driven by only one gear (3-7). Sprocket wheel sheaths and covers (3-2 and 3-3 respectively) keep the fan-fold paper clamped to the wheels.

Although they require a bit of extra mechanical support, tractor-feed paper transports are still quite popular because it is virtually impossible to walk or wrinkle a page when it is "pulled" evenly through the printer. The use of fan-fold paper also allows the printer to run for thousands of pages without stopping. This combination of reliable paper feeding and an almost inexhaustible paper supply make tractor-feed systems the preferred transport mechanism for high-volume business or industrial printing.

One of the main complaints about tractor feeding has been its unidirectional nature; you can only pull paper in one direction. However, later tractor-feed systems on high-end printers employ a bidirectional sprocket system. A bidirectional sprocket system meshes with perforated paper as it enters the printer and leaves the platen. During normal operation, this bidirectional feed pushes paper in and pulls it out simultaneously. If paper must be reversed, the mechanism will pull paper back evenly without crumpling it.

Driving the paper transport

Now that you know the basic mechanical architecture behind friction-feed and tractor-feed systems, you can understand how the systems are operated. There are two facets involved in paper transport operation: mechanical and electrical. Both are equally important for proper transport operation. The mechanical paper transport system is illustrated in figure 10-6. While the mechanical

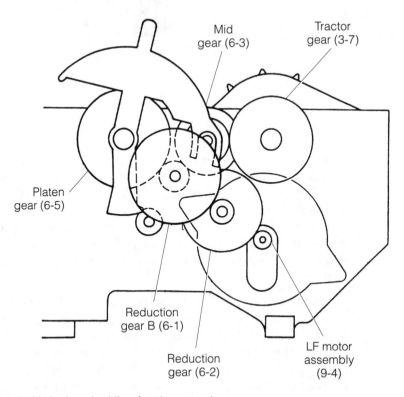

■ **10-6** *A typical line feed gear train.* Tandy Corporation

system will vary a bit from model to model, figure 10-6 offers an important glimpse of a tractor-feed gear arrangement. The heart of the mechanism is the paper feed motor (sometimes called the *line feed* or *LF* motor assembly). All of the force needed to move paper comes from this motor; if it fails, so does the paper feed.

The LF motor drives a reduction gear (marked 6-2 in figure 10-2). Force from the first reduction gear is transferred to a second reduction gear (marked 6-1). Together, these reduction gears increase the effective torque provided by the LF motor. The second reduction gear drives two elements: the platen, and a midgear (marked 6-3). The purpose of the midgear is to adjust speed and torque so that the tractor-feed speed will match the platen speed exactly. If the tractor-feed moves too slowly, paper will "pay out" of the platen too fast and bunch up. If the tractor-feed moves too quickly, it will strain and tear the paper. A midgear also allows the introduction of a clutch mechanism, which allows a user to engage and disengage the tractor gear, thus turning the tractor-feed mechanism on and off at will. When the clutch mechanism disengages the tractor gear, it also typically engages a pressure roller against the platen (effectively converting a tractor-feed paper transport into a friction-feed transport, and vice versa). As you might imagine, the entire gear train must be intact for the paper transport to function properly. Any damaged or obstructed gears can adversely affect line feeds. A slipping or otherwise damaged clutch (if one is indeed installed) can slow the tractor gear and cause paper to bunch up.

Electrically, the paper transport system revolves around the LF motor, as shown in figure 10-7. The motor itself is a four-phase stepping motor. You can tell by the depiction of each motor winding. Pins 2 and 6 represent one winding, pins 6 and 4 are a second winding (pin 6 is common), pins 3 and 5 form the third winding, and pins 5 and 1 are the fourth winding (pin 5 is also common). In this design, the common pins are tied to +24 V through Q35, which is enabled when the 'LFST signal generated by the CPU is logic 1. This configuration allows the CPU to exercise master control over the LF motor. Now, the direction and number of steps produced by the LF motor are determined by signals originating in the ECU. (Figure 10-7 illustrates U3 as the logical element.) However, logic signals alone are not enough to drive the LF motor windings. Instead, each logic signal is "amplified" through a driver transistor (depicted as Q31, Q34, Q36, and Q37). Trouble in the CPU, the gate array, or any of the driver transistors can easily disable the LF motor.

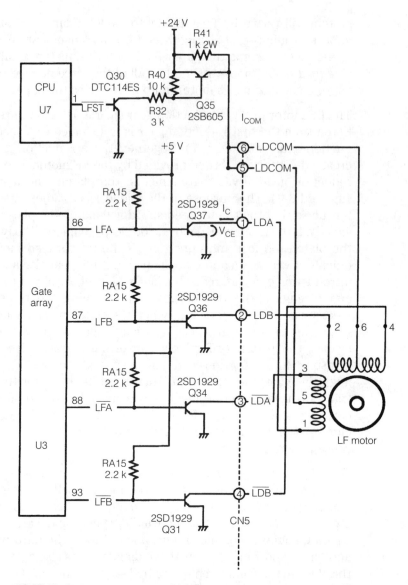

■ **10-7** *A paper transport drive system.* Tandy Corporation

The actual signal characteristics can be viewed with an oscillo-scope; you will see a display such as the one shown in figure 10-8. When a driver transistor is fired, the level of V_{ce} (shown in the upper trace) drops off, and current builds in about 3 ms. When current reaches its peak, the motor winding will exert enough force to step the rotor. When the driver transistor is turned off, its V_{ce} jumps way up because of the inductive "flyback" produced by the motor, but this voltage spike falls off and settles back to a stable

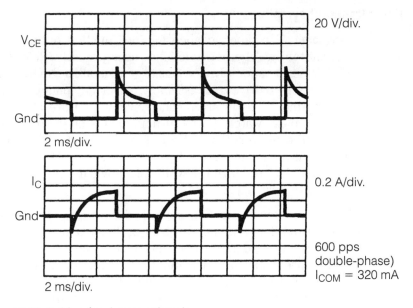

20 V/div.

2 ms/div.

0.2 A/div.

600 pps
double-phase)
$I_{COM} = 320$ mA

2 ms/div.

■ **10-8** *Line feed motor signals.* Tandy Corporation

voltage before the transistor is fired again. The actual duration of an LF motor step is about 3.33 ms.

You can see the relationship between each driver signal in the timing diagram of figure 10-9. While this diagram might seem daunting at first glance, there are several things you should notice. First, the drivers always work in pairs; that is, the LFA driver is always on when the 'LFA driver is off, and the LFB driver is always on when the 'LFB driver is off. Second, the LFA and LFB drivers are always separated in time, so the LFA signals will change first, followed closely by the LFB signals, and so on. By sequencing the signals in this way, the LF motor is stepped in the advancing direction. If the signal order were reversed, the LF motor would step in the opposite direction.

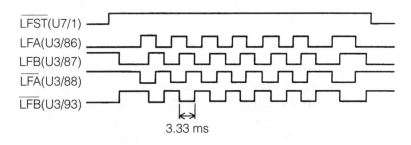

3.33 ms

■ **10-9** *Line feed system logic timing.* Tandy Corporation

Paper transport symptoms

Symptom 1 The paper advance does not function, or functions only intermittently. All other functions check properly. Refer to Chart 10-1. When a paper advance fails to work at all, begin by observing the paper feed drive train assembly. Check any pulleys or gears to ensure that all parts are meshed evenly and are able to move freely. You can watch this by turning the platen knob located outside of the printer. Remove any foreign objects or obstructions that might be jamming the drive train. Never try to force a drive train that does not turn freely! Realign any parts that appear to be slipping or misaligned.

If the gear train is intact, suspect the LF motor and driver system next. Turn off and unplug the printer, then examine the electrical connections for your paper advance motor. Make sure that all connectors are installed and seated properly. If you suspect a faulty wiring connection, use your multimeter to measure continuity across any suspicious wires. It might be necessary to disconnect the cable from at least one end to prevent false continuity readings. Replace any faulty wiring.

Warning: The following is a "hot" procedure that must be performed with the printer on, and with clear access to the ECU and LF motor. Take all precautions to protect yourself from injury.

If you have an oscilloscope, you can check the supply voltage and signals on each side of the driver transistors (as illustrated in figure 10-7). You can use the PRINTERS program to generate a series of regular line feed sequences for testing. Before you proceed, make sure that +24 V is available to the motor (you can measure it on motor pins 6 and 5). If +24 V are missing, the motor will not turn, and you should suspect a problem in the +24-V supply, the switching components (i.e., Q30 and Q35), or the CPU or other ECU logic. Once supply voltage is available, check the motor signals. If you see logic signals on the base of each transistor, you should find high-energy signals on each collector (similar to the V_{ce} trace in figure 10-8). If you see a logic signal, but the expected V_{ce} signal is absent, the corresponding driver transistor is probably defective and should be replaced. If the logic signal is missing to begin with, there is a problem in the ECU. You can replace the logic IC or the entire ECU at your discretion as discussed in Chapter 9. If +24 V and all of the V_{ce} driver signals are reaching the LF motor as expected and the motor still does not turn, the motor is probably defective and should be replaced.

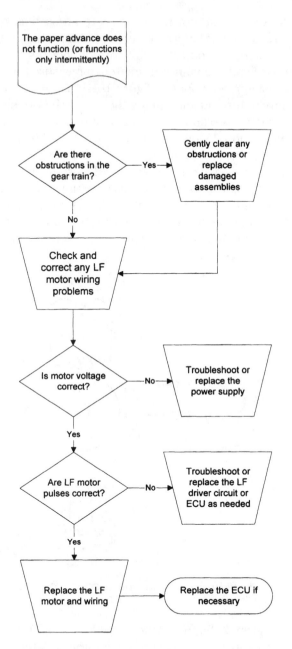

The paper advance does not function (or functions only intermittently)

Are there obstructions in the gear train? — Yes → Gently clear any obstructions or replace damaged assemblies

No

Check and correct any LF motor wiring problems

Is motor voltage correct? — No → Troubleshoot or replace the power supply

Yes

Are LF motor pulses correct? — No → Troubleshoot or replace the LF driver circuit or ECU as needed

Yes

Replace the LF motor and wiring → Replace the ECU if necessary

Chart 10-1 *Flowchart for Friction and Tractor Feed Systems Symptom 1.*

Now if you do not have access to an oscilloscope, or you do not have the time and inclination to follow the "hot" procedure outlined above, you can simply disconnect the LF motor from its PC board and use your multimeter to measure the winding resistance of each motor phase. Figure 10-10 shows you how to do this. Be sure to turn off and unplug the printer before proceeding. Measure across pins 6 and 4, 6 and 2, 5 and 3, and 5 and 1. Each measurement should yield a resistance of about 56 Ω (although your particular LF motor might have slightly different ratings). Ideally, each phase winding should yield close to the expected resistance. If any of the windings measure unusually high (open) or low (shorted), the motor is likely defective and should be replaced. If the windings all measure correctly, the motor is probably good, and your fault is in the ECU/driver circuitry, which you should simply replace outright.

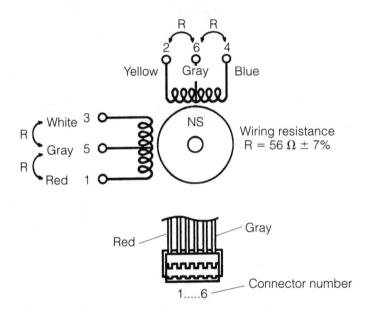

■ **10-10** *Measuring line feed motor winding resistance.*
Tandy Corporation

Symptom 2 Paper gathers or bunches up at the tractor-feed sprocket wheels. This is a symptom relatively common to older printers, or printers with a tractor-feed mechanism that was serviced improperly. The platen and sprocket wheels must transfer paper at exactly the same rate to achieve smooth paper handling. If the sprocket wheels are moving a bit too slow (relative to the platen) for any reason, the paper will eventually "bunch up" at the sprocket wheels.

As shown in figure 10-6, some printers add a clutch assembly that can engage and disengage the tractor-feed unit from the drive train. If there is a clutch mechanism present, see that it is engaged fully and properly. You might want to "pop the clutch" in and out a few times to ensure positive contact. Inspect the drive train gears very carefully, especially the teeth on the tractor gear itself. If one or more teeth are missing, the tractor assembly will "skip" a bit, and might behave quite erratically. Replace any worn or damaged gears. Another possibility is that there might be foreign matter obstructing the gear train. Dust, dirt, debris, or old and hardened lubricant can cause the tractor gear to skip or stick. Carefully remove any obstruction(s) in the gear train.

Symptom 3 Paper slips or walks around the friction-fed transport. Friction-feed paper transports are only designed to work with certain types of paper—brands within a certain range of thickness and weight. Very fine (light bond) paper or very heavy (card stock) paper will probably not advance properly. Slick or other unusual coatings can also upset a friction-feed system. Check the specifications for your particular printer to find its optimum paper type. If you find that you are using an unusual type of paper, try the printer using standard 20-lb. bond xerography-grade paper.

If the problem persists, advance the paper feed knob manually (if possible) and take careful note of each roller condition. An even, consistent paper feed depends on firm roller pressure applied evenly across its entire length. Rollers that are very dirty, or old and dry, might no longer be applying force evenly. Clean and rejuvenate your rollers with a good-quality rubber cleaning compound available from almost any comprehensive office supply store.

Caution: Rubber cleaning compounds can be dangerous, and might not be compatible with all types of synthetic roller materials. Read instructions on the chemical container carefully, and follow all safety and ventilation instructions.

Check your paper path for any debris or obstructions that might be catching part of the paper. A crumpled corner or paper jammed in the paper path or caught in the feed guide can easily interfere with subsequent sheets. Turn off and unplug the printer, and remove all obstructions (being very careful not to mark any of the rollers). A straightened paper clip can often get into spaces that your fingers and tools will not. Use your needle-nose pliers to put a small hook in the wire's end for grabbing and pulling the obstruction. Do not disassemble the rollers unless absolutely necessary.

Old rollers might also be out of alignment. Mechanical wear on shafts and bushings (or bearings) can allow some rollers to "float" around in the printer. Carefully examine the condition of each roller shaft. Use the paper loading lever (if available) to separate pressure rollers from the platen, then wiggle each shaft by hand. Ideally, each shaft should be fixed firmly within its assembly, so you should feel little or no "slack." If you feel or see a roller move within its assembly, replace its bushings, bearing, or shaft. Some pressure roller assemblies can be adjusted slightly to alter its contact force. If your particular printer uses nonadjustable pressure rollers, there is little more to be done other than to replace the mechanical paper feed assembly entirely. If you can adjust roller force (using spring tension or a screw adjustment), do so only as a last resort, and only then in small increments. Careless adjustment can easily worsen the problem. Remember to try cleaning or rejuvenating the rollers before attempting a mechanical adjustment.

Symptom 4 Paper wrinkles or tears through the printer. Tractor feed paper transport systems are remarkably reliable, so it is very rare to encounter tearing problems using the tractor approach. But many later-model printers (especially DMI printers) offer a selection of paper feed paths, as shown in figure 10-11. A mechanical lever is typically used to switch between tractor and friction feed modes. If paper suddenly seems to wrinkle or tear along its perforations during printing, the first thing to check should be the paper feed selector lever.

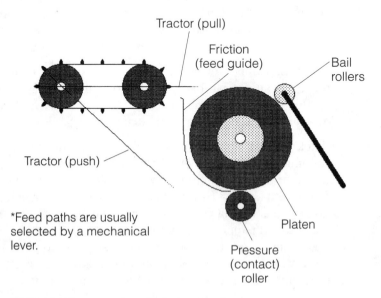

■ **10-11** *Diagram of multiple feed paths.*

If your printer's paper feed mode is set correctly, check the paper path for any debris or obstructions that might be catching the paper. Fragments of torn paper caught in the feed guide can easily jam the paper path. Carefully remove all obstructions that you might find, but use extreme caution to prevent damage to your rollers or feed guide. Do not disassemble the paper transport unless it is absolutely necessary. If problems persist, you might have to replace the entire mechanical transport assembly.

Carriage transport systems

Impact, ink jet, and many thermal printers use serial print heads. Fully formed text and graphics are formed by passing a print head left and right across a page surface. As the head moves, it places a series of vertical dots that creates the image. In this way, a complete line of text can be generated in a single pass (letter-quality text or graphics might require additional passes). As you might imagine, the process of moving a serial print head becomes a serious concern. It must move at the proper time, at the proper speed, and over the proper distance to within several thousandths of an inch—every pass. The task of moving a serial print head is handled by the carriage transport system.

Carriage transport mechanics

Figure 10-2 presents an exploded view of a typical DMI printer; you can see the individual components of the carriage transport located in the lower middle of the figure. The heart of the carriage system is the carriage transport motor or CR motor (marked 9-2). Signals from the ECU command the stepping motor to rotate clockwise or counterclockwise. As the CR motor turns, it rotates a primary pulley (7-13) that is attached to a timing belt (marked 8-1). A secondary pulley (7-11) keeps the timing belt secure. Although the illustration might not show this clearly, the timing belt and pulleys are notched, much like gears, to prevent the timing belt from slipping. The carriage platform itself (marked 8-2) attaches to a single point on the timing belt, so as the timing belt spins, it moves the carriage left or right. A carriage is also threaded through a smooth guide rod (7-3). The guide rod acts to stabilize the carriage and keep it parallel to the platen. The print head itself (such as the device marked 4-2) mounts to the carriage. As you might imagine, any problems with the CR motor, timing belt, or guide rod will have an adverse effect on carriage motion.

Of course, there are often variations to this configuration. Sometimes, a pulley system is replaced by a lead-screw type of assembly. The print head and stepping motor remain unchanged, but the carriage and drive train are modified. A CR stepping motor now drives a gear train that operates a long, coarsely threaded lead screw. This lead screw is threaded into the carriage to become one of its "rails." The second rail remains a simple, low-friction guide rod that provides stability and support for the carriage (and print head). When the stepping motor turns, it rotates the lead screw (often through a gear train). Clockwise rotation of the lead screw pushes the carriage left, while counterclockwise rotation pulls the carriage right.

For the timing belt and pulley approach, positioning precision is determined by the stepping motor's resolution and the pulley diameter. This holds true for a direct drive mechanism. If the primary pulley works through a gear train, that gear ratio must also be included. As an example, suppose a stepping motor directly drives a pulley 1 inch in diameter. The circumference of the primary pulley would be $[\pi \times d]$ $[3.14 \times 1"]$ 3.14", so one complete motor revolution (thus one complete pulley revolution) would cause the linkage to travel 3.14 inches. If the stepping motor works at 200 steps per revolution, each step would turn the pulley $\frac{1}{200}$ of a revolution, or $[3.14/200]$ 0.0157" of linear travel.

Suppose the pulley was only 0.5 inch in diameter. Its circumference would only be $[\pi \times d]$ $[3.14 \times 0.5"]$ 1.57", so a single motor rotation would only move the carriage 1.57 inches. At 200 steps per revolution, each step would drive the carriage $[1.57/200]$ 0.0079". Smaller pulleys (or motors with more steps per revolution) can achieve finer positioning, but at the cost of more steps to traverse the platen width.

A similar process holds true for the lead screw drive. For the sake of simplicity, suppose that there is an equal gear ratio, so one motor revolution will result in one lead screw rotation. In this case, positioning precision is determined by the spacing (or *pitch*) of each lead screw thread, as well as stepping motor resolution. If your lead screw has a pitch of one (one thread per one inch of screw length), one motor revolution will move the carriage one inch. At 200 steps per revolution, each motor step will turn the lead screw $\frac{1}{200}$ of a turn, which results in $[1"/200]$ 0.005" of carriage travel.

If your lead screw has a pitch of 10 (10 threads over 1" of screw length), 1 thread covers $\frac{1}{10}$" in length. With 200 steps per revolution, one motor step would still turn the lead screw $\frac{1}{200}$ of a turn, but because a thread now occupies $\frac{1}{10}$", the carriage would only turn [0.1"/200] 0.0005". The motor would now have to rotate 10 times to achieve a full inch of travel. In practice, lead screws are almost never this fine, but it demonstrates the effects of screw pitch on positioning.

Carriage transport electronics

The CR transport electronics are illustrated in figure 10-12. You might notice some similarities between the driver circuitry here, and the design of figure 10-7; the CR circuit uses a four-phase stepping motor, four high-power driver transistors, a master switching circuit to apply voltage to the motor, and logic signals generated from the ECU. However, there are some important differences that you must be aware of. The most noticeable differences are in the master motor voltage control circuit operated by the CPU. Rather than a simple on/off switching arrangement as in figure 10-7, the CR driver circuit in figure 10-12 uses a much more extensive circuit to selectively switch power to the CR motor. This approach allows the CR motor to be operated at several different speeds in order to accommodate the printer's different print modes. Table 10-1 explains the various modes.

Table 10-1 Comparison of logic conditions and carriage steps.

Motor pulses	Print mode	HCST	HCP1	HCP2
960 pps	Standard 10 cpi	low	low	high
960 pps	Standard 17 cpi	low	high	low
840 pps	Letter quality	low	low	high
700 pps	Home seeking	low	low	low
640 pps	Letter quality (reduced throughput)	low	low	high
0 pps	Standby	high	low	low

There are three logic signals being generated from the CPU. The CPU will select different combinations of these signals depending on the desired carriage speed. For example, when printing in standard 10 CPI mode, precise positioning is not required, so the carriage motor can move at a relatively fast 960 pulses per second (pps). To drive the motor at 960 pps, the CPU signals (HCST, HCP1, and HCP2) will be logic 0, logic 0, and logic 1 respectively. As another example, when the carriage is hunting for its "home" position, it will move a bit slower at 700 pps, so HCST, HCP1, and

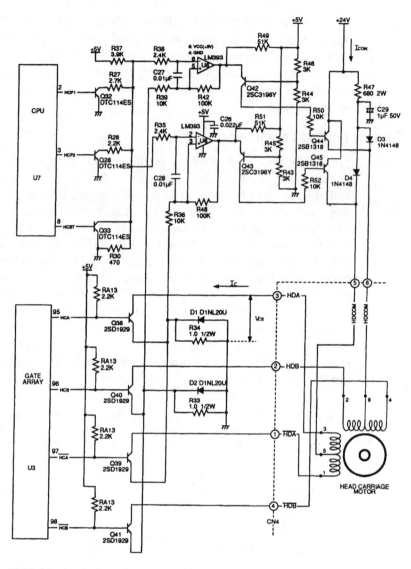

■ **10-12** *A carriage transport drive system.* Tandy Corporation

HCP2 will all be set to logic 0. The logic signals are processed through two comparators (U8) and switched through Q42, Q43, Q44, and Q45. When Q44 and Q45 are off, the CR motor is powered through a resistor (R47) and diodes (D3 and D4). The voltage drop across R47 limits the voltage available to the motor, so it can be stepped faster. If either Q44 or Q45 are on, R47 and the corresponding diode are effectively shorted out, so two of the motor windings are then connected directly to 24 V; the motor must be stepped a bit more slowly. When both Q44 and Q45 are on, R47 and both of the diodes are shorted out, and all motor phases are set to maximum voltage, and minimum step rate. An important point to remember is that the actual motor step rate and direction are determined by logic pulses generated at the gate array (U3), not by the amount of motor voltage applied to the windings.

The motor itself is a four-phase stepping motor. You can tell by the depiction of each motor winding. Pins 2 and 6 represent one winding, pins 6 and 4 are a second winding (pin 6 is common), pins 3 and 5 form the third winding, and pins 5 and 1 are the fourth winding (pin 5 is also common). In this design, the "common" pins are tied to +24 V through a diode (D3 and D4) and R47. This configuration allows the CPU to exercise master control over the CR motor and its step speed. Now, the direction and number of steps produced by the CR motor are determined by signals originating in the ECU (figure 10-12 illustrates U3 as the logical element). However, logic signals alone are not enough to drive the CR motor windings. Instead, each logic signal is "amplified" through a driver transistor (depicted as Q38, Q40, Q39, and Q41). Trouble in the CPU, the gate array, or any of the driver transistors can easily disable the CR motor. The actual motor drive signal characteristics can be viewed with an oscilloscope. By measuring the collector voltage on any one of the phase driver transistors, you will see a display similar to the one shown in figure 10-13.

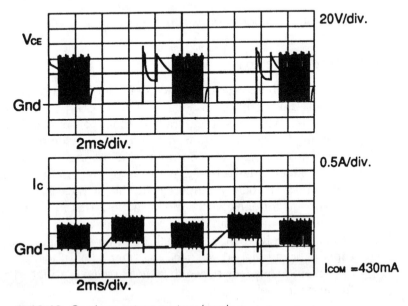

■ **10-13** *Carriage return motor signals.* Tandy Corporation

You can see the relationship between each driver signal in the timing diagram of figure 10-14. While this diagram might seem daunting at first glance, there are several things you should notice. Unlike the logical sequence for an LF system in figure 10-9, the CR logic signals do not work in pairs, but there is a clearly defined sequence in which the signals must be provided; to reverse the motor direction, the signal sequence must also be reversed. Second, each of the logical signals are always separated in time, so the 'HCB signal will change first, followed closely by the HCB signal, then the HCA signal, and the 'HCA signal. By sequencing the signals in this way, the CR motor is stepped counterclockwise, and the carriage is moved left toward its "home" position. If the signal order were reversed, the CR motor would step clockwise, and the carriage would move toward the right.

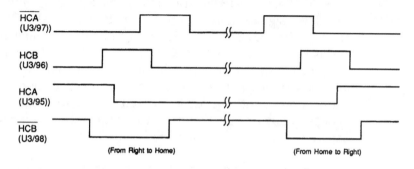

■ **10-14** *Carriage return system logic timing.* Tandy Corporation

Carriage transport symptoms

Symptom 1 The carriage advance does not function, or functions only intermittently. All other functions check properly. Refer to Chart 10-2. When a carriage transport fails to work at all, begin by observing the carriage transport drive train assembly. Check the pulleys and timing belt to ensure that all parts are meshed evenly and are able to move freely. With the printer off and unplugged, you can watch this by gently easing the carriage back and forth. Remove any foreign objects or obstructions that might be jamming the motor or drive train. Never try to force a drive train that does not move freely. Realign and tighten any parts that appear to be slipping or misaligned. Replace the timing belt if it has stretched or broken. Also check to see that the carriage is attached securely to the timing belt.

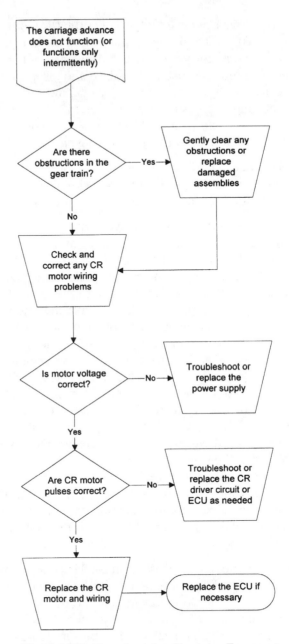

Chart 10-2 *Flowchart for Carriage Transport Systems Symptom 1.*

If the drive train is intact, suspect the CR motor and driver system next. Turn off and unplug the printer, then examine the electrical connections for your carriage transport motor. Make sure that all connectors are installed and seated properly. If you suspect a faulty wiring connection, use your multimeter to measure continuity across any suspicious wires. It might be necessary to disconnect the cable from at least one end to prevent false continuity readings. Replace any faulty wiring harnesses.

Warning: The following is a "hot" procedure that must be performed with the printer on, and with clear access to the ECU and LF motor. Take all precautions to protect yourself from injury.

If you have an oscilloscope, you can check the supply voltage and signals on each side of the driver transistors (as illustrated in figure 10-12). You can use the PRINTERS program to generate a series of regular carriage return sequences for testing. Before you proceed, make sure that +24 V is available to the motor (you can measure it at the power supply). If +24 V is missing, the motor will not turn, and you should suspect a problem in the +24-V supply. If voltage is missing at pins 6 and 5 of the motor, either resistor R47 has failed, or diodes D3 and D4 have opened. Once supply voltage is available at the motor, check the motor's phase signals. If you see logic signals on the base of each driver transistor, you should find high-energy signals on each collector (similar to the V_{ce} trace in figure 10-14). If you see a logic signal, but the expected V_{ce} signal is absent, the corresponding driver transistor is probably defective and should be replaced. If the logic signal is missing to begin with, there is a problem in the ECU. You can replace the associated logic IC or the entire ECU at your discretion as discussed in Chapter 9. If +24 V and all of the V_{ce} driver signals are reaching the CR motor as expected and the motor still does not turn, the motor is probably defective and should be replaced.

If you do not have access to an oscilloscope, or you do not have the time and inclination to follow the "hot" procedure, you can simply disconnect the CR motor from its PC board and use your multimeter to measure the winding resistance of each motor phase. Figure 10-15 shows you how to do this. Be sure to turn off and unplug the printer before proceeding. Measure across pins 6 and 4, 6 and 2, 5 and 3, and 5 and 1. Each measurement should yield a resistance of about 15 Ω (although your particular CR motor might have slightly different ratings). Ideally, each phase winding should yield close to the expected resistance. If any of the windings measure unusually high (open) or low (shorted), the CR motor is likely defective and

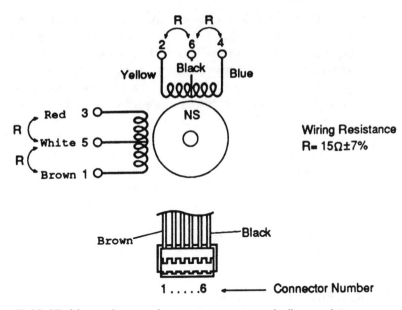

■ 10-15 *Measuring carriage return motor winding resistance.* Tandy Corporation

should be replaced. If the windings all measure correctly, the motor is probably good, and your fault is in the ECU/driver circuitry, which you should simply replace outright.

Symptom 2 Carriage operates, but it does not always position properly. The character spacing might be erratic, or the carriage might sometimes slam into side frames. Faulty mechanics are often at the heart of carriage problems. Unplug the printer and inspect the carriage drive train very carefully. Pay particular attention to the timing belt and be sure that it is seated properly and reasonably taut around each pulley. A worn or loose timing belt should be retensioned or replaced. Also check that the carriage assembly is attached securely to the timing belt; a loose carriage will certainly move erratically. Inspect any drive gears between the CR motor and primary pulley for signs of slipping or broken gear teeth. Replace any parts that appear broken or excessively worn.

If your mechanical system is intact, the trouble is probably electrical in nature. Start by checking the CR motor wiring. See that any CR motor connectors are attached properly and securely to the ECU. Faulty wiring or loose connectors can easily cause intermittent carriage operation.

Another possible problem might exist in the positioning feedback system. All moving-carriage printers need to find a "home" position. Often, this is accomplished with an optical sensor or mechanical switch. If the "home" sensor has failed, the carriage will be unable to position itself accurately. Examine the "home" sensor as outlined in the procedures of Chapter 9. While many low-end printers simply count motor pulses and approximate the carriage position, some printer designs supplement the "home" sensor with an optical encoder that generates pulses as the CR motor turns. The ECU can count these pulses and compare them with the number of pulses sent to the CR motor; if the numbers are equal, the printer knows that the carriage is positioned where it is supposed to be, otherwise, the carriage position can be adjusted "on the fly." If the optical encoder has failed, the ECU cannot determine if the carriage is where it is supposed to be, and errors in positioning can result. Check the optical position encoder as outlined in the procedures of Chapter 9.

Ribbon transport systems

All forms of printing require some sort of media. It is the media that becomes a permanent page image. Each printing technology uses its own particular type of media. For example, impact printing requires ink from a fabric or plastic ribbon, thermal printing requires heat-sensitive chemicals already in the paper, or the dry ink in a thermal transfer ribbon, ink jet printing uses a reservoir of liquid ink, and EP printing takes a supply of toner powder. Media is consumed by the printer during normal operation, so it must be fresh and available at all times. Nowhere is that process more intricate and troublesome than with ribbons.

Fabric ribbons are used almost exclusively in impact printers. A fabric ribbon is simply a long, slender length of material saturated with liquid ink, then packed or spooled into a plastic cassette, as shown in figure 10-16. Thermal transfer ribbons use a substrate (foundation) of clear plastic coated with a thin layer of dry plastic ink. Heat from a thermal print head will liquefy the dry ink and transfer corresponding points to the page. They are usually spooled in cartridges, as shown in figure 10-16.

Ribbons must be advanced during the printing process in order to keep fresh media available to the print head at all times. This mechanical operation is handled by a ribbon transport system. Transports for ribbon cartridges are often unidirectional; that is, ribbon is advanced in one direction only until it wears out. A typical fabric ribbon will survive several complete passes before wearing out. Spooled ribbon cartridges are often one-time devices used for

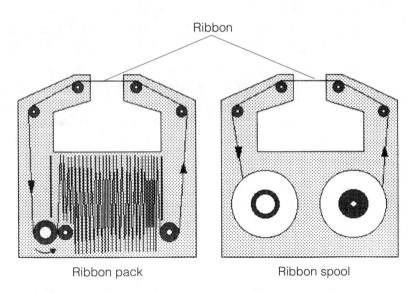

Ribbon

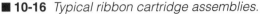

Ribbon pack   Ribbon spool

■ **10-16** *Typical ribbon cartridge assemblies.*

thermal transfer ribbon. When the supply spool is exhausted, a mechanical sensor can be tripped telling the ECU that a new ribbon is needed.

There is no independent motor or solenoid to drive the ribbon transport, so force is "stolen" from the secondary pulley of the carriage transport, as shown in figure 10-17. An intricate "transmission" of contact rollers and gears serves to keep the ribbon advancing in the same direction regardless of carriage direction. Rather than explain the drive's operation in detail, you can easily follow the arrow directions on each roller, but the key to the system is the rocker arm, which swivels based on the secondary pulley's direction. Problems with this arm (or any of the drive rollers) will cause severe problems with the ribbon transport. Be aware that ribbon transport mechanisms can vary greatly between printers, but you can easily recognize a ribbon transport mechanism from the long spindle that inserts into a ribbon cartridge.

Ribbon transport symptoms

Symptom 1 Print is light or nonexistent. All other functions appear correct. Before you actually begin to troubleshoot a ribbon transport, examine the ribbon cartridge as the printer operates. If the ribbon drive shaft advances, inspect the ribbon itself; it might simply be exhausted. Try replacing the ribbon cartridge and retest the printer. A ribbon cartridge that does not advance might be kinked or jammed within its cartridge. Install a fresh ribbon and retest. If normal operation returns, discard the defective ribbon cartridge.

If the ribbon drive shaft does not turn (or a fresh ribbon does not correct your problem), examine the ribbon transport mechanics, such as shown in figure 10-17. The PRINTERS utility can produce a set of carriage returns that you can use to observe the ribbon transport. Unplug the printer and remove the ribbon cartridge. You will observe the long ribbon drive shaft that inserts into the ribbon cartridge. Grouped just below and behind the drive shaft, you will find a series of other small gears and friction rollers that make up the ribbon transport. The mechanism can be assembled on the carriage or on the printer's mechanical frame.

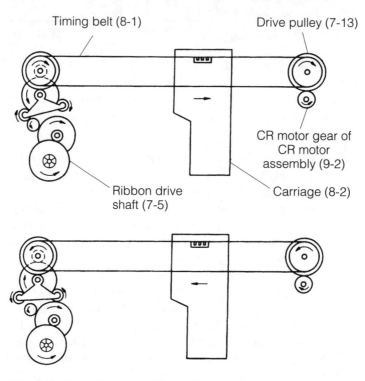

■ **10-17** *Operation of a ribbon drive mechanism.* Tandy Corporation

Although it is never desirable to operate a printer without its ribbon, it is usually safe to do for limited periods of time, as long as paper is available to absorb print wire impact or heat. Refer to your owner's manual for any specific warnings or cautions. You might have to perform some minor disassembly to observe the entire ribbon transport. While the printer is running, watch the ribbon transport mechanism for any parts that might be loose, sticking, or jammed together. Dust and debris might have accumulated to jam the mechanism. Use a clean cotton swab to wipe away any foreign matter. If the transport mechanism is severely worn, it might have to be replaced entirely.

EP mechanical systems

Where moving-carriage printers use relatively straightforward motor and gear arrangements to develop motion, the EP printer provides a much greater level of mechanical sophistication. Generally speaking, this added complexity is due to the electrophotographic process itself. As you saw in Chapter 4, there are a myriad of operations that must be executed simultaneously. This part of the chapter examines the EP printer's mechanical system in detail, and offers some troubleshooting guidance.

The overall mechanical system

Most EP printers can be broken down into three or four key mechanical areas: the laser scanning system, the development system, the paper transport/image formation system, and (in some models) a supplemental mechanical system for bin feeding (a paper feed modification). The diagram of figure 10-18 gives you some idea of the EP printer's mechanical complexity. You should note that the mechanical force that drives each of these areas is developed by its own motor, but all motors are operated under direction of the ECU.

Polygon motor (laser/scanner system)

Lasers have proven to be excellent writing mechanisms, but for the laser beam to be of any use, it must be scanned (or swept) across the photosensitive drum in a highly coordinated fashion, as illustrated in figure 10-19. The laser beam is produced by a solid-state laser diode. The beam travels past a mechanical shutter that can be closed as a safety measure to cut off the laser beam whenever the printer's enclosure is opened. A slit plate helps to narrow the beam, and a collimator lens stops what little divergence there might be, so the diameter of the laser beam will remain constant. The beam then strikes reflecting mirror A, where it is reflected by a rotating polygon mirror to reflecting mirror B. The rotation of the polygon mirror sweeps the beam. The moving beam reflected from mirror B crosses to reflecting mirror C where it is passed through a focusing/compensating lens, and finally strikes the drum. When the

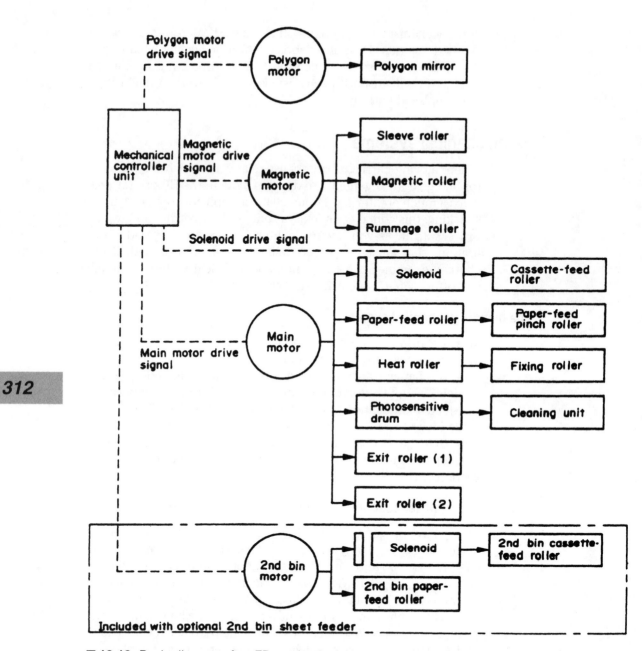

■ **10-18** *Basic diagram of an EP mechanical system.* Tandy Corporation

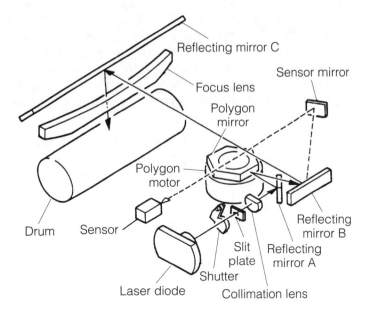

■ 10-19 *A typical laser scanning system.* Tandy Corporation

beam has reached the far end of its sweep, it is reflected to a sensor mirror, which bounces the beam to a laser sensor. This kind of sensor arrangement is used to ensure that the laser is working, and that the mechanical laser system is synchronized properly.

The heart of this mechanical sweeping system is the polygon motor. It is the motor that spins a hexagonal polygon mirror. The hexagonal shape basically causes six complete sweeps of the laser beam for every one rotation of the polygon motor; this increases the apparent speed of the scanning system (and thus increases the printer's throughput). The most important aspect of the polygon motor is that it maintains a constant speed, so the scanning rate remains constant. If the motor speed varies, even just a little, scan lines can be off significantly, and the resulting image will be extremely distorted.

Magnetic motor (development unit)

The magnetic motor (figure 10-18), also known as a developer motor, is used to drive the three main rollers in an EP development unit: the sleeve roller, the magnetic roller, and a rummage roller. The sleeve roller and magnetic roller are actually integrated into the same "development roller" (also referred to as the transfer roller) assembly located in the toner supply. The magnetic roller at the core of the assembly attracts the iron component of the toner, while the sleeve roller actually carries the toner into proximity with the EP drum. The sleeve roller receives an electric charge, which is transferred to the toner, as well as an ac bias that helps ensure proper toner transfer to the drum. If mechanical problems arise in the magnetic motor or its gear system, the transfer of toner to paper will be interrupted resulting in light print or erratic toner distribution.

The rummage roller is also part of the development system, but it works largely out of sight inside the toner reservoir itself. Over time, toner can clump and become unevenly distributed. The rummage roller works to keep toner mixed, free, and evenly distributed in the cartridge. You can see some of the mechanics for a rummage roller in figure 10-20.

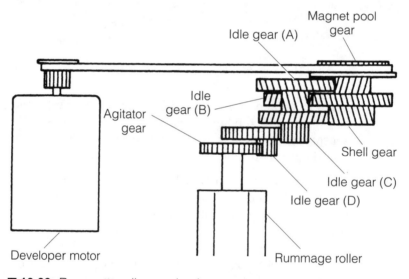

■ **10-20** *Rummage roller mechanics.* Tandy Corporation

Main motor (paper transport)

Of all the mechanisms noted in figure 10-18, you might notice that the main motor carries the biggest load in the system. The main motor is basically responsible for driving the entire EP paper transport system: the paper pickup roller, registration rollers, EP drum, fusing rollers, and exit rollers. Figure 10-21 illustrates the entire feed system.

Paper transport begins with a stack of single-sheet paper in the paper cassette. When a paper feed process begins, the main motor

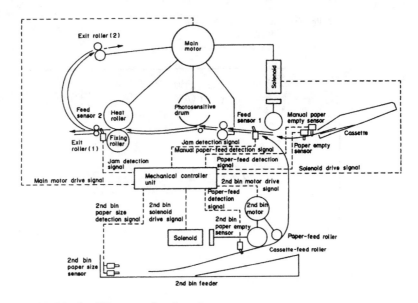

■ 10-21 *An EP paper feed system.* Tandy Corporation

starts, then engages a solenoid that starts a cassette feed roller (also known as the paper pickup roller). The roller grabs a single sheet of paper, and passes it to a set of registration rollers (marked paper feed roller and pinch roller in figure 10-18). The registration rollers will hold the paper in place until the leading edge of the drum's image is aligned with the top of the page, then the registration rollers will feed the sheet into the image formation area. Once the sheet is clear of the cassette, the solenoid disengages and the paper pick-up roller will stop.

Registration rollers continue to feed the page which obtains the image. The page then reaches the fusing assembly consisting of a heat roller and fixing roller (or pressure roller). Heat and pressure melt the toner and squeeze it into the paper rendering a permanent image. Now the completed page must be ejected from the printer. Ejection is handled by a series of exit roller assemblies. If the page is to be ejected "face up," it will be passed out of the printer from exit roller 1. If the page is to be ejected "face down," it will be passed from exit roller 1 to exit roller 2, then out of the printer. Face down is the most common ejection preference because this keeps subsequent sheets in order.

To appreciate the complexity of an EP paper transport system, you should have an understanding of the gear arrangements involved. Figure 10-22 shows the main transmission layout for the paper transport (the three gears that are not highlighted are in-

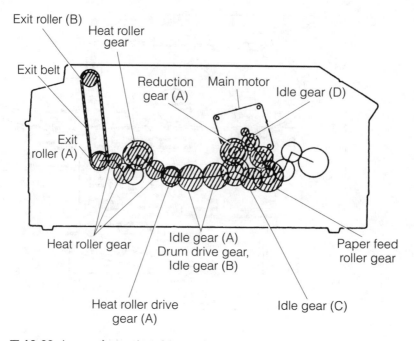

Exit roller (B)

Heat roller gear

Exit belt

Exit roller (A)

Reduction gear (A)

Main motor

Idle gear (D)

Heat roller gear

Idle gear (A)
Drum drive gear,
Idle gear (B)

Paper feed roller gear

Heat roller drive gear (A)

Idle gear (C)

■ **10-22** *Image formation drive train.* Tandy Corporation

volved in paper pickup). As you might imagine, any damage or obstructions in the main motor or its transmission will cause noticeable problems in paper transport.

Supplemental mechanics (bin feeder)

Finally, figure 10-18 illustrates a second bin motor. This is an optional assembly and is not available on all printers, but you might encounter similar designs on high-volume paper feed mechanisms. Simply speaking, the second bin feed system is intended to supplement the low-volume paper cassette. When a second bin is installed (and selected as the paper source), the second bin motor provides the mechanical force for paper pick-up. A solenoid is engaged, which starts a paper pick-up roller (cassette feed roller) to draw a single sheet from the bin supply. The sheet is moved with a paper feed roller in the bin, and passed to the registration rollers. The remainder of the paper transport process remains the same. It is important to note that when a second bin is selected as the paper source, the original paper cassette, solenoid, and pick-up system will not be used, so paper is drawn from only one source or the other. You can see the feed path for a second bin feed in figure 10-21. When there is trouble obtaining paper from the second bin, the problem is almost always mechanical in nature.

General mechanical symptoms

Symptom 1 You consistently encounter faulty image registration. Refer to Chart 10-3. Paper sheets are drawn into the printer by a pick-up roller, and held by a set of registration rollers until a drum image is ready to be transferred to paper. Under normal circumstances, the leading edge of paper will be matched (or registered) precisely with the beginning of a drum image. Poor paper quality, mechanical wear, and paper path obstructions can all contribute to registration problems.

Begin by inspecting your paper and paper tray assembly. Unusual paper weights or specially coated papers might not work properly in your paper transport system (this can also lead to paper jam problems). If you find that your paper is nonstandard, try about 50 sheets of standard 20-lb. bond xerographic-grade paper and retest the printer. Because paper is fed from a central paper tray, any obstructions or damage to the tray can adversely affect page registration (or even cause paper jams). Examine your tray carefully. Correct any damage or restrictions that you might find, or replace the entire tray outright.

If registration is still incorrect, it usually suggests mechanical wear in your paper feed assembly. Check the pick-up roller assembly first. Look for signs of excessive roller wear. Remove your printer housings to expose the paper transport system. You will have to defeat any housing interlock switches, and perhaps the EP cartridge sensitivity switches.

Warning: This will result in a hot test, so be extremely careful to avoid reaching in or around the printer while the cover interlock is defeated.

Perform a self-test or use the PRINTERS paper feed test and watch paper as it moves through the printer. The paper pick-up roller should grab a page and move it about three inches or more into the printer before registration rollers activate. If the pick-up roller clutch solenoid turns on, but the pick-up roller fails to turn immediately, your pick-up assembly is worn out. The recommended procedure is to simply replace the pick-up assembly, but you might be able to adjust the pick-up roller or clutch tension to somewhat improve printer performance.

Another common problem is wear in the registration roller assembly. If this set of rollers does not grab the waiting page and pull it through evenly at the proper time, misregistration can also occur. As you observe paper movement, watch the action of your regis-

317

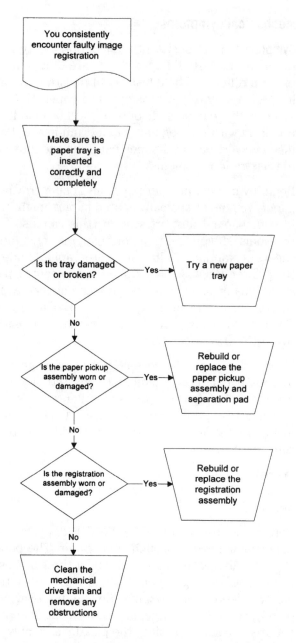

Chart 10-3 *Flowchart for EP Mechanical Systems Symptom 1.*

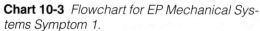

tration rollers. They should engage immediately after the pick-up roller stops turning. If the registration clutch solenoid activates, but paper does not move immediately, your registration roller assembly is worn out. The recommended procedure is to simply replace the registration assembly, but you might be able to adjust torsion spring tensions to somewhat improve printer performance.

Pay particular attention to the components in your drive train assembly. Dirty or damaged gears can jam or slip. This leads to erratic paper movement and faulty registration. Clean your drive train gears with a clean, soft cloth, but do not attempt to disassemble the gear train unless it is absolutely necessary. Use a cotton swab to clean gear teeth and tight spaces. Remove any objects or debris that might block the drive train, and replace any gears that are damaged.

Symptom 2 There are skips or gaps in the page. Images might appear similar to figure 10-23. Gaps in the print typically indicate that the page has lurched forward at some point, probably prior to the EP drum. Overlaps (or skips) suggest that instead of lurching forward, the page stalled momentarily. In either case, your problem is likely due to a fault in the drive train rather than in a paper transport roller per se. With so many gears at work in the EP printer (e.g., figure 10-22), a broken gear tooth or obstruction can easily cause erratic page transport problems. Turn off and unplug the printer, then examine each gear for signs of damage or obstruction. Make sure every gear meshes properly. Clean the gear train carefully with a lightly dampened cotton swab. If problems persist, try replacing the EP engine assembly.

An image Gap

An image Skip

■ **10-23** *An example of gaps and skips in an EP image.*

Symptom 3 Print shows regular or repetitive defects. Repetitive defects are problems that occur at regular intervals along a page (as opposed to random defects) and are most often the result of roller problems. Rollers have fixed circumferences, so as paper moves through the printer, any one point on a roller might reach a page several times. For example, if a drum has a circumference (not a diameter) of three inches, any one point on the drum will reach a standard 8.5" × 11" page up to three times. If the drum is damaged or marked at that point, those imperfections will repeat regularly in the finished image. As a consequence, it is often easy to identify the fault. The PRINTERS utility provides a drum and roller test ruler to help you check for repetitive defects, as in figure 10-24.

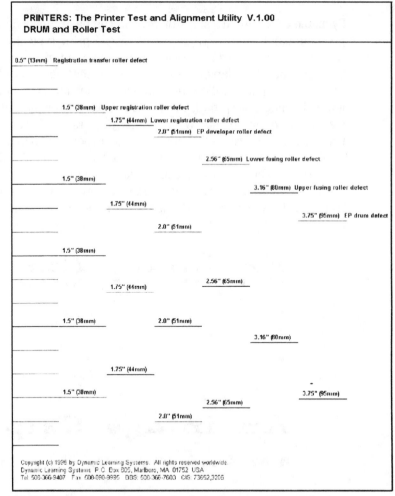

■ **10-24** *Using the defect ruler to judge repetitive defects.*

Most repetitive defects take place in the EP engine, which contains the photosensitive drum and developing roller. A typical EP drum has a circumference of about 3.75" (95 mm). Defects that occur at that interval can often be attributed to a drum defect, typically a mark on the photosensitive surface. A developing roller has a circumference of about 2" (51 mm), so problems that repeat every 2" are usually associated with the developing roller. In either case, you can replace the engine and retest the printer.

There are three key registration rollers that can cause repetitive defects. The upper registration roller is typically 1.5" (38 mm) in circumference, while the lower registration roller runs a bit bigger at 1.75" (44 mm). Trouble at either of these distances suggests a fault in the registration system. The smallest roller is the registration transfer roller at 0.5" (13 mm), and a fault there will affect the page as it travels toward the EP drum.

On the other end of the IFS, there are two fusing rollers. The upper fusing roller (the heated roller) has a circumference of about 3.16" (80 mm). Image marks or defects at that interval suggest a dirty or damaged upper fusion roller. The lower fusing roller runs a bit smaller at 2.56" (65 mm). Unplug the printer, allow at least 10 minutes for the fusion assembly to cool, then gently clean both fusing rollers, also replace the cleaning pad. If you find that the fusion rollers are physically damaged or are unable to clean them effectively, replace the fusing assembly.

Any roller that is fouled with debris or toner particles can contribute to a repeating pattern of defects. Make sure to examine each of your rollers carefully. Clean or replace any roller that you find to be causing marks.

Symptom 4 You consistently encounter paper jam conditions. As figure 10-21 shows, the electrophotographic paper transport system is much more sophisticated than those used in conventional serial or line printers. As a result of this additional complexity, main logic circuitry must be able to detect whether or not paper enters and exits the paper path as expected. As you review figure 10-21, you will notice a series of sensors spaced throughout the paper path. A paper jam condition can often be triggered from several different combinations of conditions:

☐ Paper does not reach the fusing assembly within a predefined amount of time.

☐ Paper does not leave the fusing assembly within a predefined amount of time.

☐ Paper reaches the fusing assembly, but fusing roller temperature is below a normal working temperature.

☐ If paper is exhausted, but the paper sensor arm fails to sense a paper-out condition, the printer will try to pick up a new sheet of paper. Because paper that is not there cannot possibly reach fusing rollers in any amount of time, this too can precipitate a "ghost" jam condition.

Begin by checking paper in the paper tray. If a jam condition is shown but there is no paper, it indicates that your paper sensing lever is not functioning properly. It could be broken, bent, or jammed; otherwise, the sensor itself might have failed. When there is ample paper available, take a moment to be sure that paper is the right size, texture, and weight (bond) for your printer. Unusual or special paper might not be picked up reliably. If you are uncertain as to the correct paper type, remove it and insert a good-quality 20-lb. xerography-grade paper. This type of paper usually has the weight and texture characteristics ideal for EP printing.

When a "paper jam" message is displayed, unplug the printer, open its covers, and make careful note of the paper's position. Look closely at the leading and lagging paper edges. The problem can often be isolated based on the paper's jam position. Three main trouble areas are the paper feed area, the registration/transfer area, and the exit area.

Paper feed area

The paper feed area consists of the paper tray (and paper), pick-up mechanical assembly, and electromechanical clutch, as shown in figure 10-21. If paper is not reaching your registration rollers, the trouble is probably in this area. Inspect your paper tray carefully. While the tray might seem foolproof, it actually plays an important roll in paper feed. If the plastic tray housing is cracked or damaged, replace it with a new tray and retest the printer. Note the movable metal plate in the tray's bottom. This is a lift mechanism that keeps paper positioned against the pick-up roller at all times. Remove the tray and paper. Make sure that this plate can move freely; replace your paper tray if it does not. Observe this lift plate as you insert it into the printer. The printer should lift this plate up as the tray is inserted. If this does not happen, repair or replace the printer's lift mechanism assembly. Add some fresh paper (50 sheets or so) and gently insert the paper tray. Be sure to insert the tray fully and squarely. If there is any obstruction (or the tray does not seat squarely), find and remove the obstruction.

Next, make sure that your main motor is functioning. Keep in mind that the main motor drives the entire image formation system. If this motor has failed, paper will not be drawn into the printer at all. You can observe the main motor and its gear train assembly by turning the printer on, opening an access cover, defeating the associated interlock switch (if any) and EP cartridge sensitivity switches, then initiating a printer self-test. If the fusing roller temperature is above its lower temperature limit, you should see motor operation immediately. If the main motor fails to operate, troubleshoot the motor and its driver voltage. Try a new motor. If the problem persists, replace the mechanical controller.

The main motor turns when a print signal is first generated in main logic. It continues turning as long as there is paper in the feed path. However, even though the main motor supplies the force to operate every roller, pick-up and registration rollers are operated only briefly during each print cycle. An electromagnetic clutch (also illustrated in figure 10-21) is used to switch the pick-up roller on and off at desired times. Main logic generates a clutch control signal that is amplified by driver circuits before being fed to the electromagnet. When deactivated, the plunger disengages the pick-up roller from the drive train. When activated, the plunger engages the pick-up roller, which causes the pick-up roller assembly to turn and grab the next available piece of paper. A separation pad beneath the pick-up roller prevents more than one sheet from being taken at any one time. Paper stops when it reaches the idle registration rollers. Notice how paper will tend to bow; this is a normal and harmless function in the paper path.

If the main motor operates, but the pick-up roller does not turn (you can observe this during an open-cover test print), inspect the pick-up roller clutch solenoid. Some designs might use two major solenoids: one for the pick-up roller and one for the registration rollers. When printing starts, one of the two solenoids (the pick-up solenoid) should engage immediately. If no solenoid engages, there is an electronic problem, so suspect your mechanical controller board. If the pick-up solenoid engages, but your pick-up roller does not turn (or turn properly), repair or replace the pick-up mechanical assembly. There might be a faulty clutch or other mechanical defect.

When the pick-up solenoid fails to actuate, use your multimeter to measure voltage across the solenoid. You should see voltage toggle on and off as the solenoid is switched. If voltage changes, but the solenoid does not function, replace the solenoid. If voltage remains at zero (or does not switch from some other voltage), there is probably a fault in the solenoid driver circuit. Troubleshoot the solenoid signal back into main logic, or replace the mechanical controller outright.

Finally, you can check for feed roller wear by measuring the distance between the paper's trailing edge, and the end of the paper tray just as the sheet stops. This time occurs between the point that the pick-up roller stops, and the registration rollers start. Normally, this trailing edge should advance about three inches or more. If it does not advance this far, your pick-up roller is probably worn out. Replace the pick-up roller assembly and separation pad.

Registration/transfer area

Registration rollers hold on to the page until its leading edge is aligned with the drum image. Force is supplied by the main motor, but another electromagnetic clutch is typically employed to switch the registration rollers on and off at their appropriate time. Once paper and the drum image are properly aligned, the mechanical controller sends a clutch control signal, which is amplified by driver circuits to operate the registration clutch solenoid. After the clutch is engaged, registration rollers carry the page forward to receive the developed toner image. The registration/transfer assembly usually consists of registration rollers, the drive train, a registration clutch solenoid, a transfer guide, and the transfer corona assembly. If paper does not reach the fusing rollers, your fault is probably in this area.

You can observe registration-roller operation by opening a housing, defeating any corresponding interlocks, and defeating any EP cartridge sensitivity switches, then initiating a self-test. Use extreme caution to prevent injury from high-voltage or optical radiation from the writing mechanism, especially lasers. Observe the paper path and drive train very carefully. If you observe an obstruction in the paper path or in the drive train, turn off and unplug the printer before attempting to clear the obstruction. Replace any gears or bushings that appear damaged or worn out. Pay close attention to any tension equalizing springs (called *torsion* springs) attached to the registration rollers. Reseat or replace any torsion springs that might be damaged or out of position.

Inspect the monofilament line encircling the transfer corona. Make sure that the line is intact, and not interfering with your paper path. Do not approach the corona with your hands or any metal tool. If you see signs of interference, turn off and unplug the printer, then allow high voltage to discharge for at least 10 minutes before replacing the monofilament line or transfer corona assembly.

If your main motor and drive train operate, but registration rollers do not turn (or they turn improperly), inspect the registration solenoid clutch. It will usually reside adjacent to the pick-up solenoid clutch. The solenoid should engage moments after the pick-up roller disengages. If the registration solenoid does not engage, there is an electrical problem; replace the solenoid or mechanical controller board. If the solenoid does engage, but registration rollers do not turn, your mechanical clutch or registration rollers are probably worn out. Replace the mechanical registration assembly.

When your registration solenoid fails to actuate, measure the signal driving the solenoid. You should see the signal toggle on and off with the solenoid. If the signal changes, but the solenoid does not fire, replace the jammed or defective solenoid. If voltage signal does not change, there is probably a fault in the solenoid driver circuit or mechanical controller board. Replace the solenoid or mechanical controller as required.

Exit area

At this point, a page has been completely developed with a toner powder image. The page must now be compressed between a set of fusing rollers; one provides heat, while the other applies pressure. Heat melts the toner powder, while roller pressure forces molten toner permanently into the paper fibers. This fixes the image. As a fixed page leaves the rollers, it tends to stick to the fusion roller. A set of evenly spaced separation pawls pry away the finished page, which is delivered to the output tray. A series of exit rollers direct the page to the output as in figure 10-21. Main motor force is delivered to the fusion rollers by a geared drive train. There are no clutches involved in exit area operations, so the drive train moves throughout the entire printing cycle. It is important that the printer detect when paper enters and leaves the exit area. Based on paper size and fusion roller speed, a page has only a set amount of time to enter and leave the exit area before a jam condition is initiated. To detect the flow of paper, an optoisolator is usually actuated by a weighted plastic lever. You can see these sensors in figure 10-21.

Normally, a paper flag lever protrudes down through a slot in the empty paper path. This leaves the optoisolator clear. Its resulting logic output indicates no paper. When paper reaches the lever, it is pushed up to the paper's level. This, in turn, moves the flag into the optoisolator slot, causing a logic change that indicates paper is present; a timer is started in main logic. When everything works properly, paper moves through the fusion roller assembly. As paper passes, the lever falls again, returning the optoisolator to its original logic state. If the optoisolator returns to its initial value before the timer expires, it means that paper has moved through the exit area properly. If paper remains, a "paper jam" is indicated. A long-term timer was started at the beginning of the printing cycle. If this long-term timer expires before paper first reaches the paper flag lever, a "paper jam" is also generated. As you might suspect, there are a variety of problems that can precipitate a jam error.

Begin by checking the paper path for any obstructions. Unplug the printer and expose the fusing assembly. It might be necessary to remove secondary safety guards covering the heater roller. Remove any obstructions or debris that you find to be interfering with the paper. Make sure that your separation pawls are correctly attached. Clean the pawls if they appear dirty. Inspect the paper flag lever carefully to be sure that it moves freely. Replace the flag lever assembly if it appears damaged or worn out. Also check all interconnecting cables and wiring to see that the paper lever optoisolator is attached.

The drive gears that run your fusion rollers are often attached to a door housing. In this way, fusing rollers are disengaged whenever that access door is opened. This set of gears is sometimes the "delivery coupling" assembly. If these gears are not engaging properly because of wear or damage, the fusion rollers will not operate (or operate only intermittently). Repair or replace any faulty delivery coupling components.

If the mechanics of your exit area appear to be operating correctly, you should examine the operation of the exit sensors. Turn on the printer and use your multimeter to measure voltage across the optoisolator's output. You might have to defeat any open cover interlocks to ensure proper voltage in the printer. Use extreme caution when measuring, and stay clear of the high-voltage coronas. Move the paper lever to actuate the optoisolator. You should see the output voltage toggle on and off as the optoisolator is actuated. If output voltage does not change, replace the faulty optoisolator. If voltage changes as expected, but paper jams are still indicated, troubleshoot your sensor signal back into the mechanical controller circuit.

Symptom 5 You see an error number or message indicating a polygon motor or "scanner" problem. Keep in mind that a scanner assembly is only needed for EP printers using laser writing mechanisms. LED writing mechanisms do not use scanning mirrors at all. The scanner is an optical-grade hexagonal mirror driven by a small, brushless dc motor (an induction motor). Printing will only be enabled after the scanner has reached and maintained its proper operating speed. The scanner is engaged at the beginning of a printing cycle, so the scanner turns as the main motor turns. For many "laser" printers, you will recognize the scanner motor by a somewhat distinctive, variable pitch whirring noise. Motor speed is constantly monitored and controlled by the mechanical controller board. If the motor fails to turn or maintain speed when power is applied, a scanner "error" is generated.

Turn off and unplug the printer, open its housings, and carefully inspect connectors and interconnecting wiring between the polygon motor and its driver circuits. Reseat any connectors or wiring that appears to be loose. The scanner is usually tested briefly during EP printer initialization. If you cannot hear the scanner motor, replace the entire laser/scanning assembly. A low or missing excitation voltage indicates a defect in your mechanical controller board that switches its motor voltage on and off. Troubleshoot the excitation voltage and switching circuitry back into the mechanical controller, or replace the board entirely. Remember that you might have to defeat cover interlocks to enable the printer's low-voltage power supply. Use extreme caution to protect yourself from injury any time the enclosure is opened.

Preventive maintenance

Mechanical parts are often prone to a great deal of wear and tear before a serious breakdown actually occurs. The majority of that wear is caused from the friction produced by ordinary mechanical contact, as well as additional friction generated by debris and dried lubricants in the drive trains. A routine series of cleaning and lubrication procedures can help to extend the working life of your mechanical parts. You should perform cleaning and lubrication procedures on a regular basis, and whenever your printer needs repair.

Clean the printer

Your printer is going to accumulate dust and debris around its outer case and protective coverings, as well as inside its carriage and paper handling areas. Over time, any accumulation will work its way into just about every mechanism. This can cause such troubles as sticky gears, slipping belts, or blockage across an optical sensor. A little time spent simply cleaning the printer can yield dividends later on when something really goes wrong.

Clean the outer case and coverings using a soft, clean cloth lightly dampened in water. Never spray water or cleaner directly onto the case or coverings. If cleaner is needed to remove stubborn marks, apply a little mild household detergent directly to your dampened cloth. Avoid all solvent-based cleaners, which attack the plastic in your covers. Use extra care to prevent wetting the control panel.

Remove the protective coverings from your printer and vacuum away any dust from the carriage and paper handling areas. You can loosen any adhesive materials that remain with a soft-bristled brush. You can buy a small, battery-powered electronic vacuum for this purpose, but any household vacuum with a flexible nozzle should do. While vacuuming, be careful not to change any dip switch settings or dislodge any jumpers. Replace all coverings and retest the printer.

Clean the rollers

Any dust or debris that can find its way into a carriage area can also work into the paper transport area, so it is usually a good idea to clean and rejuvenate your rollers as part of a routine maintenance regimen. Turn off and unplug your printer, then remove paper from the paper transport. Simple dust and stains can usually be removed with a soft, damp cloth and a little effort. Dampen the cloth with water only; never use detergents or harsh solvents to clean rollers. As you rub back and forth against the platen, advance the platen knob by hand. Continue cleaning the platen through several revolutions to help carry away any dirt that might have worked into adjoining contact or pressure rollers.

For a superior cleaning, use a cleaning solvent designed specifically for use on rubber rollers. Many quality stationery stores will carry this type of solvent. Not only will the correct solvent do an excellent cleaning job, but it will rejuvenate the texture and pliability of the platen and adjacent rollers. This factor is especially important in friction-feed transports where roller contact plays an important part in its operation. Use only high-quality solvents, and follow the particular manufacturer's directions exactly. Because many printers use a variety of synthetic materials on their rollers, test any specialized solvent on a small area before proceeding with general rejuvenation. Follow all safety and ventilation instructions. Never use nonapproved or caustic solvents to clean your rollers.

Clean the print head

Most paper consists of tightly interwoven fibers bonded together by chemicals, temperature, pressure, and time. Whenever paper is rubbed or struck, it liberates a small amount of fibers from its surface. These fibers settle into the carriage area just like any dust present in the open air, but some paper fibers work into the print head. Thermal print heads need almost no maintenance. You can wipe away any accumulations of dirt or dust using a soft, clean, non-abrasive cloth dampened lightly in ethyl alcohol. Regular cleanings can help prevent accumulations of debris from lifting the print head and changing its contact pressure against the paper or transfer ribbon.

Impact print heads are much more subject to problems because paper fibers will usually mix with oils deposited from ribbon ink. This forms a type of sticky "glue" that works its way into the individual print wires and causes them to jam. If you simply wipe the print head's face whenever you change the ribbon, you can prevent long-term buildup in the print wires before this "glue" becomes hard and dry. Print wires that are already jammed can be cleaned easily. Gently extend each print wire by hand (if possible) and clean around each wire with a clean cotton swab dipped lightly in ethyl alcohol or light household oil. Use extreme caution, because wires will only extend several millimeters; avoid using excessive force when extending a wire. Also be careful not to bend a wire as you work with it. Work the wire in and out, and continue to clean it until it moves freely.

Ink jet print heads pose a slightly different set of problems. An ink jet nozzle will not clog due to paper dust, but the liquid ink in its channels can begin to dry after prolonged exposure to air. Liquid ink is usually manufactured with a chemical solvent base that does not readily evaporate in air. Ink at each nozzle can often remain exposed to the open air for several days without danger of clogging. The "home" position of many ink jet printers offers a cap or sponge that helps prevent drying during periods of disuse, but aging printers will not "cap" the nozzles very well.

A replaceable ink jet cartridge is very easy to clean. Unclip the cartridge from its holder and clean the nozzles with a cotton swab or clean, soft cloth. In order to prime the cartridge, use the swab's wooden end to press gently on the ink bladder. Use enough force to form ink beads on each nozzle, but be very careful not to puncture the bladder. Repeat this several times until the head is clear. Replace the cartridge if the clogged nozzle does not clear.

Nonreplaceable ink jet heads are much more troublesome to clean. A cleaning solvent must be injected into the print head to dislodge a clog. The cleaning kit often supplied with many nonreplaceable head printers offers a syringe and a container of cleaning solvent. Fill the syringe with cleaner and attach it to the head's ink inlet. Gently force cleaner into the head, as shown in figure 10-25. Solvents reduce the ink's viscosity and it should flow easily from the nozzles. Continue injecting cleaner until all ink has been evacuated and only clear solvent flows. You might have to let the cleaner sit for a time to dissolve any stubborn clogs. Once the nozzles are cleared, reattach the ink supply and prime the head with ink until all solvent is displaced and ink beads at each nozzle. This is a very messy procedure. Wear old clothing, rubber gloves, and have towels handy.

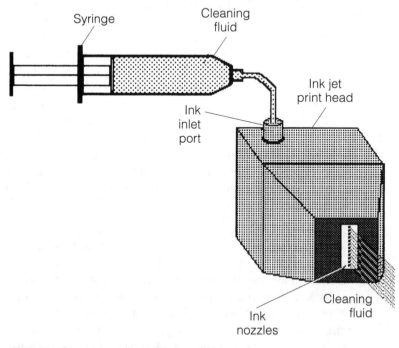

■ **10-25** *Cleaning a fixed ink jet print head.*

Clean & lubricate the paper transport

Dust and debris in your paper transport rollers and drive train can cause gears to jam or bind. It could also force rollers out of alignment. The best defense against this is regular cleaning and lubrication of the drive train. Expose the paper transport drive train. You might have to remove covers or guards that might be protecting the assembly. Clean the gear assembly carefully with a clean, soft cloth lightly dampened with water and mild household detergent. Harsh solvents or cleaners can destroy plastic gears and fittings found in many printers. Use a cotton swab to remove debris from between gear teeth. If you must disassemble a gear train to clean severe jams or realign the assembly, be certain to make very careful notes of how each part goes together.

Reapply lubricants to the drive train only if they were lubricated to begin with. Ideally, you should use the same type of lubricant, but a light household oil is usually acceptable. Place a few drops onto the teeth of one or two gears, then turn the platen knob to distribute that oil throughout the drive train. Continue adding and distributing oil until all gear teeth show a light but consistent coating. Never apply oil to wires or belts in a pulley system. Avoid heavy oils and grease; these simply attract dust and debris that will be even harder to clean the next time around.

Clean & lubricate the head transport

For proper operation, a carriage must be able to ride freely on its guide rail(s) at all times. Through a combination of humidity, dust, and dirt, rust can emerge along steel rails (stainless steel rails are sometimes used, but these are expensive), but plastic is frequently used as a substitute. Rust causes a dramatic increase in friction between the rails and carriage. This can slow the carriage down and cause rough positioning. In extreme cases, it can actually bind the carriage.

Turn off and unplug the printer and wipe your guide rails with a clean, soft cloth to remove any loose dust or rust particles. Most other rust can be swept away with a light touch of steel wool. Use caution when removing rust. Steel wool fibers and rust particles are conductive, so keep your electronic circuits covered and be sure to vacuum any rust or wool that might have been shed. Place a few drops of light household oil onto each rail, then move the carriage back and forth to distribute oil evenly. A light coat of oil will reduce friction, and help protect rails from further rust. Never use heavy oils or grease to lubricate rails.

331

Electrophotographic
printer service techniques

11

ELECTROPHOTOGRAPHIC (EP) PRINTERS USE A COMPLEX combination of light, static electricity, heat, chemistry, and pressure, all guided by a fairly complex ECU (figure 11-1). EP printing is not a "one-step" action as it is with impact, thermal, or ink jet systems. There is no one single part responsible for applying print; EP printers use a series of individual assemblies that make up its Image Formation System (IFS). Because EP image formation uses a "process" rather than a "print head," there are many more variable conditions that will affect the print's ultimate quality and appearance. Now that you have seen how EP printers work and learned the operation of basic power supplies, electronic controls, and mechanical systems, this chapter will present detailed explanations and troubleshooting procedures specifically for EP printers.

Hewlett-Packard Co.

■ **11-1** *An HP LaserJet IIID printer.*

Inside the EP printer

Before proceeding with specific symptoms, you should have a thorough understanding of the EP printer's internal components. Figure 11-2 illustrates a complete cross-section of an image formation system. Each of the major components and subassemblies are clearly pointed out. Paper starts in the paper tray. When a printing cycle begins, a feed roller (also called a *paper pick-up roller*) grabs the next sheet in the tray. A separation pad below the feed roller helps to ensure that only one page at a time actually enters the printer. The feed roller positions the page against a set of registration rollers that hold the sheet in place until the drum image is synchronized with the paper's top margin. A transfer roller carries the page past the transfer corona where high voltage places a large charge on the page. As the page continues to move near the EP drum, the latent image developed in the drum will be transferred to the page. A feed guide carries the developed page to the fusing assembly where the upper fusing roller applies heat, and the lower fusing roller applies pressure, fixing the image to the page. The page moves up the printer where it is grabbed by the delivery assembly and ejected to the face-down tray.

334

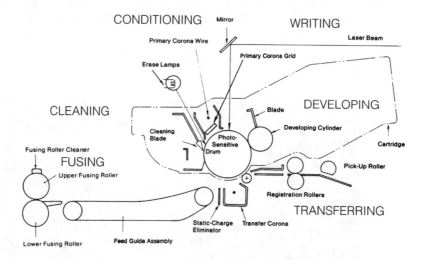

■ **11-2** *Cross-section of an EP image formation system.* Hewlett-Packard Co.

There are additional assemblies that you should be familiar with in the SX-type engine. A set of erase lamps (3) cleans any charges off the drum, while a primary corona (4) applies an even conditioning charge to the drum surface. Once the drum is cleaned and conditioned, it is ready to receive an image from the writing mechanism.

Writing is accomplished with a laser/scanning assembly (6). The laser beam is modulated by the printer's logic and swept by a rotating polygon mirror. The swept beam leaves the assembly and is reflected from a beam-to-drum mirror (5) to the drum surface. It is this latent image that is developed with toner and transferred to the page at the transfer station (13).

As you might imagine, trouble in any part of the printer can seriously impair the printer's operation. For the purposes of this chapter, trouble can be broken down into eight major areas: controller problems, registration problems, fusing problems, laser/scanner problems, drive and transmission problems, high-voltage power supply (HVPS) problems, transfer problems, and a small assortment of miscellaneous symptoms. This chapter shows you the symptoms and solutions for over 60 EP printer problems.

Controller (logic) systems

Most EP printers use an ECU consisting of two parts: a main board and a mechanical controller. The main board provides the core logic for the printer: CPU, memory, an interface for the control panel, the communication circuits, and other processing elements. The mechanical controller provides an interface between the pure logic and the electromechanical components of the printer. For example, a mechanical controller holds the driver circuitry controlling the printer's motors and solenoids. Some printers integrate these functions onto a single PC board, while other printer designs employ two separate boards. While controller circuitry is generally quite reliable, it does fail from time to time, so it is important that you recognize the signs of trouble.

Warning: Some of the procedures outlined below will suggest that you replace ICs or PC boards as part of the corrective process. Remember that ICs and circuit boards are highly sensitive to damage from electrostatic discharge (ESD). Be sure to use all available static controls to prevent additional damage.

Symptoms

Symptom 1 The printer's LCD shows a "CPU Error." Some printer designs might show this error as a series of blinking LEDs, or a sequence of beeps. The CPU is the heart of your printer's logical operation. When you first start the printer, the CPU and its associated core logic (figure 11-3) is tested, much like the BIOS of

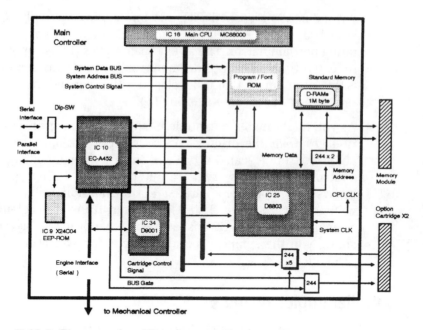

11-3 *Diagram of an EP main controller board.* Tandy Corporation

a computer will execute a self test. If the CPU fails to pass all of its test requirements, an error will be generated.

As you might imagine, a CPU failure is catastrophic; that is, the printer simply will not work without it. Start by turning off and unplugging the printer, then examine each of the connectors on the main controller. Each connector should be installed properly and completely. If problems persist, you will have to replace the CPU. Replacing the CPU can be either cheap or expensive depending on how it is mounted. If the CPU is socket-mounted on the main controller, you can often just remove the old CPU and plug in a new one. However, if the CPU is soldered to the main controller board, you will have to desolder and resolder the CPU (if you have the proper surface-mount soldering tools), or replace the entire main controller.

Symptom 2 The printer's LCD shows a "ROM Checksum Error." Your particular printer might use an error number (e.g., ERROR 11) to represent the condition. As in a computer, all of the printer's on-board instructions and programming are held in a ROM on the main controller board. It is the ROM that provides key instructions and data to the CPU for processing. When the printer starts, a checksum test is run on the ROM to verify the integrity of its contents. If the resulting checksum does not match the checksum reference number stored in the ROM, an error is generated. You can see the "program/font ROM" in figure 11-3.

First, check to see that any supplemental font or option cartridges are installed properly; you might try removing the cartridge(s) to find if the problem disappears. If there are no option cartridges, or the problem persists, you must replace the ROM IC. In many cases, ROM ICs are socket-mounted devices because they must be programmed outside of the logic board's assembly process. When this is the case, you might be able to replace the ROM IC directly. If the ROM IC is soldered to the main logic board (or a replacement ROM IC is simply not available), you will have to replace the entire main logic board.

Symptom 3 The printer's LCD shows a "RAM R/W Error," a "Memory Error," or other memory defect. Your particular printer might use an error number (e.g., ERROR 12 or ERROR 30) to represent the condition. Dynamic RAM (or DRAM) serves as the workspace for an EP printer. Where moving-carriage printers typically offer buffers of 8KB or 16KB, the EP printer can easily offer 1MB or more; some high-end printers can accommodate 16MB or more. This volume of memory is necessary because the EP printer must be able to construct the data needed to form an entire page at a time. For an 8.5" × 11" page at high resolutions, this can be a phenomenal volume of data. Unfortunately, trouble in any part of the DRAM can adversely affect the image, especially PostScript images. Memory is tested when the printer is first initialized. Like PCs, the more memory that is installed, the longer it takes the printer to initialize. A typical test involves writing a known byte to each address, then reading those bytes back. If the read byte matches the written byte, the address is considered good; otherwise, a RAM error is reported. You can see the default 1MB of DRAM in figure 11-3.

It is rare that a RAM error message will indicate the specific location of the error, but you can easily isolate the fault to a bad memory module or the standard (resident) memory. Turn off and unplug the printer, then remove any expansion memory modules that might be installed. You might have to set jumpers or DIP switches to tell the printer that memory has been removed. If the problem disappears, one or more of your expansion memory modules has failed. Try reinstalling one module at a time until the problem reoccurs; the last module to be installed when the error surfaced is the faulty module. If the problem persists when memory modules are removed, you can be confident that the fault is in your resident memory. Although memory modules often take the form of SIMMs or other plug-in modules, resident RAM is typically hard-soldered to the main controller board. You might attempt to replace the RAM if you have the proper desoldering tools and replacement RAM ICs on hand. Otherwise, simply replace the main controller board.

Symptom 4 Your printer's LCD shows a "Memory Overflow" error. Your particular printer might use an error code (e.g., ERROR 20) to represent the condition. When data is sent from the computer to the printer, part of that data consists of "user information," such as soft-fonts and macro commands. If the amount of "user information" exceeds the amount of RAM set aside for it, a "Memory Overflow" (or similar error) will be generated. While this error is not directly related to the image size or complexity, complex images typically carry a larger overhead of "user information," so you might find that simplifying the image can sometimes clear the problem even though the image itself is not really at fault. Generally speaking, you can eliminate this error by adding optional memory, or reducing the amount of data that must be downloaded to the printer (often a function of the application doing the printing).

Symptom 5 Your printer's LCD shows a "Print Overrun" error. Your particular printer might use an error code (e.g., ERROR 21) to represent the condition. Unlike the last error, "print overrun" problems almost always indicate that the page to be printed is too complicated for the printer; there is just not enough memory to hold all of the data required to form the image. To overcome this type of problem, try simplifying the image (e.g., use fewer fonts or try using solid shading instead of dithering). You might also try making the printed area smaller. For example, instead of printing an image at $8 \times 8"$, try printing it at $5 \times 5"$. The smaller image requires less raw data. The ideal way to correct this problem over the long term is to add memory to the printer.

Symptom 6 The printer reports an "I/O Protocol Error." Refer to Chart 11-1. Your particular printer might use an error code (e.g., ERROR 22) to represent the condition. This is a communication fault. The term *protocol* basically means agreement or rules. So when a protocol error arises, it suggests that the computer and printer are not communicating "by the rules." The most blatant protocol error is connecting a serial port to a parallel printer, or a parallel port to a serial printer, but this is an extremely rare over-

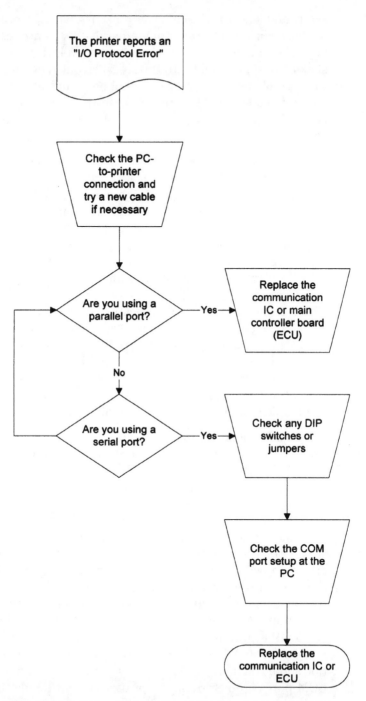

Chart 11-1 *Flowchart for Controller (Logic) Symptom 6.*

sight, and one that usually only occurs when a printer is first installed. Protocol errors among parallel ports are also very rare because parallel port operation is very well defined with hand-shaking designed right into the signal layout. The most likely protocol problems can arise with serial communication; there are so many variables in the serial process that must be matched between the printer and computer, that even the slightest error can cause problems.

Start by checking the connections between the computer and printer. See that the communication link is parallel-to-parallel or serial-to-serial. Also try a new, high-quality cable (serial or parallel as appropriate) between the printer and computer. When parallel communication is being used, a protocol error suggests a failure in the communication interface IC (IC10 in figure 11-3). You can try replacing the communication IC, or replace the main controller board entirely. When serial communication is being used, you should examine any DIP switches or jumpers inside the printer. Check to see that the communication speed and framing bits are all set as expected, then see that the corresponding COM port in the PC is configured similarly (through the printing application or the Windows Printer Control Panel). If problems persist, even when the serial communication link is set properly, suspect trouble in the communication interface IC. You can try replacing the communication IC, or replace the main controller outright.

Symptom 7 The image is composed of "garbage" and disassociated symbols. Your printer might also generate a "Parity/Framing Error," or use an error code (e.g., ERROR 40) to represent the condition. This error indicates that there is a problem with serial data framing. As detailed in Chapter 9, serial data must be "framed" with the proper number of start, data, parity, and stop bits. These bits must be set the same way at the printer and the computer's COM port. If either end of the communication link is set improperly, data passed from the computer to the printer will be misinterpreted (resulting in highly distorted printout). Check the printer first and note any DIP switch or jumper settings that affect the data frame. Next, check the COM port settings at the computer (under the printing application or the Windows Printer Control Panel). The COM port's start, data, parity, and stop bit configuration should all match the printer's settings. If not, adjust either the COM port parameters or the printer DIP switch settings so that both ends of the communication link are set the same way. If problems persist, there might be a fault in the printer's communication

IC (e.g., IC10 in figure 11-3). You might try replacing the IC, or you might replace the main controller entirely.

Symptom 8 The image appears "stitched." Stitching is an image distortion where points in the image appear to have been "pulled" in the horizontal direction (typically to the right). Figure 11-4 illustrates the "stitching" effect, along with some manifestations of other controller errors. Images are formed by scanning a laser beam repetitively across the drum. Pixels are formed by turning the laser beam on and off while scanning, a function performed by the mechanical controller board, as shown in figure 11-5. If there is an intermittent fault in the mechanical controller logic, beam modulation might fail during one or more scanning passes, resulting in random "pulls" in the image.

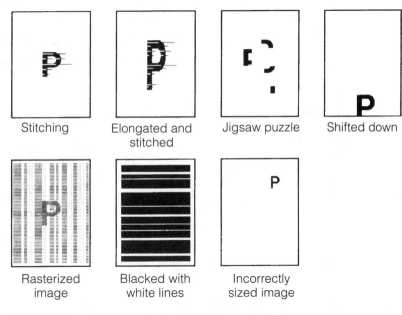

■ **11-4** *Recognizing main controller board faults.*

Start your examination by checking all of the cables between the laser/scanner assembly and the mechanical controller. Loose wiring might result in intermittent laser fire. Turn off and unplug the printer, then try removing and reinstalling each of the connectors. Check any other wiring on the mechanical controller as well. If problems persist, chances are very good that the mechanical controller has failed. You might attempt to troubleshoot the mechanical controller, as described in Chapter 9, but it is often more efficient to simply replace the mechanical controller board out-

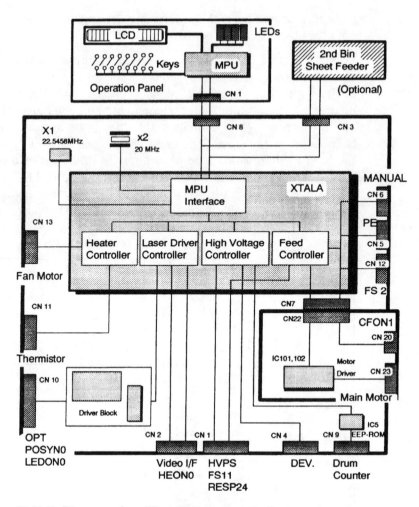

■ **11-5** *Diagram of an EP mechanical controller board.* Tandy Corporation

right. If a new mechanical controller fails to resolve the problem, replace the laser/scanner assembly.

Symptom 9 The image appears elongated and "stitched." This is a variation of Symptom 8. Not only is the laser beam misfiring intermittently, but the image is being "stretched" along the page. Under most circumstances, there has been a logic failure on the mechanical controller. Before you attempt work on the controller, however, try reseating each of the connectors on the mechanical controller. Be sure to turn off and unplug the printer before fiddling with any connectors. If problems persist, there is a serious fault in the printer's ECU. You might attempt to troubleshoot the ECU as outlined in Chapter 9, but this type of fault can be very difficult to track down. As a result, it is often better to try replacing the mechanical controller board first. If that fails to correct the problem, try a new main logic board. If your particular printer design integrates all of the logic and controlling circuitry on a single ECU board, replace that board outright.

Symptom 10 Portions of the image are disassociated like a jigsaw puzzle. Of all the controller failures, this is perhaps one of the most perplexing. You might notice that some elements of the printed image are fine, but other (larger) areas of print seem jumbled around. To make matters worse, the problem is often intermittent, so some printed pages might appear fine. Under most circumstances, there has been a logic failure on the main controller board. Before you attempt work on the main controller, however, try reseating each of the connectors on the main and mechanical controller boards. Be sure to turn off and unplug the printer before working with any connectors. If problems persist, there is a serious fault in the printer's ECU. You might attempt to troubleshoot the ECU as outlined in Chapter 9, but this type of "jigsaw puzzle" operation can be very difficult to track down. As a result, it is often better to try replacing the main controller board first. If that fails to correct the problem, try a new mechanical controller board. If your particular printer design integrates all of the logic and controlling circuitry on a single ECU board, replace that board outright.

Symptom 11 The image appears to be shifted down very significantly. You can see this type of problem illustrated in figure 11-4. At first glance, you might be tempted to think that this is a registration problem (and cannot be ruled out), but it is also possible that fault on the mechanical controller (probably the "feed control" circuit as in figure 11-5) is passing the page through far too soon before the developed image is aligned. Chances are that the pick-up and registration mechanics are working correctly; otherwise, the page would likely lose its top margin or appear smudged. When the top margin is excessive, suspect a logic fault. Specifically, you should suspect that a logic error is firing the registration system too soon after a printing cycle starts. You should address this type of problem by troubleshooting or replacing the mechanical controller board.

Symptom 12 The image appears "rasterized" with no intelligible information. A "rasterized" image is a complete distortion; there is rarely any discernible information in the printed page. Instead, the image is composed of broken horizontal lines, such as in figure 11-4. The trick with this type of fault is that it is not always easy to determine the problem origin. Turn off and unplug the printer. Open the printer and check each cable and wiring harness at the controller board(s) and laser/scanning assembly. Try reseating each of the connectors. If the problem persists, the fault is almost certainly in the main controller board. Chapter 9 offers some essential ECU troubleshooting information, but this type of logical troubleshooting can be extremely challenging and time-consuming. So it is often easier to just replace the main controller board and retest the printer. If a new controller board fails to correct the problem, you should troubleshoot or replace the mechanical controller board.

Symptom 13 The image is blacked out with white horizontal lines. This type of problem creates a page that is blacked out except for a series of white horizontal bars, and will typically eradicate any discernible image on the page. As it turns out, connector problems can readily cause this type of problem, so start your examination there. Turn off and unplug the printer, then check the wiring harnesses and reseat each connector on the main and mechanical controller boards. Be extremely careful to replace each connector carefully, and avoid bending any of the connector pins.

If the problem continues, your fault is likely to be in the main controller board. Chapter 9 offers some essential ECU troubleshooting information, but this type of logical troubleshooting can be extremely challenging and time-consuming, especially under these symptoms. So it is often easier to just replace the main controller board and retest the printer. If a new controller board fails to correct the problem, you should troubleshoot or replace the mechanical controller board.

Symptom 14 The image is incorrectly sized along the vertical axis. Ideally, an image should be sized according to the size of whatever paper tray is installed. When the image size is significantly smaller than expected, you should first check to see that the proper paper tray is installed, and that the printing application is set to use the correct paper size (especially under Windows). If everything is configured properly, you should examine the paper tray sensors as described in Chapter 9. Replace any defective tray sensor microswitches. If problems persist, you should also inspect any wiring harnesses and connectors at the mechanical controller. Loose or defective wiring can cause erroneous page sizing. If the connectors check properly, you should suspect a logical problem in the mechanical controller board. You might attempt to troubleshoot the mechanical controller if you wish, or you might simply choose to replace the mechanical controller outright. If that should fail to resolve the problem, try a new main controller board.

Registration symptoms

The "registration" process involves picking up a sheet of paper and positioning it for use. As a result, any problems in the paper tray, pick-up roller, separation pad, registration rollers, or the related drive train can result in any one of the following problems. Registration problems are quite common, especially in older printers, where age and wear can affect the rollers, gears, and critical mechanical spacing. In very mild cases, you might be able to correct a registration problem with careful cleaning and a bit of readjustment. For most situations, however, you will need to replace a defective mechanical assembly, or a failing electromechanical device (such as a clutch).

Symptoms

Symptom 1 The print contains lines of print, usually in the lower half of the page, that appear smudged. You can see a simple example of the problem in figure 11-6. This symptom is almost always the result of a problem with your registration rollers. Uneven wear can allow the registration rollers to grip the page firmly at one point, then loosely at another point. When the grip tightens, the page jerks forward just a fraction, but enough to smudge (or blur) the print at that point. Turn off and unplug the printer, then expose the registration assembly and examine it closely. Look for any accumulations of debris or obstructions that might force the registration rollers apart at different points. Remove any obstructions and try cleaning the roller pair. Examine the registration drive train and look for any gears that might be damaged or obstructed. Clean the drive train and replace any damaged gears. If the problem persists, you should consider replacing the registration assembly.

Symptom 2 There is no apparent top margin. The image might run off the top of the page, as in figure 11-6. In just about every case. There is a fault in the pick-up assembly. As a consequence, the page is not being passed to the registration assembly in time to be aligned with the leading edge of the drum image, so the image appears cut off at the page top. Start by examining your paper tray. Make sure that the "lift mechanism" is not jammed or otherwise interfering with paper leaving the tray. If you're not sure, try a dif-

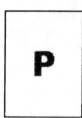   

Smudged print   No top margin   Top smudging   Too much top margin

Image skew

■ **11-6** *Recognizing pick-up/registration faults.*

ferent paper tray. Also consider your paper itself. Unusually light or specially coated papers might simply not be picked up properly. Try a standard 20-lb. bond xerography-grade paper. If problems persist at this point, chances are that your pick-up assembly is failing. Turn off and unplug the printer, then examine your pick-up system closely. Check for any accumulations of debris or obstructions that might interfere with the pick-up sequence. Remove any obstructions, then clean the pick-up roller and separation pad. Also examine the pick-up drive train and clutch. Any jammed or damaged gears should be replaced. If the solenoid clutch is sticking or failing, you should replace the solenoid clutch or clutch PC board. If that fails to resolve the problem, you will have little alternative but to replace the pick-up assembly and separation pad.

Symptom 3 There is pronounced smudging at the top of the image (generally near the top margin). This symptom suggests that the registration assembly is failing, or it has not been installed correctly. Turn off and unplug the printer, then expose the registration system. Carefully inspect the system to see that the rollers and drive drain are installed properly. Try reinstalling the registration assembly. If problems persist, try a new registration assembly.

Symptom 4 There is too much margin space on top of the image. When there is excessive margin space at the top of the image, it generally indicates that the paper has been allowed into the IFS too soon; paper is traveling through the printer before the drum image was ready. This fault is usually related to the registration roller clutch. You see, the registration rollers are supposed to hold the page until the drum image is aligned properly. This means that the registrations rollers must be engaged or disengaged as required, typically through a clutch mechanism. If the clutch is jammed in the engaged position, the registration rollers will always run (passing each new page through immediately). Turn off and unplug the printer, then examine your registration clutch closely. If the clutch is jammed, try to free it and clean surrounding mechanics to remove any accumulations of debris. If the clutch fails to re-engage or remains jammed again, replace the registration clutch entirely, or replace the clutch solenoid PC board.

Symptom 5 The image is "skewed" (not square with the page). Refer to Chart 11-2. Skew occurs when the page is passed through the printer at an angle (rather than straight). Typically, paper must enter the printer straight because of the paper tray, so the page must shift due to a mechanical problem. In actual practice, however, a loose or bent paper guide tab can often shift the paper as it enters the printer. Start your examination by checking the paper tray, specifically the paper cassette guide tab. If the tab is loose or bent, replace it or try a new paper tray. If the paper tray is intact, consider the paper itself. Unusually light or specially coated papers can skew in the pick-up and registration mechanics. If you are using an unusual paper, try some standard 20-lb. xerography-grade paper. If the problem should continue, turn off and unplug the printer, then examine the pick-up and registration mechanics. Check for obstructions, or any accumulations of foreign matter that might interfere with the paper path and cause the page to skew. If there is nothing conclusive, try replacing the pick-up assembly and separation pad, then the registration assembly (in that order).

Laser/scanner problems

A laser beam must be modulated (turned on and off corresponding to the presence or absence of a dot) and scanned across the conditioned drum. Both modulation and scanning must take place at a fairly high rate in order to form an image, up to 12 pages per minute and more. However, the process of writing with a laser beam is not a simple task, as you see in figure 11-7. Variations in laser output power (often due to age), variations in polygon motor speed (also due to age and wear), and the accumulation of dust

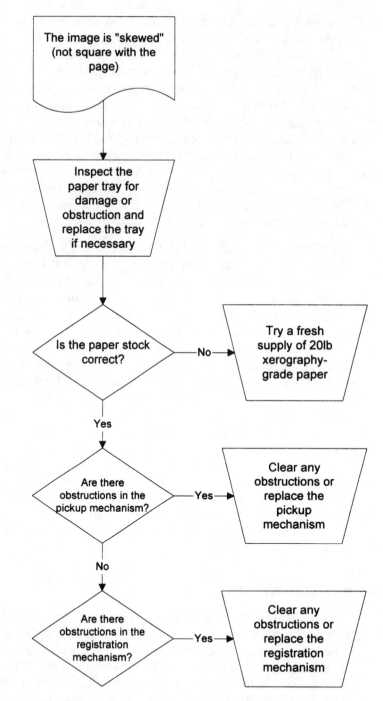

Chart 11-2 *Flowchart for Registration/Pickup Symptom 5.*

348

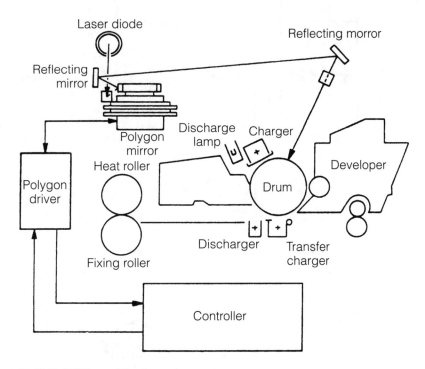

■ **11-7** *Writing with a laser beam.* Tandy Corporation

and debris on the polygon mirror and other optical components will all have an adverse impact on the final image. Faults can even creep into the laser sensor and affect beam detection and alignment. EP printer designers responded to the problems associated with such a delicate assembly by placing all of the laser, control, and scanning components into a single "laser/scanner" assembly. Today, the laser/scanner is an easily replaceable module, and that is how we will treat it.

Symptoms

Symptom 1 Right-hand text appears missing or distorted. Figure 11-8 illustrates a typical example. In many cases, this is simply a manifestation of low toner in your EP/toner cartridge. If any area of the development roller receives insufficient toner, it will result in very light or missing image areas. Turn off and unplug the printer, remove the EP/toner cartridge, and redistribute the toner. Follow your manufacturer's recommendations for toner redistribution. If you see an improvement in image quality (at least temporarily), replace the EP/toner cartridge.

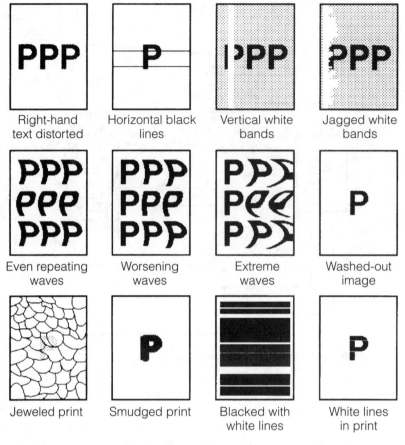

Right-hand text distorted	Horizontal black lines	Vertical white bands	Jagged white bands
Even repeating waves	Worsening waves	Extreme waves	Washed-out image
Jeweled print	Smudged print	Blacked with white lines	White lines in print

■ **11-8** *Recognizing laser/scanner problems.*

Examine the shock mountings that support your laser/scanner assembly. If the laser/scanner assembly is loose or not mounted correctly, scan lines might not be delivered to the proper drum locations. Try remounting the laser/scanner assembly. If the problem persists, replace the writing mechanism entirely. If you are using a laser writing mechanism, pay special attention to the installation and alignment of the laser beam sensor.

Symptom 2 You encounter horizontal black lines spaced randomly through the print. Remember that black areas are the result of light striking the drum. If your printer uses a laser/scanner assembly, a defective or improperly seated beam detector could send false scan timing signals to the main logic. The laser would then make its scan line while main logic waits to send its data. At the beginning of each scan cycle, the laser beam strikes a detector. The detector carries laser light through an optical fiber to a circuit, which converts light into an electronic logic signal that is compatible with the mechanical controller's logic. Circuitry interprets this "beam detect signal" and knows the polygon mirror is properly aligned to begin a new scan. The mechanical controller then modulates the laser beam on and off corresponding to the presence or absence of dots in the scan line.

Positioning and alignment are critical here. If the beam detector is misaligned or loose, the printer's motor vibrations might cause the detector to occasionally miss the beam. Printer circuitry responds to this by activating the laser full-duty in an effort to synchronize itself again. Reseat the laser/scanner assembly, or try reseating the beam detector and optical fiber. If the problem persists, try replacing the beam detector and cable, or replace the laser/scanner unit outright.

Symptom 3 Your printer's LCD reports a "polygon motor synchronization error." The printer might also display an error code (e.g., ERROR 31) to represent the condition. The polygon mirror is the heart of the laser scanning system. The motor's speed must be absolutely steady. If the motor fails to rotate, or fails to synchronize at a constant rate within a few seconds of power-up, scanning will fail. In early EP printers, the polygon motor and mirror were implemented as discrete devices. In today's EP printers, however, the laser, scanner motor, and polygon mirror are all integrated into a replaceable laser/scanner assembly. When a scanner error is reported, you should first shut down the entire printer, let it rest for several minutes, then turn it back on to see if the error clears. If the fault persists, your best course is simply to replace the laser/scanner assembly entirely.

Symptom 4 There are one or more vertical white bands in the image. At first glance, this symptom might appear to be a problem with the transfer corona. However, you will notice that the white band(s) appearing here are thick and well-defined (and cleaning the transfer corona will have no effect). A hard white band such as this suggests that the laser beam (or LED light) is being blocked. This is not as uncommon as you might imagine. Dust, foreign matter, and debris can accumulate on the focusing lens and obstruct the light path. It is also possible that there is a chip or scratch in the lens.

Turn off and unplug the printer. Start your examination by checking and cleaning the transfer corona; the trouble is probably not here, but perform a quick check just to eliminate that possibility. If the transfer corona should prove dirty, certainly retest the printer.

If the problem persists, expose the "beam-to-drum" mirror and focusing lens, and examine both closely. Look for dust, dirt, toner, paper fragments, or any other foreign matter that might have accumulated on the optics. If you find foreign matter, you should not just blow it out with compressed air; it will make a mess, and the dust will eventually resettle somewhere else. Take the nozzle of a vacuum cleaner and hold it in proximity of the optical area, then blow the optics clean with a canister of compressed air. This way, the foreign matter loosened by the compressed air will be vacuumed away rather than resettle in the printer. The key idea to remember here is do not touch the optics.

For stains or stubborn debris, clean the afflicted optics gently with a high-quality lens cleaner fluid and wipes from any photography store. Be very careful not to dislodge the "beam-to-drum" mirror or lens from its mounting. Never blow on a lens or mirror yourself; breath vapor and particles can condense and dry on a lens to cause even more problems in the future. Allow any cleaner residue to dry completely before reassembling and retesting the printer.

If the problem should persist, suspect a problem with the laser/scanner assembly. There might be some foreign matter on the laser aperture that blocks the scanned beam as it leaves the scanner. Check the laser/scanner's beam aperture and clean away any foreign matter. If the material is inside the laser/scanner assembly, it should be replaced.

Symptom 5 There is a white jagged band in the image. This symptom is similar in nature to the previous symptom; foreign matter is interfering with the laser beam path. The major difference is that instead of a solid white band, you see a random jagged white band. A major difference, however, is that the obstruction is random (drifting in and out of the laser path unpredictably). This suggests that you are dealing with a loose obstruction, such as a paper fragment, which is able to move freely. Turn off and unplug the printer, then check for obstructions around the transfer corona. While the transfer corona itself is probably not fouled, a paper fragment stuck on the monofilament line can flutter back and forth resulting in the same jagged appearance.

Next, check the optical path for any loose material that can obstruct the laser beam. Be particularly concerned with paper fragments or peeling labels. Fortunately, such obstructions are relatively easy to spot and remove. When removing an obstruction, be careful to avoid scratching or moving any of the optical components. If the problem persists, you might have an obstruction inside of the laser/scanner assembly, so it should be replaced.

Symptom 6 There are repetitive waves in the image. You can see a simple example of this fault in figure 11-8. All of the image elements are printed, but there is a regular "wave" in the image. This kind of distortion is typically referred to as *scanner modulation*, where scanner speed oscillates up and down just a bit during the scanning process. In virtually all cases, the fault lies in your laser/scanner assembly. Turn off and unplug the printer, then try reseating the cables and wiring harnesses connected to the laser/scanner unit. Try the printer again. If problems persist, replace the laser/scanner assembly.

Symptom 7 There are worsening waves in the image. This type of problem is a variation of the "scanner modulation" fault, shown in the previous symptom. In this case, however, the modulation is relatively mild on the left side of the page, and gradually increases in magnitude toward the right side. These "worsening waves" can take several forms, as shown in figure 11-8; typical manifestations can be heavy or light. Regardless of the modulation intensity, all of these symptoms can often be traced to a connector problem at the laser/scanner assembly. Turn off and unplug the printer, then carefully reseat each connector and wiring harness between the laser/scanner unit and the mechanical controller board. If problems persist, replace the laser/scanner assembly.

Symptom 8 The image appears washed out; there is little or no intelligible information in the image. Typically, you will see random dots appearing over the page, but there are not enough dots to form a coherent image. Light images might suggest a problem with the high-voltage power supply or the transfer system, but in many such circumstances, some hint of an image is visible. You might also suspect the toner supply, but toner that is too low to form an image will register a "low toner" error. Still, a quick check is always advisable. Remove the EP cartridge and try redistributing the toner, then try darkening the print density wheel setting. If the image improves, check the EP/toner cartridge and suspect the HVPS. Otherwise, you should suspect a failure at the laser diode itself. Although solid-state lasers tend to run for long periods with little real degradation in power, an aging laser diode might produce enough energy to satisfy the laser sensor, but not nearly enough to discharge the EP drum. Try replacing the laser/scanner assembly.

Symptom 9 The print appears "jeweled." You can see this kind or print in figure 11-8. This is caused when the laser beam is totally unable to synchronize with the printer; the laser sensor is failing to detect the beam. In many cases, the fiber-optic cable carrying the laser signal has been detached or broken. When the optical cable is a stand-alone component, it is a relatively easy matter to replace the cable and sensor. If the cable and sensor are integrated into the laser/scanner assembly, your best course is to reseat the cables and wiring harnesses between the laser/scanner and the mechanical controller board. If problems persist, replace the laser/scanner assembly outright.

Symptom 10 You see regular "smudging" in the print. When dirt, dust, and other foreign matter accumulate on the "beam-to-drum" mirror or compensating lens, they tend to block laser light at those points, resulting in vertical white bars or lines down the image. However, mild accumulations of dust or debris that might not be heavy enough to block laser light might be enough to "scatter" some of the light. This "scattered light" spreads like shrapnel resulting in unwanted exposures. Because each point of exposure becomes dark, this often manifests itself as a "dirty" or "smudged" appearance in the print. Your best course is to clean the printer's optical deck. Turn off and unplug the printer, then expose the optical area. Place the nozzle from your vacuum cleaner in the immediate area, and blow away any dust and debris with a can of photography-grade compressed air. Do not attempt to vacuum inside of the printer! Just let it remove any airborne contaminants dislodged by the compressed air.

Symptom 11 The print is blacked out with white horizontal lines. In order to modulate the laser beam to form dots, the data must be synchronized with the position of the laser beam. This synchronization is accomplished by the beam detector, which is typically located in the contemporary laser/scanner assembly. If the detector fails to detect the laser beam, it will fire full-duty in an attempt to reestablish synchronization. When the beam fires, it will produce a black line across the page. Multiple subsequent black lines will effectively black-out the image. The white gaps occur if the beam is sensed, or if a time-out/retry period has elapsed. In most cases, the beam sensor in the laser/scanner assembly has failed or become intermittent. Turn off and unplug the printer, then try reseating each of the cables from the laser/scanner. If the problem persists, your best course is simply to replace the laser/scanner assembly entirely.

Symptom 12 The image forms correctly except for random white gaps that appear horizontally across the page. This is another manifestation of trouble in the laser beam detection process. A kink in the fiber-optic cable can result in intermittent losses of laser power. In older printers with a discrete fiber-optic cable, it was a simple matter to replace the cable outright. Now that beam detection is accomplished in the laser/scanner assembly itself, your best course is simply to replace the unit outright.

Drive or transmission problems

With so much emphasis placed on the key electronic and mechanical subassemblies of an EP printer, it can be easy to forget that each of those mechanical assemblies are coupled together with a comprehensive drive train of motors, gears, and (sometimes) pulleys. A failure, even an intermittent, at any point in the drive or transmission will have some serious consequences in the printed image. Figure 11-9 outlines the printer's mechanical system.

The mechanical controller (sometimes referred to as the dc controller) starts the main motor. Once the main motor starts, a gear train will operate the EP drum, the transfer (or feed) roller(s), and the fusing rollers. In some designs, the main motor will also operate a set of exit rollers that direct the page to an output tray. Of course, there must also be a provision to pick up and register each page, but those assemblies cannot run full-duty; instead, they must be switched on and off at the proper time. To accomplish this timing, a solenoid-driven clutch (marked "Solenoid") is added to the pick-up roller and registration roller assemblies. For the system in figure 11-9, a separate motor (marked "Magnetic motor") is used to drive the development system.

Symptoms

Symptom 1 There are gaps and overlaps in the print. Figure 11-10 shows you an example of this symptom. The problem here is a slipping gear or failing drive motor. Unfortunately, this is not a simple problem to spot; gear assemblies such as the one in figure 11-11 are generally quite fine, and an intermittent gear movement can easily go unnoticed. Start with a careful inspection of the gear train. Make sure that all gears are attached and meshed securely. It is not uncommon for older gear assemblies to loosen with wear. Also check that the main motor is mounted securely and meshed properly with other gears. Be especially careful to check for ob-

355

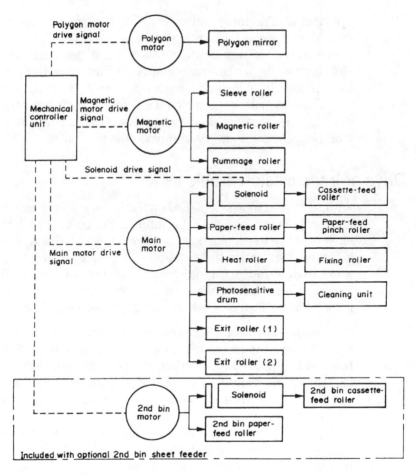

■ **11-9** *An EP mechanical system.* Tandy Corporation

structions or foreign matter that might be lodged in the gear train. Finally, you will need to check each gear for broken teeth, which is a time-consuming and tedious process, but it is preferable to dismantling the entire drive train. A high-intensity pen light will help to highlight broken gear teeth. Replace any gears that might be damaged. If the problem persists and the drive train is flawless, try replacing the main motor assembly.

Symptom 2 The print has a "roller-coaster" appearance. This type of roller coaster distortion is typically the result of a fault in the gear train. Start with a careful inspection of the gear train. Make sure that all gears are attached and meshed securely. It is not uncommon for older gear assemblies to loosen with wear. Also check that the main motor is mounted securely and meshed properly with other gears. Be especially careful to check for obstructions or

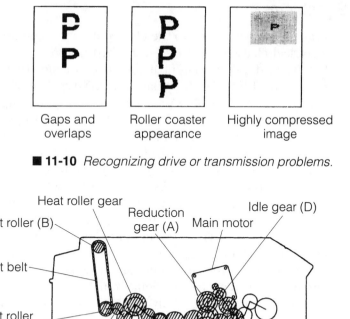

Gaps and overlaps

Roller coaster appearance

Highly compressed image

■ **11-10** *Recognizing drive or transmission problems.*

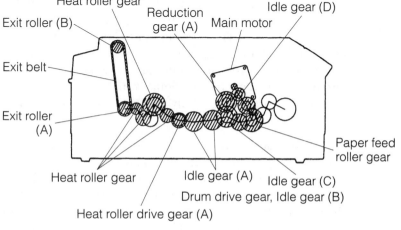

Heat roller gear

Exit roller (B)

Reduction gear (A)

Main motor

Idle gear (D)

Exit belt

Exit roller (A)

Heat roller gear

Idle gear (A)

Heat roller drive gear (A)

Drum drive gear, Idle gear (B)

Idle gear (C)

Paper feed roller gear

■ **11-11** *A typical image formation gear assembly.* Tandy Corporation

foreign matter that might be lodged in the gear train. Finally, you will need to check each gear for broken teeth, which is a time-consuming and tedious process, but it is preferable to dismantling the entire drive train. A high-intensity pen light will help to highlight broken gear teeth. Replace any gears that might be damaged.

Symptom 3 The image is highly compressed in the vertical axis. A highly compressed image can indicate a failing main motor, especially when the amount of "compression" varies randomly from page to page. Because the main motor is responsible for driving the entire system, a fault can interrupt the page transport. Check the main motor to see that it is mounted securely to the frame and meshed properly with other gears. Also check the connector and wiring harness at the main motor and mechanical controller board to be sure that everything is attached properly. If problems persist, try replacing the main motor. If that fails to correct the problem, replace the mechanical controller board.

HVPS problems

High voltage is the key to the electrophotographic process. Huge electrical charges must be established in order to condition the EP drum, develop a latent image, and transfer that image to a page. An HVPS for the classical "SX-type" engine develops –6,000, +6,000, and –600 V. The newer "CX-type" engine requires far less voltage (–1,000, +1,000, and –400 V). Still, high voltages impose some important demands on the power supply and its associated wiring. First, high-voltage supplies require precise component values that are rated for high-voltage operation. While ordinary circuits might easily tolerate a "close" component value, HV supplies demand direct replacements. Installing a "close" value (or a part with a loose tolerance) in an HVPS can throw the output(s) way off. The other factor to consider is the wiring. Most commercial wire is only insulated to 600 V or so; higher voltages can jump the inexpensive commercial insulation and arc or short-circuit; they can even electrocute you. So HVPS wiring harnesses and connectors are specially designed to operate safely at high voltages.

As a technician, these factors present some special problems. Replacement components are expensive and often difficult to find. Installing those components can be tedious and time-consuming. And even when things are working perfectly, you can not measure the outputs directly without specialized test leads and equipment. When all of this is taken into account, it is almost always preferable to replace a suspected HVPS outright rather than attempt to troubleshoot it.

Symptoms

Symptom 1   Your printer's LCD displays a "high-voltage error." The printer might also use an error code (e.g., ERROR 35) to represent the condition. This indicates that one or more outputs from the HVPS are low or absent. The preferred technique is to replace the HVPS outright. Before replacing an HVPS, turn off and unplug the printer, and allow at least 15 minutes for charges in the HVPS to dissipate. When replacing the HVPS, be very careful to route any wiring away from logic circuitry, and pay close attention when installing new connectors. It is also important that you bolt the new HVPS securely into place; this ensures proper grounding.

Symptom 2   The image is visible, but the print out is darkened. You can see this type of symptom in figure 11-12. In order for an image

358

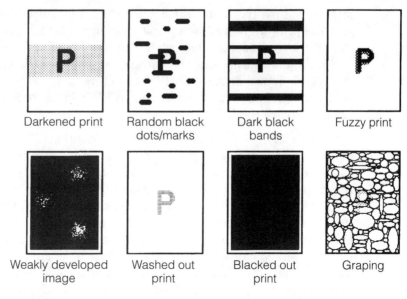

Darkened print	Random black dots/marks	Dark black bands	Fuzzy print
Weakly developed image	Washed out print	Blacked out print	Graping

■ **11-12** *Recognizing a high-voltage power supply problem.*

to be developed, the EP drum must be discharged. This can also happen if the primary corona fails to place a conditioning charge on the drum. At first, you might suspect that the –6,000-V source is low, but in actual practice, this type of symptom is typically the result of bad HVPS grounding. Turn off and unplug the printer, then allow at least 15 minutes for the HVPS to discharge. Open the printer and inspect the mounting bolts holding the HVPS in place. Chances are that you'll find one or more grounding screws loose. Gently tighten each of the mounting/grounding hardware (you don't want to strip any of the mounting holes). Secure the printer and retest.

Symptom 3 There are random black splotches in the image. Generally, the image will appear, but it will contain a series of small black marks spaced randomly throughout the page. This type of image problem suggests that the HVPS is arcing internally (and is probably close to failure). Turn off and unplug the printer, then allow at least 15 minutes for charges in the HVPS to dissipate. Check the high-voltage connectors and high-voltage wiring harness. Try reseating the connectors to check for failing contacts. If problems persist, replace the HVPS.

Symptom 4 There is "graping" in the image. The "graping" effect places small, dark, oval-shaped marks on the page, usually along one side of the page. Graping is often due to a short-circuit in the primary corona HV connector; HV is arcing out. Turn off and un-plug the printer, then allow at least 15 minutes for any charges in the HVPS to dissipate. Inspect the primary corona wiring, and check for any shorts along the corona, or along the HV lead from the HVPS. Try reseating the HV connectors and wiring. If problems persist, replace the primary corona HV lead (if possible); otherwise, replace the HVPS.

Symptom 5 The image appears, but it contains heavy black bands. You can see an example of this in figure 11-12. Although this symptom might look quite different from Symptom 2, it is really quite similar. If the HVPS ground is loose, the image can be darkened, but if the HVPS ground is simply intermittent, portions of the image might be exposed just fine. As the grounding cuts out, however, primary voltage fails, and the lack of conditioning voltage causes a black band to form. When the ground kicks in again, the image formation resumes, and so on. Turn off and unplug the printer, and allow at least 15 minutes for the HVPS to discharge. Gently tighten or reseat each of the grounding screws holding the HVPS in place (be careful not to strip the threaded holes). Also check the HV wiring harness to see that it is not crimped or shorted by other assemblies. If problems persist, replace the HVPS.

Symptom 6 The image appears fuzzy, letters and graphics appear "smudged" or "out of focus." This type of problem suggests a fault in the ac bias voltage. Toner is heavily attracted to the exposed drum. When toner jumps to the drum, some toner lands in nonex-posed areas near the exposed points. By using an ac developer bias, the developer voltage varies up and down. As developer voltage increases, more toner is passed to the exposed drum areas. As developer voltage decreases, toner is pulled back from the nonex-posed drum areas. This action increases image contrast while cleaning up any "collateral" toner that might have landed improperly. If the ac component of your developer voltage fails, that contrast-enhancing feature will go away, resulting in fuzzy print. Because developer voltage is generated in the HVPS, try replacing the HVPS outright.

Symptom 7 There are weakly developed areas in the image. The image appears, but various areas of the image are unusually light. This can be attributed to moisture in the paper. Try a supply of fresh, dry paper in the paper cassette. Also make sure that the paper does not have a specialized coating. If the problem persists, you should suspect that the HVPS is weak and nearing failure. For a symptom such as this, your best course is usually just to replace the HVPS outright. Be sure to turn off and unplug the printer, and allow at least 15 minutes for the HVPS to discharge before attempting replacement.

Symptom 8 The image appears washed out. This type of symptom can often be the result of several causes. Before proceeding, check the print density wheel and try increasing the density setting. If problems continue, check your paper supply. Specially or chemically coated papers might not transfer very well. If the problem continues, you should suspect that the transfer voltage is weak or absent, or the primary grid voltage might be failing. In either case, you should replace the HVPS. Be sure to turn off and unplug the printer, and allow at least 15 minutes for the HVPS to discharge before attempting replacement.

Symptom 9 The page is blacked out. This symptom suggests that the primary corona voltage has failed. Without a conditioning charge, the EP drum will remain completely discharged; this will attract full toner, which will result in a black page. Before attempting to replace the HVPS, check the primary corona to see that it is still intact, and check the wiring between the primary corona and the HVPS. If the primary corona is damaged, replace the EP/toner cartridge; otherwise, replace the HVPS. Be sure to turn off and unplug the printer, and allow at least 15 minutes for the HVPS to discharge before attempting replacement.

Fusing assembly problems

The fusing assembly is another focal point for many printer problems. In order to fix toner to the page surface, a combination of heat and pressure is applied with a set of fusing rollers. The upper roller provides heat while the lower roller provides pressure. In order for the fusing assembly to work properly, there are several factors that must be in place. First, the heating roller must reach and maintain a constant temperature; that temperature must be consistent across the roller's surface. Second, pressure must be constant all the way across the two rollers, so the two rollers must be aligned properly. Third, not all melted toner will stick to the page; some will adhere to the heating roller. So there must be some provision for cleaning the heating roller. Finally, there must be a reliable method for protecting the printer from overheating.

The fusing unit design shown in figure 11-13 addresses these concerns. Heat is generated by a bar heater or a long quartz lamp mounted inside of the upper fusing roller. Power to operate the heater is provided from the dc power supply (typically 24 V). A separate thermistor in the roller changes resistance versus temperature, so it acts as a temperature detector. The thermistor's resistance is measured by a circuit on the mechanical controller board, which in turn, modulates the power feeding the heater. This process "closes the loop" to achieve a stable operating temperature. If a failure should occur that allows the heater to run continuously, a thermal fuse will open and cut off voltage to the heater. Although figure 11-13 does not show it, the upper and lower fusing rollers are held together with torsion springs; the springs keep both rollers together with the right amount of compression, and can adjust for slight variations in paper weight and system wear. Toner that sticks to the upper fusing roller can transfer off the roller elsewhere on the page, resulting in a "speckled" appearance. Therefore, the upper roller is coated with Teflon to reduce sticking, and a cleaning pad rubs any toner off the roller. You will find that temperature, alignment, and cleaning problems are some of the most frequent fusing troubles.

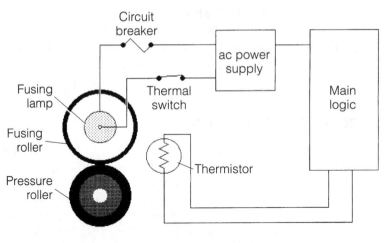

■ **11-13** *Fusing unit temperature control loop.*

Symptoms

Symptom 1 The printer's LCD indicates a "Heater Error" or other type of fusing temperature malfunction. Refer to Chart 11-3. Your particular printer might use an error number (e.g., ERROR 32) to represent the condition. Fusing is integral to the successful operation of any EP printer. Toner that is not fused success-

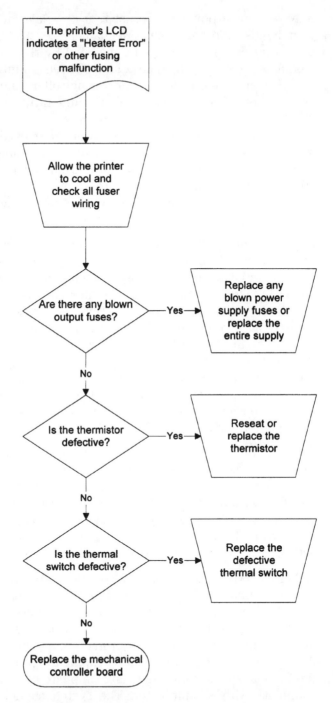

The printer's LCD indicates a "Heater Error" or other fusing malfunction

Allow the printer to cool and check all fuser wiring

Are there any blown output fuses? —Yes→ Replace any blown power supply fuses or replace the entire supply

No

Is the thermistor defective? —Yes→ Reseat or replace the thermistor

No

Is the thermal switch defective? —Yes→ Replace the defective thermal switch

No

Replace the mechanical controller board

Chart 11-3 *Flowchart for Fusing Assembly Symptom 1.*

fully remains a powder or crust that can flake or rub off onto your hands or other pages. Mechanical controller logic interprets the temperature signal developed by the thermistor and modulates power to the quartz lamp. Three conditions will generate a fusing malfunction error: fusing roller temperature falls below about 140°C; fusing roller temperature climbs above 230°C; or fusing roller temperature does not reach 165°C in 90 seconds after the printer is powered up. Your particular printer might use slightly different temperature and timing parameters. Also note that a fusing error will often remain with a printer for 10 minutes or so after it is powered down, so be sure to allow plenty of time for the system to cool before examining the fusing system.

Begin by examining the installation of your fusing assembly. Check to see that all wiring and connectors are tight and seated properly. The quartz heater power supply is often equipped with a fuse or circuit breaker that protects the printer (this is not the thermal switch shown in figure 11-13). If this fuse or circuit breaker is open, replace your fuse or reset your circuit breaker, then retest the printer. Remember to clear the error, or allow enough time for the error to clear by itself. If the fuse or breaker trips again during retest, you have a serious short circuit in your fusing assembly or power supply. You can attempt to isolate the short circuit, or simply replace your suspected assemblies (fusing assembly first, then the dc power supply).

Turn off and unplug the printer, allow it to cool, and check your temperature sensing thermistor by measuring its resistance with a multimeter. At room temperature, the thermistor should read about 1 kΩ (depending on the particular thermistor). If the printer has been at running temperature, thermistor resistance might be much lower. If the thermistor appears open or shorted, replace it with an exact replacement part and retest the printer.

A thermal switch (sometimes called a *thermoprotector*) is added in series with the fusing lamp. If a thermistor or main logic failure should allow temperature to climb out of control, the thermal switch will open and break the circuit once it senses temperatures over its preset threshold. This protects the printer from severe damage and possibly a fire hazard. Unplug the printer, disconnect the thermal switch from the fusing lamp circuit, and measure its continuity with a multimeter. The switch should normally be closed. If you find an open switch, it should be replaced. Check the quartz lamp next by measuring continuity across the bulb itself. If you read an open circuit, replace the quartz lamp (or the entire fusing assembly). Be sure to secure any disconnected wires. If the printer still does not reach its desired temperature, or continuously opens the thermal switch, troubleshoot your thermistor signal conditioning circuit and the fusing lamp control signal from the mechanical controller, or replace the mechanical controller board entirely.

Symptom 2 Print appears smeared or fused improperly. Refer to Chart 11-4. Temperature and pressure are two key variables of the EP printing process. Toner must be melted and bonded to a page in order to fix an image permanently. If fusing temperature or roller pressure is too low during the fusing operation, toner might remain in its powder form. Resulting images can be smeared or smudged with a touch. You can run the PRINTERS fusing test to check fusing quality by running a series of continuous prints. Place the first and last printout on a firm surface and rub both surfaces with your fingertips. No smearing should occur. If your fusing level varies between pages (one page might smear while another might not), clean the thermistor temperature sensor and repeat this test. Remember to wait 10 minutes or so before working on the fusing assembly. If fusing performance does not improve, replace the thermistor and troubleshoot its signal conditioning circuit at the mechanical controller. If smearing persists, replace the fusing assembly and cleaning pad.

Static teeth just beyond your transfer corona are used to discharge the paper once toner has been attracted away from the EP drum. This helps paper to clear the drum without being attracted to it. An even charge is needed to discharge paper evenly, otherwise, some portions of the page might retain a local charge. As paper moves toward the fusing assembly, remaining charge forces might shift some toner resulting in an image that does not smear to the touch, but has a smeared or pulled appearance. Examine the static discharge comb once the printer is unplugged and discharged. If any of its teeth are bent or missing, replace the comb.

A cleaning pad rubs against the fusing roller to wipe away any accumulations of toner particles or dust. If this cleaning pad is worn out or missing, contamination on the fusing roller can be transferred to the page, resulting in smeared print. Check your cleaning pad in the fusing assembly. Worn out or missing pads should be replaced immediately.

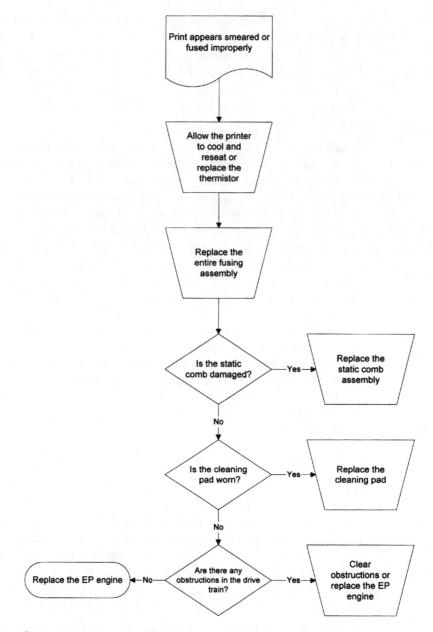

Chart 11-4 *Flowchart for Fusing Assembly Symptom 2.*

Inspect your drive train for any gears that show signs of damage or excessive wear. Slipping gears could allow the EP drum and paper to move at different speeds. This can easily cause portions of an image to appear smudged; such areas would appear bolder or darker than other portions of the image. Replace any gears that you find to be defective. If you do not find any defective drive train components, try replacing the EP cartridge. Finally, a foreign object in the paper path can rub against a toner powder image and smudge it before fusing. Check the paper path and remove any debris or paper fragments that might be interfering with the image.

Symptom 3 There are narrow horizontal bands of smudged print. Figure 11-14 shows you an example of this problem. While smudging usually suggests a fusing problem, its occurrence in relatively narrow bands actually points to a problem with the paper feed; the registration or transfer rollers are not moving evenly, so they are jerking the paper. When the paper jerks, the toner immediately being transferred from the EP drum becomes smudged.

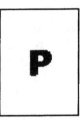

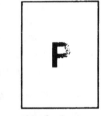

Smudged print   Smudged zones   Pencil line(s)   No fusing on part of page

■ **11-14** *Recognizing fusing system problems.*

Check your paper stock first. Unusually light or specially coated papers might slip periodically, resulting in a slight jerking motion. Try a standard 20-lb. xerography-grade paper. If the paper is appropriate, there are three causes for this kind of paper jerk: either your rollers are worn (allowing loose contact at some point in their rotation); the rollers are obstructed (effectively jamming the paper at some point in their rotation); or there is a fault in the drive train. Unfortunately, observing the paper path while the printer is running will rarely reveal subtle mechanical defects, so turn off and unplug the printer, then inspect your registration rollers for signs of wear or accumulations of foreign matter. If the registration rollers appear damaged or worn, replace the registration assembly. If there is a buildup of foreign matter, carefully clean the registration rollers.

If the problem persists, inspect the drive train carefully. Check each gear to see that they are meshed properly, and see if there are any broken gear teeth. A small, high-intensity pen light will make this inspection easier. Replace any gears that are worn or damaged. If there are obstructions in the gear train, clean them away carefully with a cotton swab lightly dampened in isopropyl alcohol. If this still fails to correct the problem, the fault is probably in the EP engine mechanics. Try replacing the EP/toner cartridge (the "engine").

Symptom 4 There are wide horizontal areas of smudged print. You can see this type of symptom in figure 11-14. Although this problem might sound quite similar to the previous symptom, the fault is almost always in the fusing assembly. Excessive pressure from the lower fusing roller squeezes the page so tightly that the print is smudged, typically across a wide area. Turn off and unplug the printer, and allow 15 minutes or so for the fusing assembly to cool. Inspect the fusing assembly carefully. If there are torsion springs holding the upper and lower fusing rollers together, you can probably reduce the tension to relieve some of the pressure. You might have to work in small increments to get the best results. Also check the lower fusing roller itself; if the roller is worn or damaged, it should be replaced. As an alternative, you can replace the entire fusing assembly outright.

Symptom 5 There are dark creases in the print. These are visible creases (also referred to as *pencil lines*) in the page itself, not just in the printed image. In virtually all cases, pencil lines are the result of a bloated lower fusing roller. The way in which it applies pressure on the page causes a crease in the page. First, check your paper supply. Light bond or specially coated papers might be especially susceptible to this kind of problem. Try a standard 20-lb. xerography-grade paper. If the problem persists, you will need to inspect the fusing assembly. Turn off and unplug the printer, and allow 15 minutes for the fusing system to cool before opening the printer. Check the lower fusing roller for signs of bloating, excessive wear, or other damage. Try replacing the lower fusing roller; otherwise, you should replace the entire fusing assembly.

Symptom 6 There is little or no fusing on one side of the image. However, the other half is fused properly. This problem occurs when there is a gap in the fusing rollers. Even if the upper fusing roller is producing the correct amount of heat, it will not fuse toner without pressure from the lower fusing roller. Gaps are often caused from physical damage to the fusing assembly, or an accumulation of foreign matter that forces the rollers apart. Turn off and unplug the printer, then wait about 15 minutes for the fusing assembly to cool. Inspect the rollers carefully. You can expect to find your problem on the side that does not fuse. For example, if the right side of the page is not fusing, the problem is likely on the right side of the fusing rollers. Check for mechanical alignment of the rollers. You might be able to restore operation by adjusting torsion spring tension. If problems continue, you should replace the entire fusing assembly.

Corona (charge roller) problems

There are two high-voltage charge areas in the EP printer: the primary area and the transfer area. Classical "SX-type" engines use corona wires, so the primary area will use a primary corona, and the transfer area will use a transfer corona. The newer "CX-type" engines replace the corona wires with charge rollers, so the primary area will use a primary charge roller, and the transfer area will use a transfer charge roller. Although these areas very rarely fail, there are a suite of problems that plague the coronas. This part of the chapter shows you some of the more pervasive faults.

Symptoms

Symptom 1 Pages are completely blacked out, and might appear blotched with an undefined border. You can see an example of this problem in figure 11-15. Turn off and unplug the printer, remove the EP cartridge, and examine its primary corona wire. Remember from Chapter 4 that a primary corona applies an even charge across a drum surface. This charge readily repels toner, except at those points exposed to light by the writing mechanism, which discharge those points and attract toner. A failure in the primary corona will prevent charge development on the drum. As a result, the entire drum surface will tend to attract toner (even if your writing mechanism works perfectly). This creates a totally black image. If you find a broken or fouled corona wire, clean the wire or replace the EP cartridge.

If your blacked-out page shows print with sharp, clearly defined borders, your writing mechanism might be running out of control. LEDs in a solid state print bar or laser beam might be shorted in an ON condition, or receiving erroneous data bits from its control circuitry (all logic 1s). In this case, the primary corona is working just fine, but a writing mechanism that is always on will effectively expose the entire drum and discharge whatever charge was applied

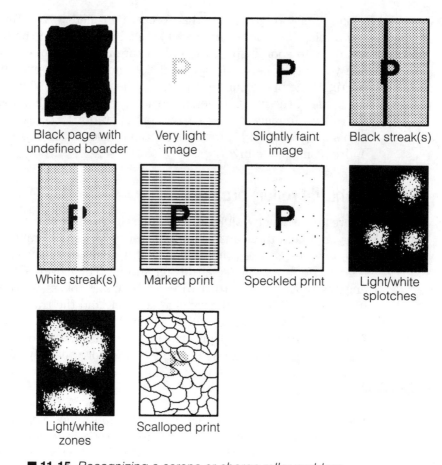

Black page with undefined boarder | Very light image | Slightly faint image | Black streak(s)

White streak(s) | Marked print | Speckled print | Light/white splotches

Light/white zones | Scalloped print

■ **11-15** *Recognizing a corona or charge roller problem.*

by the primary corona. The net result of attracting toner would be the same, but whatever image is formed would probably appear crisper, more deliberate.

Use your oscilloscope to measure the data signals reaching your writing mechanism during a print cycle. You should find a semi-random square wave representing the 1s and 0s composing the image. If you find only one logic state, troubleshoot your main logic and driving circuits handling the data, or replace the mechanical controller board. If data entering the writing mechanism appears normal, replace your writing mechanism (LED bar or laser/scanner assembly). You might wish to cross-reference this symptom with an HVPS problem earlier in this chapter.

Symptom 2 Print is very faint. Refer to Chart 11-5. Turn off and unplug the printer, remove the EP cartridge, and try redistributing

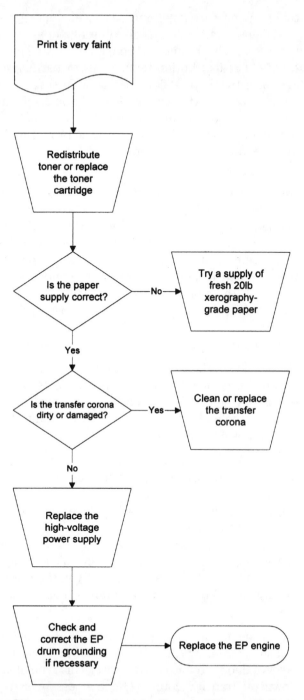

Chart 11-5 *Flowchart for Corona (Charge Roller) Symptom 2.*

toner in the cartridge. Your user's manual probably offers preferred instructions for redistributing toner. Keep in mind that toner is largely organic; as such, it has only a limited shelf and useful life. If redistribution temporarily or partially improves the image, or if the EP cartridge has been in service for more than six months, replace the EP cartridge. If you are using a paper with a moisture content, finish, or conductivity that is not acceptable, image formation might not take place properly. Try a standard 20-lb. xerography-grade paper.

Check your transfer corona or transfer charge roller. It is the transfer corona that applies a charge to paper that pulls toner off the drum. A weak transfer corona or charge roller might not apply enough charge to attract all the toner in a drum image. This can result in very faint images. Turn off and unplug the printer, allow ample time for the high-voltage power supply to discharge completely, then inspect all wiring and connections at the transfer corona. If the monofilament line encircling the transfer corona is damaged, replace the transfer corona assembly, or attempt to re-thread the monofilament line. If faint images persist, repair or replace the high-voltage power supply assembly.

Finally, check the drum ground contacts to be sure that they are secure. Dirty or damaged ground contacts will not readily allow exposed drum areas to discharge. As a result, very little toner will be attracted and only faint images will result. If the problem persists, replace the EP engine.

Symptom 3 Print is just slightly faint. Print that is only slightly faint does not necessarily suggest a serious problem. There are a series of fairly simple checks that can narrow down the problem. Check the print density control dial. Turn the dial to a lower setting to increase contrast (or whatever darker setting there is for your particular printer). Check your paper supply next. Unusual or specially coated paper may cause fused toner images to appear faint. If you are unsure about the paper currently in the printer, insert a good-quality, standard-weight xerographic paper and test the printer again.

Over time, natural dust particles in the air will be attracted to the transfer corona and accumulate there. This eventually causes a layer of debris to form on the wire. This type of accumulation cuts down on transfer corona effectiveness, which places less of a charge on paper. Less toner is pulled from the drum, so the resulting image appears fainter. Turn off and unplug the printer, allow ample time for the high-voltage power supply to discharge, then gently clean the transfer corona with a clean cotton swab or corona cleaning tool. Be very careful not to break the monofilament line wrapped about the transfer corona assembly. If this line does break, the transfer corona assembly will have to be rewrapped or replaced.

Check your toner level. Unplug the printer, remove the EP cartridge, and redistribute toner. Follow all manufacturer's recommendations when it comes to redistributing toner. The toner supply might just be slightly low at the developing roller. Unplug your printer and examine the EP cartridge sensitivity switch settings. These microswitches are actuated by molded tabs attached to your EP cartridge. This tab configuration represents the relative sensitivity of the drum. Main logic uses this code to set the power level of its writing mechanism to ensure optimum print quality. These switches also tell main logic whether an EP cartridge is installed at all. If one of these tabs is broken, or if a switch has failed, the drum might not be receiving enough light energy to achieve proper contrast. Check your sensitivity switches as outlined for a "No EP Cartridge" error, shown later in this chapter. If the problem persists, your high-voltage power supply is probably failing. Replace your high-voltage power supply.

Symptom 4 There are one or more vertical black streaks in the print. Black streaks might range from narrow lines to wide bands depending on the severity of the problem. In most cases, this fault is due to foreign matter accumulating on the primary corona. Foreign matter will prevent charges from forming on the drum. In turn, this will invariably attract toner, which creates black streaks. Typically, the edges of these streaks are fuzzy and ill-defined. Your best course is simply to clean the primary corona; most printers enclose a cleaning tool for just this purpose. The process takes no more than a minute. If the problem should persist (very unlikely), replace the EP engine.

Symptom 5 There are one or more vertical white streaks in the print. Begin by checking your toner level. Toner might be distributed unevenly along the cartridge's length. Turn off and unplug the printer, remove the EP cartridge, and redistribute the toner. Follow your manufacturer's recommendations when handling the EP cartridge. If this improves your print quality (at least temporarily), replace the nearly exhausted EP cartridge.

Next, examine your transfer corona for areas of blockage or extreme contamination. Such faults would prevent the transfer corona from generating an even charge along its length; corrosion acts as an insulator that reduces the corona's electric field. Uncharged page areas will not attract toner from the drum, so those page areas will remain white. Clean the transfer corona very carefully with a clean cotton swab. If your printer comes with a corona cleaning tool, use that instead. When cleaning, be sure to avoid the monofilament line wrapped around the transfer corona assembly. If the line breaks, it will have to be rewrapped, or the entire transfer corona assembly will have to be replaced.

Check the optical assembly for any accumulation of dust or debris that could block out sections of light. Because EP drums are only scanned as fine horizontal lines, it would take little more than a fragment of debris to block light through a focusing lens. Gently blow off any dust or debris with a can of high-quality, optical-grade compressed air available from any photography store. For stains or stubborn debris, clean the afflicted lens gently with a high-quality lens cleaner and wipes from any photography store. Be very careful not to dislodge the lens from its mounting. Never blow on a lens or mirror yourself; breath vapor and particles can condense and dry on a lens to cause even more problems in the future.

Symptom 6 The print appears "scalloped." You can see an example of "scalloping" in figure 11-15. The scalloping effect has a unique and unmistakable appearance, and almost always indicates that the primary corona has broken. The image that forms is then expressly the result of random discharge from the erase lamps. In many cases, the failure of a primary corona will simply blacken the page. In some circumstances, however, the erase lamps will leave a latent image that will develop into the scalloped pattern. You should immediately suspect a failure in the primary corona. Your best course is simply to replace the primary corona by exchanging the EP/toner cartridge.

Symptom 7 The print suggests a shorted transfer corona, as shown in figure 11-15. An image appears as expected, but it is marked with vertical swatches of small horizontal tics. Experience has demonstrated that this type of symptom is frequently due to a short-circuited transfer corona. Turn off and unplug the printer, then allow at least 15 minutes for the printer to cool and discharge. Inspect the transfer corona carefully, as well as any wiring at the corona. Gently clear away any foreign material, especially conductive material, from the transfer area, and try the printer again. If the problem persists, try replacing the transfer corona assembly.

Symptom 8 Print appears speckled. In almost all cases, speckled print is the result of a fault in your primary corona grid. A grid is essentially a fine wire mesh between the primary corona and drum surface. A constant voltage applied across the grid serves to regulate the charge applied to the drum to establish a more consistent charge distribution. Grid failure will allow much higher charge levels to be applied unevenly. A higher conditioning charge might not be discharged sufficiently by the writing mechanism; toner might not be attracted to the drum even though the writing is working as expected. This results in a very light image (almost absent except for some light speckles across the page). Because the primary grid assembly is part of the EP cartridge, replace the EP/toner cartridge and retest the printer. If speckled print persists, you should suspect a fault in the HVPS.

Symptom 9 There are light/white splotches in the image. When you see a symptom such as this, your first suspicion should be moisture in the paper supply, which is a common occurrence in humid summer months. When the paper becomes damp (even just from the air's humidity), charges do not distribute properly across the page. As a result, paper will not charge in the damp areas, so toner is not attracted from the drum. Damp areas then remain very light or white. Paper that is unusually coated can have similar problems. In virtually all cases, a supply of fresh, dry, 20-lb., xerography-grade paper should correct the problem. To correct the problem over the long term, consider adding a dehumidifier or air conditioner in the work area to keep paper dry.

Symptom 10 There are light/white zones spread through the image. At first glance, you might think that this symptom is similar to the previous one. In practice, however, random white zones in the printed page are much larger and more distinct than simple light splotches; in effect, the white areas have just disappeared. This symptom is indigenous to the CX-type EP engine that uses charge rollers rather than coronas. In most cases, you will find that the transfer charge roller has failed or is missing. Even without a working transfer corona, the CX engine can transfer portions of the latent image to the page, but you can see from figure 11-15 that the transfer is very unstable. Check and replace the transfer charge roller or replace the EP engine outright.

Miscellaneous problems

This chapter has focused on problems that plague key areas of the EP printer. However, there are some symptoms that cannot easily be associated with any particular area of the printer. As a consequence, these problems can be difficult to track down and correct. This part of the chapter illustrates some of the printer's miscellaneous problems.

Symptoms

Symptom 1 Your printer never leaves its warm-up mode. There is a continuous "Warming Up" status code. EP printers must perform two important tasks during initialization. First, a self-test is performed to check the printer's logic circuits and electromechanical components. This usually takes no more than 10 seconds from the time power is first applied. Second, its fusing rollers must warm up to a working temperature. Fusing temperature is typically acceptable within 90 seconds from a cold start. At that point, the printer will establish communication with the host computer and stand by to accept data, so its "Warming Up" code should change to an "On-Line" or "Ready" code.

When the printer fails to go on-line, it might be the result of a faulty communication interface, or a control panel problem. Turn the printer off, disconnect its communication cable, and restore power. If the printer finally becomes ready without its communication cable, check the cable itself and its connection at the computer. You might have plugged a parallel printer into the computer's serial port, or vice versa. There might be a faulty communication interface in your host computer.

If the printer still fails to become ready, unplug the printer and check that the control panel cables or interconnecting wiring is attached properly. Check the control panel to see that it is operating correctly. Also check the control panel interface circuit (sometimes called an *interface/formatter* circuit). Repair or replace your faulty control panel or interface/formatter circuit. Depending on the complexity of your particular printer, the interface/formatter might be a separate printed circuit plugged into the main logic board, or its functions might be incorporated right into the main logic board itself.

Symptom 2 You find a "Paper out" message. When the printer generates a "Paper out" message, it means that either paper is exhausted, or the paper tray has been removed. When a paper tray is inserted, a series of metal or plastic tabs make contact with a set of microswitches, as shown in figure 11-16. The presence or absence of tabs will form a code that is unique to that particular pa-

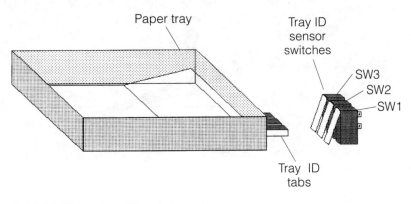

Paper tray

Tray ID
sensor
switches

SW3
SW2
SW1

Tray ID
tabs

■ **11-16** *Paper tray ID switch system.*

per size. Microswitches are activated by the presence of tabs. Main logic interprets this paper type code, and knows automatically what kind of media (paper, envelopes, etc.) that it is working with. This allows the printer to automatically scale the image according to paper size. Table 11-1 shows a typical paper code table.

■ **Table 11-1 Typical tray switch configurations.**

Paper tray	SW1	SW2	SW3
Executive	1	1	1
A4	1	1	0
Legal	0	0	1
Envelope	0	1	1
Letter	1	0	0
* No tray	0	0	0

1 = on (engaged)
0 = off (disengaged)

The presence of paper is detected by a mechanical sensing lever, as shown in figure 11-17. When paper is available, a lever rests on the paper. A metal or plastic shaft links this lever to a thin plastic flag. While paper is available, this flag is clear of the paper-out sensor. If the tray becomes empty, this lever falls through a slot in the tray, which rotates its flag into the paper-out sensor. This indicates that paper is exhausted. The paper-out sensor is usually mounted on an auxiliary PC board (known as the *paper control* board), and its signal is typically interpreted by the mechanical controller board.

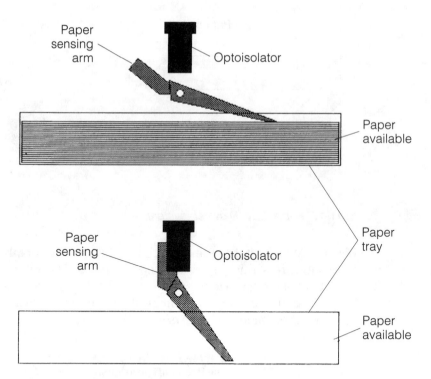

■ **11-17** *Operation of a paper sensing arm.*

Begin your check by removing the paper tray. Be sure that there is paper in the tray, and that any ID tabs are intact, especially if you have just recently dropped the tray. Reinsert the filled paper tray carefully and completely. If the "Paper out" message continues, then there is either a problem with your paper ID microswitches, paper-sensing lever, or the paper-out optoisolator.

You can check the paper ID microswitches by removing the paper tray and actuating the paper sensing lever by hand (so the printer thinks that paper is available). Refer to Table 11-1 and actuate each switch in turn using the eraser of a pencil. Actuate one switch at a time and observe the printer's display. The "Paper out" error should go away whenever at least one microswitch is pressed. If the error remains when a switch is pressed, that switch is probably defective. Unplug the printer and use your multimeter to check continuity across the suspect switch as you actuate it. Replace any defective switch. If the switches work electrically, but the printer does not register them, troubleshoot or replace the main logic board. Inspect the paper-out lever and optoisolator next.

When paper is available, the paper-out lever should move its plastic flag clear of the optoisolator. When paper is empty, the lever should place its flag into the optoisolator slot. Note: This logic might be reversed depending on the particular logic of the printer. This check confirms that the paper sensing arm works properly. If you see the lever mechanism jammed or bent, repair or replace the mechanism. Check the paper-out optoisolator, as shown in the procedures of Chapter 9. Replace the optoisolator if it appears defective. If the sensors appear operational, replace the mechanical controller board.

Symptom 3 You see a "Printer open" message. Printers can be opened in order to perform routine cleaning and EP cartridge replacement. The cover(s) that can be opened to access your printer are usually interlocked with the writing mechanism and high-voltage power supply to prevent possible injury from laser light or high voltages while the printer is opened. A simplified interlock assembly is shown in figure 11-18. The top cover (or some other cover assembly) uses a pushrod to actuate a simple electrical switch. When the top cover is opened, the interlock switch opens, and the printer's driver voltage (+24 Vdc is shown) is cut off from all other circuits. This effectively disables the printer's operation. When the top cover is closed again, the interlock switch is reactivated, and printer operation is restored.

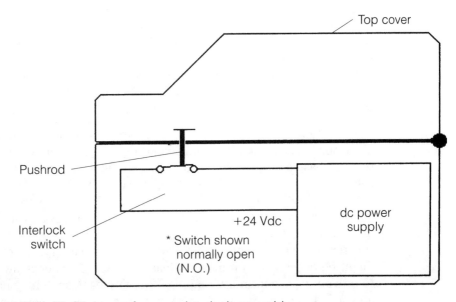

■ **11-18** *Diagram of a cover interlock assembly.*

Make sure that your cover(s) are all shut securely (try opening and reclosing each cover). Inspect any actuating levers or pushrods carefully. Replace any bent, broken, or missing mechanical levers. Unplug the printer and observe how each interlock is actuated (it might be necessary to disassemble other covers to observe interlock operation). Adjust the pushrods or switch positions if necessary to ensure firm contact.

Turn off and unplug the printer, then use your multimeter to measure continuity across any questionable interlock switches. It might be necessary to remove at least one wire from the switch to prevent false readings. Actuate the switch by hand to be sure that it works properly. Replace any defective interlock switch, reattach all connectors and interconnecting wiring, and retest the printer. If the switch itself works correctly, check the signals feeding the switch. For figure 11-18, check the dc voltage at the switch. If the voltage is low or absent, trace the voltage back to the power supply or other signal source. If signals are behaving as expected but a "Printer open" message remains, trace the interlock signal into the mechanical controller board and troubleshoot your electronics, or replace the mechanical controller outright.

Symptom 4 You see a "No EP cartridge" message. An electrophotographic engine assembly uses several tabs (known as *sensitivity tabs*) to register its presence, as well as inform the printer about the drum's relative sensitivity level. The ECU regulates the output power of its writing mechanism based on these tab arrangements (e.g., high-power, medium-power, low-power, or no-power, no cartridge). Sensitivity tabs are used to actuate microswitches located on a secondary PC board. The sequence of switch contacts forms a "sensitivity" code that is interpreted by the mechanical controller.

Begin by checking the installation of your current EP engine. Make sure that it is in place and seated properly. Check to be sure that at least one sensitivity tab is actuating a sensor switch. If there are no tabs on the EP engine, replace it with a new or correct-model EP engine having at least one tab. Retest the printer. If your "No EP cartridge" error persists, check all sensitivity switches in the printer. Turn off and unplug the printer, then use your multimeter to measure continuity across each sensitivity switch. It might be necessary to remove at least one wire from each switch to prevent false continuity readings. Actuate each switch by hand and see that each one works properly. Replace any microswitch that appears defective or intermittent. Replace any connectors or interconnecting wiring, and retest the printer. If the sensitivity switches are working properly, troubleshoot or replace the mechanical controller board or replace it outright.

Symptom 5 You see a "Toner low" message constantly, or the error never appears. A toner sensor is located within the EP/toner cartridge itself. Functionally, the sensor is little more than an antenna receiving a signal from the high-voltage ac developer bias, as shown in figure 11-19. When toner is plentiful, much of the electromagnetic field generated by the presence of high-voltage ac is blocked. As a result, the toner sensor only generates a small voltage. This weak signal is often conditioned in the mechanical controller by an amplifier using some type of operational amplifier circuit that compares sensed voltage to a preset reference voltage. For the sensor in figure 11-19, sensed voltage is normally below the reference voltage; its output is a logic 0. Main logic would interpret this signal as a satisfactory toner supply. As toner volume decreases, more high-voltage energy is picked up by the toner sensor, in turn developing a higher voltage signal. When toner is too low, sensed voltage will exceed the reference, and the comparator's output will switch to a logic 1. This is handled in main logic, and a "Toner low" warning is produced.

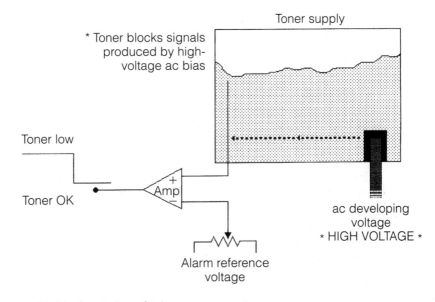

■ **11-19** *Operation of a low-toner sensor.*

Unfortunately, there is no good way to test the toner sensor. High voltage is very dangerous to measure directly without the appropriate test probes, and the signal picked up at the receiving wire is too small to measure without a sensitive meter or oscilloscope. Turn off and unplug the printer and begin your check by shaking the toner to redistribute the toner supply (or insert a fresh EP/toner cartridge). Refer to the user's manual for your particular printer to find the recommended procedure for redistributing toner, then retest the printer. If the problem persists, there might be a fault in the mechanical controller board's detection circuit. Troubleshoot the mechanical controller or replace it entirely.

Symptom 6 Your printer's LCD displays a "fan motor error" or similar fault. The printer might also use an error code (e.g., ERROR 34) to represent the condition. The typical EP printer uses two fans: a high-voltage cooling fan, and an ozone venting fan. In most cases, the ozone venting fan runs off-line and is not detected by the printer. The power supply cooling fan, however, is vital for the supply's reliability; if the cooling fan fails, the power supply will quickly overheat and fail. To prevent this from happening, the fan's operation is often monitored. If the fan quits, an error will be produced. In most cases, the fan is simply defective and should be replaced. Check the voltage available at the fan. If fan voltage is available (but the fan does not spin), the fan is defective and should be replaced. If fan voltage is missing, you will need to work back into the printer to find where fan power was lost; check any loose wiring or connectors.

Symptom 7 There is ghosting in the image. The expected image prints normally, but upon inspection, you can see faint traces of previous image portions, as shown in figure 11-20. This is a case of poor housekeeping; ordinarily, a cleaning blade should scrape away any residual toner remaining on the EP drum prior to erasing and conditioning. If the cleaning blade is worn out, or the scrap toner reservoir is full, cleaning might not take place as expected, and toner will remain on the EP drum for one or more subsequent rotations. If the residual toner comes off on another rotation, it will often appear as the "ghost" of a previous image. Unfortunately, the only way to really correct this problem is to replace the entire EP engine. Cleaning blades are hardly replaceable parts, and if the scrap toner bin is full, there is no way to recycle the toner back into the reservoir. Turn off and unplug the printer, replace the EP engine, and try the printer again.

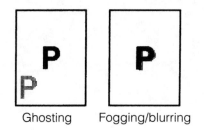

Ghosting   Fogging/blurring

■ 11-20
Recognizing miscellaneous problems

Symptom 8 The print appears fogged or blurred. This might appear somewhat like smudging in previous symptoms, but where smudging was generally limited in other symptoms, it occurs throughout the page here. This is a situation where you should examine the paper transfer guide, the passage between the static discharge comb (the transfer area) and the fusing assembly. Most transfer guides are coated with Teflon or similar material to reduce static. Over time and use, the antistatic coating can wear off revealing the static-prone plastic below the coating. If the plastic of the transfer guide causes a static charge, it might be strong enough to "drag" the toner image just slightly, resulting in a blurred of fogged image. If you find wear in the transfer guide, replace the transfer guide assembly.

383

Printing problems under Windows & Windows 95

12

This book was originally designed to deal with hardware-based printer problems. But as the PC world has shifted from DOS to Windows (and now to Windows 95), we cannot ignore the setup and configuration of the operating system itself as a source of printing problems (figure 12-1). It is no longer enough to just "fix the printer." Today's technicians should also be able to recognize

NEC Technologies, Inc.

■ **12-1** *An NEC SuperScript Color 3000 printer.*

print quality and performance issues that arise outside of the printer. This chapter provides you with that extra measure of information. Before proceeding, you should understand that this chapter focuses on symptoms and solutions. It does not discuss the design of Windows, or the architecture of its printing system; those topics are quite lengthy, and are detailed in many of the fine Windows and Windows 95 books currently available.

Windows 3.1x

When working under DOS, printing functions had to be incorporated into each application. Each DOS application printed with its own distinctive "look and feel," and some were better than others. Windows designers envisioned a world where printing is a function performed independently of the particular Windows applications. This relieves the need to specify printers and ports for each application as you do under DOS; printers and ports need only be selected once through a single, centralized printing applet (e.g., a Print Manager). By loading specific printer drivers for each printer attached to the PC, it is possible to handle the detailed font and image data typically present under Windows. Ultimately, centralized Windows provides a uniform printing "engine" capable of producing high-quality text and graphics on a vast array of printers, yet it does not require Windows applications to drive the printers directly. For most everyday users, centralized printing is an effective and problem-free process. Unfortunately, there are some serious limitations and flaws to Windows printing. If you test the printer and find it to be working properly, suspect a Windows problem.

General Windows troubleshooting

Symptom 1 The printer does not work at all. When the printer fails to work at all, you can spend quite a bit of time chasing the wrong leads. Start your examination with the printer itself. Make sure the printer is plugged in and turned on (this might sound silly, but it really happens). Check the cable between the printer and computer. A loose or faulty printer cable can wreak havoc on your printing. A good sanity check might be to try leaving Windows and printing through a DOS application, such as EDIT or another simple word processor. If you can get a DOS application to print, you know the printer, cable, and computer (the hardware) is working together, and the trouble lies in the Windows configuration. If a DOS application will not print, suspect a hardware fault. Concentrate your search on the printer's particular DIP switch, jumper, or front panel control settings, then refer to other troubleshooting tactics outlined in this book. Once you achieve printing through a DOS application, you can concentrate on the Windows setup and connection parameters.

Back in the Windows environment, access your Print Manager and take a look at the Printer control window. Make sure the correct printer is selected (you might have to install a new printer driver through the Add button) and connected to the proper printer port. If the wrong printer driver or port is active, your printer will probably not work at all. When your new printer comes with its own printer driver, you should install that driver instead of the original Windows equivalent because the driver accompanying the new printer is typically better than the native Windows version. If you are using a serial printer, check the serial parameters through the Settings function in the Connect window. Keep in mind that the Settings selection is only active once a COM port has been chosen for connection. If everything looks right so far, take a look at the printer Setup window. If the paper size, resolution, or memory setting is invalid, the printer might hang up or fail to print. Double check the installation of any memory or font cartridges in the printer (you should power down the printer before doing this), and make sure the correct memory and cartridge(s) are selected in the Setup window.

Symptom 2 The page prints, but the format is incorrect. Your printer hardware is most likely set up properly, but one or more Windows selections are probably wrong. Begin by considering your particular Windows application. Most applications (e.g., Microsoft Publisher) provide a comprehensive set of layout and format parameters that direct the formation of the printed page. If the application is setting up the page wrong, the page will certainly be printed incorrectly. As a sanity check, try printing through another Windows application if possible. If another application produces proper print, the problem is likely in the original application and not in the Print Manager configuration. If, however, the trouble persists in another Windows application, the trouble might be in the Print Manager.

If you suspect trouble in the Print Manager, start by checking the printer selection in the Printer control window. If the wrong printer is selected, the erroneous printer driver can upset your printing. Select the correct printer and try printing again. You might have to add the correct printer driver if it is not yet available. If your new printer came with its own printer driver, you should install that driver instead of the original Windows equivalent because the driver accompanying a new printer is typically better than the Windows version. Next, check the parameters in the printer Setup window. Paper size, graphics resolution, memory setting, and page orientation can all affect the page formatting. A bad setting here can also cause printing problems. If you cannot find an error in the Print Manager, take a look at your printer itself. Refer to the printer's user manual and check any DIP switches, jumpers, or front panel controls that affect page formatting. If any font cartridges are installed, power down the printer and see that they are all installed correctly.

Symptom 3 Paper advances, but no print appears on the page. Chances are good that communication is taking place properly between the printer and computer, but there is a glitch in the Windows setup. Begin by checking the paper source specified in the Setup window. An incorrect paper source might allow paper to advance from an incorrect source, but no printing can be completed. Next, check the selected printer in the Printer control window. If the wrong printer is currently selected, the printer command codes cannot be recognized by the printer (except perhaps for the page feed). Select the correct printer and try printing again. If your printer came with its own printer driver, you should install that driver instead of the original Windows equivalent because the driver accompanying the printer typically provides better performance than the Windows version.

Windows printing speed & performance

Windows printing speed is related to a variety of factors including the printer itself (i.e., its capacity in pages per minute), the amount of memory it contains, and the information that must be printed. For example, a full-page graphic image generated at 600 × 600 dpi will take a bit longer to print than a text-only document at 300 × 300 dpi. This part of the chapter addresses problems related to general performance issues under Windows 3.1x.

Symptom 1 Printing seems far too slow. Because Windows prints everything in graphics mode (text as well as images), printing tends to be inherently slower when compared to text printing under DOS. However, there are some tactics that you can use to boost printing speed. If you are using the Windows Print Manager (or another third-party print manager), try increasing the priority of the print job. Increasing the print priority increases the amount of time Windows allocates to the job. You might also wish to shut down any unneeded applications running in the background before starting a new print job; this also frees time for Windows to concentrate on the print job.

When printing images and other graphics, Windows produces temporary files, and these files can sometimes be quite large. If your hard drive is heavily fragmented, the printing spool might be delayed by excessive hard drive seeks. Leave Windows and try defragmenting your hard drive. If you are printing to a serial printer, try printing to a parallel printer (if possible). Parallel data transfers tend to be a bit faster than serial transfers. For additional speed, try printing your documents using only printer fonts; these are the font names denoted with a printer symbol next to them rather than a TT (TrueType) symbol. You might also consider lowering the resolution of your printer. For example, if you are printing at 600×600 resolution, try lowering the resolution to 300×300. Finally, make sure that you are using the correct printer driver for your specific printer, and check that the driver is the latest version. If drivers came bundled with your printer, try using them instead of native Windows drivers because custom or manufacturer's specific drivers often provide better performance than native Windows drivers.

Symptom 2 PostScript graphics appear to print slowly from an LPT port. In some PC configurations and driver combinations, graphics might seem to print unusually slowly. Often, this is not a problem per se, but simply the result of your system setup. However, you can often speed printing by disabling the Compress Bitmaps check box in the printer's Advanced Options dialog. While this will speed printing, it will increase the amount of time needed for an application to regain control (e.g., the hourglass symbol will remain longer). Also, do not disable the Compress Bitmaps check box if you are printing through a serial port (COM1 through COM4). Note that this tactic will only affect graphics printing; it will not affect the printing speed of text-only documents.

Symptom 3 You see an error message, such as "Offending Command: MSTT #### Undefined" when trying to print to a PostScript printer. This is a problem that occurs frequently when the printer runs out of virtual memory. Because Windows transfers fonts to the PostScript printer using an MSTT #### numbering scheme (where #### is the four-digit number), the printer might drop fonts, causing an error when the font is needed in the printing process. Start your correction by checking that the virtual memory setting (in the Advanced Options dialog under the PostScript driver) is appropriate for the amount of virtual memory in the printer. You can also reduce the amount of printer memory used by Windows by disabling the Clear Memory Per Page box in the Advanced Options dialog. This might slow printing a bit, but will reduce the demands on printer memory.

Symptom 4 Your bidirectional dot-matrix or ink jet printer is not printing in bidirectional mode under Windows. This is not the fault of Windows, the driver, or the printer, but is necessary because moving-carriage printers cannot achieve the precision necessary for high-resolution printing when operating in bidirectional mode, so Windows printer drivers are designed to be unidirectional. If you are using an unsupported printer that is set to operate in bidirectional mode, you might see lines and text appear jagged. Try setting the printer to unidirectional (or "graphics") mode.

Windows setup & system configuration problems

Automatic installation routines frequently make changes to CONFIG.SYS and AUTOEXEC.BAT files when new applications are added to the Windows platform. Unfortunately, those routines are not always perfect, and sometimes an unexpected error can result in unusual Windows problems. Even editing the startup (.INI) files manually can result in problems if a needed line is deleted or altered unexpectedly. In addition, drivers might conflict with other drivers (especially video drivers) and hardware in the system. The symptoms below illustrate these effects.

Symptom 1 After installing or upgrading an application, you encounter problems printing from one or more Windows applications. You (or the installation program) might have accidentally inserted a space in the SET TEMP= line in AUTOEXEC.BAT. The extra space causes a problem when Windows attempts to locate the TEMP directory, and prevents temporary printing files from being produced; as a result, you cannot print. The same problem can occur when you forget to include a backslash character in the TEMP directory path (e.g., SET TEMP=C:TEMP). To correct this problem, use a text editor to edit your AUTOEXEC.BAT file. Locate the SET TEMP= line and be sure that it is entered correctly, and see that there are no extra spaces added after the line. After you make your changes, save the corrected AUTOEXEC.BAT file and restart the system so that those changes can take effect.

Symptom 2 Some of the text (usually on a dot-matrix printer) is cut off or missing in the printout. Often, this problem is related to an incompatibility between the video board (e.g., an ATI Mach 32 board) and the printer driver. Because most new Gateway 2000 systems are shipped with ATI video cards and drivers, there is a higher probability of this problem occurring on Gateway systems. You might try an alternate or generic printer driver. If this is not acceptable, reducing the video mode to VGA or SVGA through the Windows Setup can alleviate this problem, at least temporarily. Ink jet and laser jet printers are typically free of this problem.

Symptom 3 Fine lines (under ¼ pt) appear randomly across the printed page. This occurs primarily with older HP LaserJet printers, and is often the result of a conflict between the printer driver and the Orchid video driver (such as for the Orchid Fahrenheit 1280 video board). Orchid has addressed these problems with an upgraded video BIOS and video driver. Your best course is to contact the video board manufacturer for upgrade information.

Symptom 4 The printing is garbled or missing. Printing is taking place, but it is distorted or missing in areas. Begin by inspecting the printer cable. Make sure the cable is intact and connected properly. You can also try turning the printer off for a few moments to clear its internal memory (also remember to delete any jobs outstanding in the Print Manager). Restart the printer and try printing again. Open the printer control window and check the selected printer. An incorrect printer driver can result in all types of garbled or intermittent printing. Select the proper printer (add a printer if necessary) and try printing again. If your printer came with its own printer driver, you should install that driver instead of the original Windows equivalent because the driver accompanying the printer is typically superior to the native Windows version.

If you are using a serial printer, there might be a serious problem with your handshaking (the way the flow of information is controlled between computer and printer). There are two types of handshaking: XON/XOFF (or software handshaking), and CTS/RTS (or hardware handshaking). If handshaking is inoperative, a high baud rate can "spill" data that overflows the printer's buffer. Try reducing the baud rate to a slow crawl of about 300 and try printing again. If print is correct or greatly improved at a low baud rate, check the flow control (perhaps try a known-good cable). Make sure the printer's flow control method matches the Windows Program Manager settings.

Symptom 5 Only part of the page is printed correctly. The remainder of the page is either garbled or missing. Turn your attention directly to the printer Setup window. Check the paper size and page orientation settings. Faulty information there can turn your printing in strange directions that confuse your desired format. Once page size and orientation are set correctly, consider your printer memory. Make sure the memory selection is set properly for your particular printer. If memory is set properly but part of the printed image is still being lost, there simply might not be enough memory to hold the entire image (also known as a *buffer overflow*). Refer to your application and try condensing or resizing the print to a smaller area of the page. You can also try adding memory to the printer (be sure to update the memory setting in the Setup window). Finally, you might consider reducing the printer's resolution.

Issues with third-party products

The Windows Print Manager is adequate for many general-purpose applications, but it has its limitations. Third-party applications have attempted to provide superior print management tools that are typically more flexible and powerful than Print Manager, but they also suffer from incompatibilities under some circumstances. The following section outlines some printer issues you might encounter when using third-party print manager products.

Symptom 1 After upgrading LaserMaster's WinPrint 1.0 to WinPrint 1.5, you receive an error message, such as "Illegal Function Call." The WinPrint utility is attempting to reference an earlier WinPrint font that might not be available. Unfortunately, you will need to remove the fonts from your system, then reinstall WinPrint. Exit Windows and reboot the computer. Restart Windows and open the WinPrint program group. Under the program group, you will see the Fontman icon. Select Fontman, choose all installed fonts, then select Remove. This will remove the WinPrint fonts from your system. If you see fonts remaining, choose Verify, then select NO when asked if you want to use those fonts under WinPrinter. Next, choose YES when asked if you want to delete references to these fonts for WinPrint and WinSpool. Exit Windows and then restart it. When Windows is running again, restart the WinPrint installation program. Once the reinstallation is complete, you can add fonts in the WinPrint Font Manager.

Symptom 2 The PC hangs when using SuperPrint 2.0 under Windows 3.1. There are basically two variations of this problem. First, the system might hang when SuperPrint 2.0 is installed. Second, you cannot print to an HP LaserJet printer once SuperPrint 2.0 is installed. In both cases, the reason for this trouble is that the Zenographics SuperPrint product requires version 2.2 in order to be fully compatible with Windows 3.1. Your best course is to contact Zenographics for an upgrade. If you must go ahead and install the older version, do not perform a default setup. Of course, if Super-Print is already installed and there is difficulty printing to an HP LaserJet, disable the SuperQue (print spooler) under SuperPrint.

Symptom 3 When using the LaserMaster WinSpool or WinJet accelerators with a PostScript printer, you see an error message such as "This Postscript job is not supported by WinSpool." This is a problem with the PostScript driver (versions 3.52 and 3.53). Try accepting the error message and continuing; in many cases, the printing will still occur properly. If problems persist, try reinstalling the LaserMaster driver on LPT2, which will also enable printing from LPT1. If LaserMaster is allowed to install on LPT1 during its default installation, there might be a conflict. As a more long-term fix, you should consider trying version 3.5 or 3.51 of the LaserMaster PostScript driver.

Symptom 4 When you switch from a LaserMaster WinJet 300 to a WinJet 800, you see an error message, such as "WinPrint Manager Printer Error: VPD cannot register frame buffer." This might then be followed by other system error messages. When you switch from a LaserMaster WinJet 300 to the WinJet 800, you need to edit the SYSTEM.INI file. If you fail to do this, errors will occur. When a WinJet 300 is installed, you will need to add the following lines to the [386Enh] section of SYSTEM.INI:

```
device=LMLPV.386 ;WinPrint
device=LMCAP.386 ;WinPrint
device=LMMI.386 ;WinPrint
```

When you install a WinJet 300 over a WinJet 800, you must remove the following line from the [386Enh] section of SYSTEM.INI such as:

```
device=LMLPV.386 ;WinPrint
;device=LMHAROLD.386 ;WinPrint
device=LMCAP.386 ;WinPrint
device=LMMI.386 ;WinPrint
```

The WinJet 800 installation adds the following lines to the [386Enh] section of the SYSTEM.INI file:

```
device=LMHAROLD.386 ;WinPrint
device=LMCAP.386 ;WinPrint
device=LMMI.386 ;WinPrint
```

When you install a WinJet 800 over a WinJet 300, you must remove the following line from the [386Enh] section of SYSTEM.INI such as:

```
;device=LMLPV.386 ;WinPrint
device=LMHAROLD.386 ;WinPrint
device=LMCAP.386 ;WinPrint
device=LMMI.386 ;WinPrint
```

Symptom 5 The printer does not use its resident font cartridge(s) or soft font(s). When the printer uses one or more font cartridges, turn off the printer and reseat all cartridges to be sure they are installed correctly. Restart the printer and try printing again. If the problem persists, check the printer Setup window and be sure that all appropriate font cartridges are selected. Windows 3.1 can support a maximum of two cartridge listings. If the font cartridge you are using is not shown in the cartridge selector list, you will probably have to identify the cartridge's fonts through the soft font installer. If you are using a new font cartridge that was developed after the printer driver was written, you might need a .PCM (printer cartridge metrics) file to tell Windows how to handle those "new" fonts. Install the .PCM file included with the font cartridge the same way you would install a soft font.

For soft fonts, make sure the desired fonts are listed in the HP Font Installer window. Reset the printer and try reloading the soft fonts to the printer. Printer memory must hold the soft fonts, so a printer with limited memory might run out of memory. Try loading only one or two soft fonts and attempt printing again. If a limited soft font download works, you might want to expand the printer's memory capacity.

Windows & printer drivers

Windows is constantly printing in the graphics mode, so proper printing of text and images relies on printer drivers. Printing performance is tied to using the right driver with the right printer. While there are generic drivers available under Windows that will often work to support unusual or specialized printers, such drivers are typically limited in resolutions and features; even though you might get the printer to "work," it probably will not work properly. As a consequence, you should always match the printer and its latest driver. The following symptoms address problems often related to printer drivers.

Symptom 1 The Generic/Text-Only printer driver does not use the multiple paper trays and various paper sizes of a printer. This is a direct limitation of the printer driver. The "generic" printer driver was designed to support a wide variety of different printers, but in order to be compatible across a wide range of printers, it is necessary to abandon particular features that one printer might have but another might not. If you cannot find a specific printer driver on the Windows installation disks, contact the printer manufacturer to obtain the latest copy of the Windows printer driver(s). Virtually all laser and ink jet printer manufacturers are now bundling Windows drivers with the printer. The alternative is to continue using the generic driver without the benefit of your printer's special features.

Symptom 2 You see an error message, such as "Control Panel cannot perform the current operation because <filename> is not a valid printer-driver file." This error can often occur when choosing Setup from the Printers dialog under the Control Panel. Usually, there are three causes for this error: the EXPAND.EXE or LZEXPAND.DLL files are corrupted or from a different version of Windows; the printer driver is invalid or corrupted; or you are installing a Windows 3.11 printer driver under Windows 3.1. When an invalid printer driver is specified, you might see the Print command of most Windows applications dimmed (grayed out).

Return to MS-DOS. Check the file dates on EXPAND.EXE and LZEXPAND.DLL. Verify that their file dates are consistent with other Windows or WFWG files on your system. If not, you should manually expand the proper versions of these files from installation disks into their appropriate directories. A typical command to copy EXPAND.EXE and reinstall the LZEXPAND.DLL file from the installation disk might be:

```
copy a:\expand.exe c:\windows\expand.exe
```

then

```
expand a:\lzexpand.dl_ c:\windows\system\lzexpand.dll
```

Now that you have the correct versions of EXPAND.EXE and LZEXPAND.DLL on the system, rename the suspect printer driver file(s), and check the path statement in AUTOEXEC.BAT to be sure that the Windows directory is the first directory in the path. Restart Windows, then reinstall the suspect printer drivers by choosing the Printers icon from the Control Panel. You can then select the Add button, and Install the correct driver(s) again. You could also run the Print Manager, then choose Printer Setup, Add, then Install. Finally, if you have been installing Windows 3.11 drivers in Windows 3.1, install the proper drivers for the version of Windows you are using, or download the driver from the Microsoft Download Service (MSDL) at (206) 936-6735.

Symptom 3 When printing on an ink jet printer (e.g., the HP Desk-Jet 500), you see garbage lines, blocks, and unwanted characters when printing using TrueType fonts. Chances are that the printer driver being used with the ink jet printer is outdated or corrupt. Obtain the latest version of the Windows driver from Microsoft or the printer manufacturer. Until a new driver is installed, try reducing the printer's resolution. In some cases, the unwanted printing will go away (at the expense of resolution).

Symptom 4 When printing on a dot-matrix printer (e.g., a Panasonic KX-P1124 or KX-P2123), you see garbled, missing, or misaligned text. Either the current printer driver is missing or corrupt, or the driver is incompatible with the printer's internal ROM. Start by checking the printer driver to see that the latest version is installed correctly. If problems persist, consider upgrading the printer's internal ROM. Note that this symptom typically occurs only with older dot-matrix printers, and a new ROM is often not available for obsolete or discontinued products. When this situation develops, try an alternate compatible or generic printer driver.

Symptom 5 The printer always prints in the highest quality mode available, even though a "draft" quality mode has been selected. This is typically due to an error or oversight in the printer driver, especially prevalent in older drivers for dot-matrix and ink jet printers. The preferred method of correcting this problem is to update the printer drivers with newer versions, or choose an alternate compatible or generic printer driver that does provide draft-mode printing.

Symptom 6 When printing a scaled TIFF image to an HP LaserJet III or IV, a vertical line prints on the left side of the image. The line might also appear on the right side of the image, or in a corner. This is due to problems with some HP printer drivers (such as HP-PCL5A.DRV). If you are encountering this problem printing to an HP LaserJet III, obtain the newest printer driver from Microsoft. If you are encountering the problem with an HP LaserJet IV, get the latest driver from HP. Until a new driver can be obtained, try scaling the image to a different size. The problem seems to occur only at particular horiz/vert size ratios, so altering this ratio even slightly might circumvent the problem.

Symptom 7 When using a color ink jet printer, the black output appears somewhat green. This problem appears typically in HP ink jet printers, such as the HP 500C. When using the color ink cartridge with the print mode set to All Color, only the color cartridge is used to print. As a result, dark colors, such as black, are actually made up of yellow, magenta, and cyan. It is the printer driver that decides how to mix the colors, but an incorrectly tested printer driver might allow less-than-ideal mixing, resulting in a slightly "greenish" black. The way to correct this problem is to set the print mode to Black and Color rather than All Color. You can make this adjustment under the Control Panel function by selecting the Printer icon, choosing Setup, then changing the print mode. This feature will cause the ink cartridge to use true black ink when printing blacks. In all cases, you should be sure that you are using the latest driver for your ink jet printer.

Symptom 8 You encounter GPF errors when attempting to print very small bitmap images. This problem occurs most frequently with HP LaserJet II printers operating at low resolutions (e.g., 75 dpi), but it might occur on any EP printer using the UNIDRV.DLL driver. This is typically the result of a problem in older versions of the Microsoft Universal Printer driver UNIDRV.DLL. Your best course here is to update UNIDRV.DLL by downloading a new version from the Microsoft Windows Driver Library (WDL) on CompuServe.

Windows 95 troubleshooting

As with earlier versions of Windows, the Windows 95 platform provides all of the printing resources needed by Windows applications. But the Windows 95 printing system makes some significant advances over older printing systems, incorporating a suite of new (and badly needed features). A 32-bit printing engine and enhanced parallel port support promise to produce smoother printing, while returning control to the application sooner. Image color matching capabilities allow the screen image colors to better match the colors generated by a color printer. Usability features such as point-and-print support, deferred printing, and print services for NetWare try to streamline the printing process. Enhanced font support allows you to install an unlimited number of fonts, and use up to 100 fonts in a document.

Still, with all of these enhancements, printing under Windows 95 is not always as foolproof as users and technicians like to believe. Printer installation/driver problems, network problems, and general printer errors can all occur. This part of the chapter is intended to illustrate some of the troubleshooting techniques used to identify and isolate printing problems, then explain the solutions for a selection of typical Windows 95 printing faults. First, there are some tactics you can use to help isolate problems.

The safe mode

Windows 95 can crash if an incorrect or corrupt driver is selected, or if there is a conflict between two or more system drivers. There might also be problems if Windows 95 fails to detect the correct video board in the PC. The symptoms of such a fault might range from poor or erratic video performance to complete system failures. Under Windows 3.1x, you would probably address such problems by trying the PC in "standard VGA mode" ($640 \times 480 \times 16$), but this tactic required you to switch drivers in Windows Setup.

Windows 95 provides you with a "safe mode" option that you can select during start-up. Restart the computer. When you see "Starting Windows 95" displayed on the screen, press the F8 key. Choose the Safe Mode startup option to run the PC in VGA mode. If your problem disappears in VGA mode, you can safely suspect that the video driver is corrupt, outdated, incorrect, or conflicting with another driver.

Check & correct the printer driver

In order for the printer to run efficiently, a properly written 32-bit printer driver must be loaded under Windows 95. If the driver is outdated, incorrect, or "buggy," the printer will not run correctly (if at all). You can check your printer driver by double-clicking on the My Computer icon, then double-click on the Printers icon. You can then add a printer or select one of your currently installed printers. Right-click on the desired printer, then click on Properties. This will bring up the printer properties dialog. Select the Details "page." The controls on this page allow you to adjust port settings, drivers, time-outs, and spool settings. Next, select the Paper "page," and click on the About button. This will tell you which driver version is in use. If this is not the latest version, try a newer driver.

Printing directly to the printer port

Often, it can be difficult to tell whether the source of trouble is in printer hardware or system software. Printing directly to the printer port is one way to verify the printer hardware without the clutter of

398

Windows 95 or its applications. Exit Windows 95 to the DOS prompt, and use the following command to print a file:

```
copy /b filename lpt1:
```

This command takes the binary file filename, and sends that file directly to the selected port. If the printer responds and prints the file correctly, the printer's hardware is working fine, and your trouble is in Windows 95 or its application(s). If the printer does not run under DOS, the printer, the PC, or the communication cable might be defective.

Controlling bidirectional support

Windows 95 is designed to support EPP printer ports conforming to the IEEE 1284 standard. However, not all PCs, printers, and parallel cable assemblies are designed to accommodate the added demands of IEEE 1284, and printing problems can result. You can disable bidirectional support in the Spool Settings dialog. You can reach the Spool Settings dialog by clicking the Spool Settings button in the Details page you saw previously. If problems occur with bidirectional printing on, then disappear once bidirectional control is turned off, you should leave bidirectional printing off until you can arrange more compatible hardware.

Clearing spooler files

There are some instances when errors in the print spool can cause print faults. When this occurs, the spool will not always clear, causing the error to persist even when Windows 95 is restarted. Fortunately, it is not too difficult to clear the spool files. Leave Windows 95 to MS-DOS, and switch to the \SYSYEM\SPOOL\PRINTERS directory, then delete all .SPL files. Next, switch to the \TEMP directory and erase any .TMP files. You should then shut down and restart the computer to finish cleaning up the .SPL files.

Windows 95 symptoms

Symptom 1 You cannot print to a printer (local or network). Start with the basics. Check to see that your printer is plugged in and turned on. It might sound silly, but this is a frequent oversight. Also see that the printer's communication cable is secure between the printer and host PC. Check the printer for adequate paper, and address any error messages that might be present. If the printer is on, connected, and on-line, but no printing is taking place, turn the printer off and wait about 10 seconds before turning it on again. This clears the printer's internal buffer.

If problems persist, try printing to a file (rather than to the printer). Then go to the MS-DOS prompt and print the file directly to the printer port as explained above. If this works, the problem is in the Windows 95 setup. If this does not work, you might have a fault in the printer cable or communication circuitry at the printer or PC.

Symptom 2 You cannot print due to a printer driver problem. Printer drivers are the key to successful printer operation. If the wrong driver is installed (or if the correct driver is outdated or corrupt), the printer will simply not work properly (if at all). Under Windows 95, printer driver problems are most significant when an older 16-bit driver (e.g., from Windows 3.1x) must be used under Windows 95 because no 32-bit driver exists. While this condition should not exist for long, this "mismatch" can often cause the printer to freeze or produce irregular print. Errors will also occur if the incorrect 32-bit driver is selected for a printer. Check the printer driver and see that it is a current 32-bit driver. If not, install an appropriate driver (you might need to download an updated driver from the printer manufacturer's BBS or on-line forum). If problems persist, try reinstalling the suspect driver.

Symptom 3 You cannot print due to an application problem. Start by checking the setup and configuration of your printing application. Make sure that the desired printer is selected, along with the correct tray, page orientation, resolution, and other printing parameters. If problems persist, try saving and closing other applications that might be open; this means you will be printing from the only running application. If the problem disappears, you might have a conflict between one or more applications under Windows 95. Try restarting each of the closed applications individually, and try printing after each application is started. The point at which problems return will reveal the conflict. If isolating the application does not correct the problem, try printing from a different application. When you are able to print successfully from a different application, your original application might be corrupt or contain a software bug. You might try reinstalling the suspect application, or contact the application's manufacturer for a work-around or software patch.

Symptom 4 You cannot print due to a print spooler problem. This problem occurs fairly infrequently, and might not be intuitively obvious from any outward symptoms. Try shutting down the print spooler (as described above) and print directly to the printer port. If the problem disappears, you might have a printer spooler problem.

Use SCANDISK to examine your disk space and integrity. Repair any disk problems that you might encounter, then try restoring the print spooler. If problems persist, switch "EMF spooling" to "RAW spooling" in the Spool Settings dialog, or leave spooling disabled.

Symptom 5 You cannot print with a bidirectional printer setup. Recent PC designs and multi-I/O boards make use of the IEEE 1284-compliant advanced parallel ports (also known as *ECP* or *EPP ports*). When faced with printing problems, especially for a newly installed printer, try disabling bidirectional support. If the problem disappears, the PC's parallel port might not be IEEE 1284-compatible, or your printer cable might not support IEEE 1284 communication. In that case, you will need to leave bidirectional printing disabled until you have the hardware in place to support advanced printing.

Symptom 6 Graphic images are garbled or otherwise printed incorrectly. There are many reasons why a printer might not receive a graphic image correctly. First, check the printer configuration in the printing application to be sure that the tray, orientation, resolution, and other image-related parameters are set properly. Next, check the printer to see that it has enough memory to support the graphic image size. Large images can require substantial amounts of memory. If you are attempting to print a large document, or deal with a large number of documents over a network, try printing shorter or fewer jobs; the spooler may be overloaded. You might also try shutting the spooler off. Finally, start the computer in its "safe mode" and try printing again. If the problem disappears, there might be a printer driver problem or conflict that you will have to identify.

Symptom 7 Only partial pages are printed. There are several reasons why a printer might not receive a complete page correctly. Setup issues should always be suspected first. Check the printer configuration in the printing application to be sure that the tray, orientation, resolution, and other image-related parameters are set properly. Also see if the printer has enough memory to support the printed image size; complex page layouts typically require substantial amounts of memory, and incomplete pages are often the result of high complexity in the printed page. Try simplifying the page layout or content (i.e., remove some objects). If the printed page is missing certain text styles, check that the corresponding font (or a suitable substitute) is installed. If problems persist, enable TrueType fonts as graphics.

Symptom 8 Printing is noticeably slower than normal. Slow printing under Windows 95 is typically the result of poor hardware performance, or a driver conflict. Begin by checking your available disk space. Printing can be very demanding of disk space for temporary files, so a drive nearing maximum capacity might not have enough space to create the needed temporary files. If your drive is down to a few MB, you should free additional space by backing up and eliminating unneeded files. Another printing speed factor is drive fragmentation. Excessive fragmentation will make the drive work much harder when reading and writing files, so printing from highly fragmented files will be correspondingly slower. Run a disk defragmenter to examine and correct file fragmentation on the drive. Suspect your system resources, especially if you are running several complex applications simultaneously. Next, verify that print spooling is enabled, and that EMF (Enhanced Metafile) spooling support is selected. Also check to see that the current and correct printer driver is selected for your printer. If problems persist to this point, try starting the system in its "safe mode," then try printing again. If this corrects the problem, there might be an application or driver conflict that you will need to isolate.

Symptom 9 The computer stalls during the printing process. This can sometimes occur if there is insufficient disk space to develop adequate temporary files. Check the amount of free disk space, and free additional space if necessary. If disk space is adequate, clear the spool files and try printing again. If problems persist, you probably have a conflict in your Windows 95 drivers. Start Windows 95 in its "safe mode." When printing runs correctly, there might be a conflict between the video driver and the printer driver. Check and reinstall the video driver, or check and reinstall the printer driver. Even if you choose not to reinstall the drivers, check that they are all up to date.

Using PRINTERS for diagnostics

13

One of the major limitations of printer troubleshooting has been testing. Traditionally, a technician was limited to the "self test" of each unique printer, or printing simple documents from a text editor or other basic application. There are two problems with this haphazard approach. First, self-tests and simple printouts do not always test every feature of the printer in a clear fashion. Second, such testing is hardly ever uniform; the quality and range of testing can vary radically from printer to printer. Dynamic Learning Systems has addressed this problem by developing PRINTERS, which is a PC-based utility designed to provide you with a suite of standardized printer tests. This chapter introduces you to PRINTERS. It details how to get your own copy of PRINTERS, how to install it on your PC, and how to use it productively in a matter of minutes.

Before going any further, you should understand that you do not have to purchase PRINTERS to troubleshoot your printer effectively; that would certainly not be fair to you. The majority of this book is written without relying on the use of a companion disk. However, PRINTERS provides you with a reliable platform for testing dot-matrix, ink jet, and EP printers. PRINTERS not only exercises a printer's main functions (e.g., carriage, line feed, print head, and so on), but it also allows you to test printer-specific functions through the use of escape sequences. You will find PRINTERS to be an inexpensive but handy addition to your toolbox.

All about PRINTERS

PRINTERS is a stand-alone DOS utility designed to drive virtually any commercial impact, ink jet, and Laser/LED (EP) printer through a series of exercises and test patterns specially tailored to reveal faults in the printer's major subassemblies. By reviewing the printed results, you will be able to estimate the source of a printer's problems with a high degree of confidence. On-line help and tutorial modes provide additional information about each test, and help you to understand the printed results. A variety of options allow you configure PRINTERS for over 220 unique printers, and tailor performance for speed and print quality. A handy "Manual Code" section allows you to enter Escape Code Sequences and text that can test specific functions of any printer. The system requirements for PRINTERS is shown in Table 13-1.

■ 13-1 System requirements for PRINTERS.

PC platform:	IBM PC/AT or 100% compatible
DOS:	DOS 5.0 or later
Mouse or Trackball:	Microsoft Mouse or 100% compatible
Memory:	2MB of RAM
Video:	VGA or SVGA display (640×480×16)
Floppy drive:	3.5" or 5.25" floppy disk drive (for HDD installation or FDD operation)
Hard drive:	1MB of hard disk drive space (for HDD installation)

Obtaining your copy of PRINTERS

You can buy a copy of PRINTERS directly from Dynamic Learning Systems. An advertisement and order form are included at the end of this book. Feel free to photocopy the order form (or tear out the page), and fill out the requested information carefully. Please remember that all purchases must be made in United States funds. When you fill out the order form, you can select the companion disk alone, a one-year subscription to the premier newsletter *The PC Toolbox*, or take advantage of a very special rate for the disk and subscription. Subscribers also receive extended access to the Dynamic Learning Systems BBS that allows you to exchange e-mail with other PC enthusiasts, and download hundreds of DOS and Windows PC utilities. Be sure to specify your desired disk size (3.5" or 5.25").

Installing & starting from the floppy drive

Your first task should be to make a backup copy of PRINTERS on a blank floppy disk. You can use the DOS DISKCOPY function to make your backup. For example, the command line:

```
C:\> diskcopy a: a: <ENTER>
```

will copy the original disk. Keep in mind that you will have to do a bit of disk swapping with this command. If you wish to use a floppy drive besides A:, you should substitute the corresponding letter for that drive. If you are uncomfortable with the DISKCOPY command, refer to your DOS manual for additional information. PRINTERS is designed to be run directly from the floppy disk, so you can keep the original disk locked away while you run from the copy. This allows you to take the disk from machine to machine so that you will not clutter your hard drive, or violate the licensee agreement by loading the software onto more than one machine simultaneously.

The following instructions will help you through the process:

1. To start PRINTERS from the floppy drive, insert the floppy into the drive and type the letter of that drive at the command prompt and press enter. The new drive letter should now be visible. For example, you can switch to the A: drive by typing:

```
C:\> a:  <ENTER>
```

the system will respond with the new drive letter:

```
A:\>_
```

2. Then, type the name of the executable file:

```
A:\> printers  <ENTER>
```

3. If you are using a floppy drive other than A:, you should substitute that drive letter (such as B:) in place of the A:. PRINTERS will start in a few moments and you will see the title screen and disclaimer. Press any key to pass the title screen and disclaimer, and you will then see the main menu.

Installing & starting from the hard drive

If you will only be using one PC and you have an extra 1.0 MB or so, you should still go ahead and make a backup copy of the PRINTERS disk as described in the previous section, but it would probably be more convenient to install the utilities to your hard drive. There is no automated installation procedure to do this, but the steps are very straightforward:

1. Boot your PC from the hard drive and when you see the command prompt, switch to the root directory by typing the cd\ command:

```
C:\> cd\  <ENTER>
```

the system should respond with the root command prompt:

```
C:\>_
```

2. Use the DOS md command to create a new subdirectory that will contain the companion disk's files. One suggestion is to use the name PRINTERS such as:

```
C:\> md printers  <ENTER>
```

3. Switch to the new subdirectory using the cd\ command:

```
C:\> cd\printers  <ENTER>
```

The system should respond with the new subdirectory label such as:

```
C:\PRINTERS>_
```

You might of course use any DOS-valid name for the subdirectory, or nest the directory under other directories if you wish. If you are working with a hard drive other than C:, substitute that drive label for C:.

4. Insert the backup floppy disk into the floppy drive. Use the DOS COPY command to copy all of the floppy disk files to the hard drive such as:

```
C:\PRINTERS> copy a:*.* c:  <ENTER>
```

This instructs the system to copy all files from the A: drive to the current directory of the C: drive. Because PRINTERS is not distributed in compressed form, uncompressing (or unzipping) is not needed.

5. After all files have been copied, remove the floppy disk and store it in a safe place. Then, type the name of the utility you wish to use such as:

```
C:\> printers  <ENTER>
```

The title screen and disclaimer for PRINTERS should appear almost immediately. Press any key to pass the title screen and disclaimer, and you will then see the main menu.

The work screen

After you pass the title screen and disclaimer, you will see the work screen, as illustrated in figure 13-1. The top of the work screen contains the title bar and main menu bar. The bottom of the work screen contains the message bar and copyright bar. Most of the work screen is empty now. There are six entries in the main menu bar; these are the essential areas that you will be concerned with while using PRINTERS.

☐ Configure   Allows you to select the program's operating parameters.

☐ Impact   Allows you to run a selection of tests for Impact printers.

☐ Ink Jet   Allows you to run a selection of tests for Ink Jet printers.

☐ Laser/LED   Allows you to run a selection of tests for EP printers.

☐ About   Shows you more information about PRINTERS.

☐ Quit   Leaves PRINTERS and returns to DOS

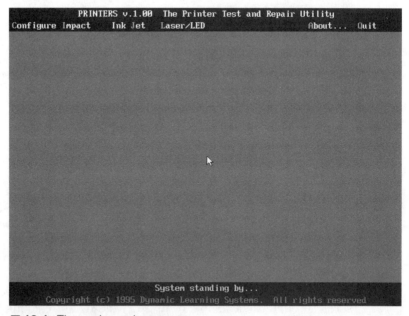

■ **13-1** *The main work screen.*

Configuring the program

To configure the various options available in PRINTERS, click on "Configure" in the main menu bar (or press C). The CONFIGURE menu will appear, as shown in figure 13-2. You can return to the work screen at any time by pressing the <ESC> key or right clicking anywhere in the display.

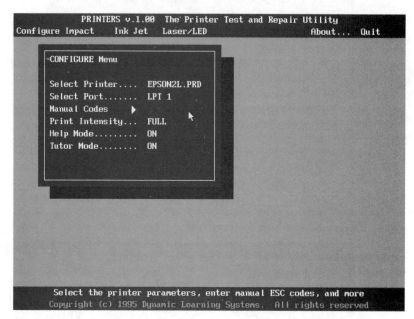

■ **13-2** *The "Configure" menu.*

Select printer

PRINTERS is compatible with the vast majority of Epson and Hewlett-Packard compatible printers now in the market, but the utility provides an extensive library of over 220 specific printer drivers. These drivers allow PRINTERS to produce the detailed graphic test patterns used in the program. Table 13-2 shows a comprehensive listing of supported printers and their corresponding driver entries. By default, PRINTERS is set to use the EPSON2L.DRV driver (a generic Epson 24-pin impact printer driver). Refer to Table 13-2 to select the appropriate driver, then scroll through the available printer drivers by left clicking on "Select Printer" (or press "p"). As you scroll through each driver, you can see details about each driver in the message bar.

■ Table 13-2 Printer driver index.

Manufacturer/printer model	Definition	Resolution	B&W/CL
Adobe Postscript—All models	PS.PRD	300×300	B&W
Color Postscript—All models	PSC.PRD	300×300	COL
Alps			
ALPS DMX800	EPSON9L.PRD	60×72	B&W
ALPS DMX800	EPSON9M.PRD	120×72	B&W
ALPS DMX800	EPSON9H.PRD	120×216	B&W
ALPS DMX800	EPSON9VH.PRD	240×216	B&W
LSX 1600	HPLSRL.PRD	75×75	B&W
LSX 1600	HPLSRM.PRD	100×100	B&W
LSX 1600	HPLSRH.PRD	150×150	B&W
LSX 1600	HPLSRVH.PRD	300×300	B&W
AMT			
Accel, Office Printer	AMTVL.PRD	60×60	B&W
Accel, Office Printer	AMTL.PRD	120×60	B&W
Accel, Office Printer	AMTM.PRD	120×120	B&W
Accel, Office Printer	AMTH.PRD	240×120	B&W
Accel, Office Printer	AMTVH.PRD	240×240	B&W
Accel, Office Printer	AMTVVH.PRD	480×240	B&W
Accel, Office Printer	AMTCVL.PRD	60×60	COL
Accel, Office Printer	AMTCL.PRD	120×60	COL
Accel, Office Printer	AMTCM.PRD	120×120	COL
Accel, Office Printer	AMTCH.PRD	240×120	COL
Accel, Office Printer	AMTCVH.PRD	240×240	COL
Accel, Office Printer	AMTCVVH.PRD	480×240	COL
TracJet	HPLSRL.PRD	75×75	B&W
TracJet	HPLSRM.PRD	100×100	B&W
TracJet	HPLSRH.PRD	150×150	B&W
TracJet	HPLSRVH.PRD	300×300	B&W
Anadex			
DP Series	ANDXDPL.PRD	72×72	B&W
DP Series	ANDXDPH.PRD	144×144	B&W
WP Series	ANDXWPL.PRD	72×72	B&W
WP Series	ANDXWPH.PRD	144×144	B&W
WP Series	ANDXWPCL.PRD	72×72	COL
WP Series	ANDXWPCH.PRD	144×144	COL
Anatex Data Systems			
ADS 2000	EPSON9L.PRD	60×72	B&W
ADS 2000	EPSON9M.PRD	120×72	B&W
ADS 2000	EPSON9H.PRD	120×216	B&W
ADS 2000	EPSON9VH.PRD	240×216	B&W
Apple			
Imagewriter II	APPLECL.PRD	60×72	COL
Imagewriter II	APPLECM.PRD	120×72	COL

Manufacturer/printer model	Definition	Resolution	B&W/CL
Imagewriter II	APPLEL.PRD	60 × 72	B&W
Imagewriter II	APPLEM.PRD	120 × 72	B&W
Laserwriter, IIf, IIg, Personal	PS.PRD	300 × 300	B&W
AT&T			
Model 475	CITOHVL.PRD	80 × 72	B&W
Model 475	CITOHL.PRD	96 × 72	B&W
Model 475	CITOHM.PRD	136 × 72	B&W
Model 475	CITOHH.PRD	160 × 72	B&W
Model 475	CITOHVH.PRD	160 × 144	B&W
Model 570	EPSON9L.PRD	60 × 72	B&W
Model 570	EPSON9M.PRD	120 × 72	B&W
Model 570	EPSON9H.PRD	120 × 216	B&W
Model 570	EPSON9VH.PRD	240 × 216	B&W
Model 583	EPSON2L.PRD	60 × 60	B&W
Model 583	EPSON2M.PRD	120 × 60	B&W
Model 583	EPSON2H.PRD	180 × 180	B&W
Axonix			
LiteWrite, MilWrite	EPSON9L.PRD	60 × 72	B&W
LiteWrite, MilWrite	EPSON9M.PRD	120 × 72	B&W
LiteWrite, MilWrite	EPSON9H.PRD	120 × 216	B&W
LiteWrite, MilWrite	EPSON9VH.PRD	240 × 216	B&W
Bezier			
BP4040	PS.PRD	300 × 300	B&W
Blue Chip			
M 200	EPSON9L.PRD	60 × 72	B&W
M 200	EPSON9M.PRD	120 × 72	B&W
M 200	EPSON9H.PRD	120 × 216	B&W
M 200	EPSON9VH.PRD	240 × 216	B&W
Brother			
1824L, 2024L	BRO24H.PRD	180 × 180	B&W
1550, 1809, HL-8e	BRO9L.PRD	60 × 72	B&W
1550, 1809, HL-8e	BRO9M.PRD	120 × 72	B&W
1550, 1809, HL-8e	BRO9H.PRD	120 × 216	B&W
1550, 1809, HL-8e	BRO9VH.PRD	240 × 216	B&W
Twinriter 5 WP mode	BROTWNL.PRD	60 × 72	B&W
Twinriter 5 WP mode	BROTWNM.PRD	120 × 72	B&W
Twinriter 5 WP mode	BROTWNH.PRD	120 × 216	B&W
Twinriter 5 WP mode	BROTWNVH.PRD	240 × 216	B&W
M-4309A	EPSON9L.PRD	60 × 72	B&W
M-4309A	EPSON9M.PRD	120 × 72	B&W
M-4309A	EPSON9H.PRD	120 × 216	B&W
M-4309A	EPSON9VH.PRD	240 × 216	B&W
HL-8V, -10V, -4Ve	HPLSRL.PRD	75 × 75	B&W
HL-8V, -10V, -4Ve	HPLSRM.PRD	100 × 100	B&W

Manufacturer/printer model	Definition	Resolution	B&W/CL
HL-8V, -10V, -4Ve	HPLSRH.PRD	150 × 150	B&W
HL-8V, -10V, -4Ve	HPLSRVH.PRD	300 × 300	B&W
HL-4PS, HL-8PS	PS.PRD	300 × 300	B&W
HT-500PS	PSC.PRD	300 × 300	COL
Bull HN Information Systems			
Compuprint 970	EPSON9L.PRD	60 × 72	B&W
Compuprint 970	EPSON9M.PRD	120 × 72	B&W
Compuprint 970	EPSON9H.PRD	120 × 216	B&W
Compuprint 970	EPSON9VH.PRD	240 × 216	B&W
Camintonn			
TurboLaser PS-Plus 3	PS.PRD	300 × 300	B&W
CAL-ABCO			
Legend 1385, CP-VII	EPSON9L.PRD	60 × 72	B&W
Legend 1385, CP-VII	EPSON9M.PRD	120 × 72	B&W
Legend 1385, CP-VII	EPSON9H.PRD	120 × 216	B&W
Legend 1385, CP-VII	EPSON9VH.PRD	240 × 216	B&W
CalComp			
ColorMaster Plus	PSC.PRD	300 × 300	COL
Canon			
BJ 130 Inkjet	CANONBJH.PRD	180 × 180	B&W
BJ 130 Inkjet	CANONBJV.PRD	360 × 360	B&W
LBP-8	CANONLL.PRD	75 × 75	B&W
LBP-8	CANONLM.PRD	100 × 100	B&W
LBP-8	CANONLH.PRD	150 × 150	B&W
LBP-8	CANONLVH.PRD	300 × 300	B&W
PW-1156A	EPSON9L.PRD	60 × 72	B&W
PW-1156A	EPSON9M.PRD	120 × 72	B&W
PW-1156A	EPSON9H.PRD	120 × 216	B&W
PW-1156A	EPSON9VH.PRD	240 × 216	B&W
BJ-800, BJ-830, BJ-20	EPSON2L.PRD	60 × 60	B&W
BJ-800, BJ-830, BJ-20	EPSON2M.PRD	120 × 60	B&W
BJ-800, BJ-830, BJ-20	EPSON2H.PRD	180 × 180	B&W
BJC-800, BJC-830	EPSON2CH.PRD	180 × 180	COL
BJC-800, BJC-830	EPSON2CV.PRD	360 × 360	COL
PJ1080A Inkjet	CANONPJ.PRD	84 × 84	COL
Centronics			
All Models	CENTRONL.PRD	60 × 60	B&W
CIE			
CI-250, CI-500	EPSON9L.PRD	60 × 72	B&W
CI-250, CI-500	EPSON9M.PRD	120 × 72	B&W
CI-250, CI-500	EPSON9H.PRD	120 × 216	B&W
CI-250, CI-500	EPSON9VH.PRD	240 × 216	B&W
Citizen			
MSP-10/25, 200GX	CITZN9L.PRD	60 × 72	B&W

411

Manufacturer/printer model	Definition	Resolution	B&W/CL
MSP-10/25, 200GX	CITZN9M.PRD	120 × 72	B&W
MSP-10/25, 200GX	CITZN9H.PRD	120 × 216	B&W
MSP-10/25, 200GX	CITZN9VH.PRD	240 × 216	B&W
MSP-10/25, 200GX	CITZN9CL.PRD	60 × 72	COL
MSP-10/25, 200GX	CITZN9CM.PRD	120 × 72	COL
MSP-10/25, 200GX	CITZN9CH.PRD	120 × 216	COL
MSP-10/25, 200GX	CITZN9CV.PRD	240 × 216	COL
GSX-140/130/145/240, PN48	CITZN24L.PRD	60 × 60	B&W
GSX-140/130/145/240, PN48	CITZN24M.PRD	120 × 60	B&W
GSX-140/130/145/240, PN48	CITZN24H.PRD	180 × 180	B&W
GSX-140/130/145/240, PN48	CITZN24V.PRD	360 × 360	B&W
GSX-140/130/145/240, PN48	CITZ24CH.PRD	180 × 180	COL
GSX-140/130/145/240, PN48	CITZ24CV.PRD	360 × 360	COL
Compaq			
PageMarq 15/20	HPLSRL.PRD	75 × 75	B&W
PageMarq 15/20	HPLSRM.PRD	100 × 100	B&W
PageMarq 15/20	HPLSRH.PRD	150 × 150	B&W
PageMarq 15/20	HPLSRVH.PRD	300 × 300	B&W
C.Itoh			
8510, 8600, Prowriter	CITOHVL.PRD	80 × 72	B&W
8510, 8600, Prowriter	CITOHL.PRD	96 × 72	B&W
8510, 8600, Prowriter	CITOHM.PRD	136 × 72	B&W
8510, 8600, Prowriter	CITOHH.PRD	160 × 72	B&W
8510, 8600, Prowriter	CITOHVH.PRD	160 × 144	B&W
C-310, 5000	EPSON9L.PRD	60 × 72	B&W
C-310, 5000	EPSON9M.PRD	120 × 72	B&W
C-310, 5000	EPSON9H.PRD	120 × 216	B&W
C-310, 5000	EPSON9VH.PRD	240 × 216	B&W
C-610, C-610II, Prowriter	EPSON2L.PRD	60 × 60	B&W
C-610, C-610II, Prowriter	EPSON2M.PRD	120 × 60	B&W
C-610, C-610II, Prowriter	EPSON2H.PRD	180 × 180	B&W
ProWriter CI-4/CI-8/CI-8e	HPLSRL.PRD	75 × 75	B&W
ProWriter CI-4/CI-8/CI-8e	HPLSRM.PRD	100 × 100	B&W
ProWriter CI-4/CI-8/CI-8e	HPLSRH.PRD	150 × 150	B&W
ProWriter CI-4/CI-8/CI-8e	HPLSRVH.PRD	300 × 300	B&W
Dataproducts			
8050/8070	DATAPM.PRD	168 × 84	B&W
8050/8070	DATAPCM.PRD	168 × 84	COL
8052C	IBMCLRL.PRD	60 × 72	B&W
8052C	IBMCLRM.PRD	120 × 72	B&W
LX-455	EPSON9L.PRD	60 × 72	B&W
LX-455	EPSON9M.PRD	120 × 72	B&W
LX-455	EPSON9H.PRD	120 × 216	B&W
LX-455	EPSON9VH.PRD	240 × 216	B&W

Manufacturer/printer model	Definition	Resolution	B&W/CL
LZR 1555/1560	HPLSRL.PRD	75×75	B&W
LZR 1555/1560	HPLSRM.PRD	100×100	B&W
LZR 1555/1560	HPLSRH.PRD	150×150	B&W
LZR 1555/1560	HPLSRVH.PRD	300×300	B&W
LZR-960	PS.PRD	300×300	B&W
Datasouth			
All Models	DATASL.PRD	72×72	B&W
All Models	DATASH.PRD	144×144	B&W
XL-300	EPSON9L.PRD	60×72	B&W
XL-300	EPSON9M.PRD	120×72	B&W
XL-300	EPSON9H.PRD	120×216	B&W
XL-300	EPSON9VH.PRD	240×216	B&W
DEC			
LA50, LA100, LN03, DECwriter	DECLAL.PRD	144×72	B&W
LA50, LA100, LN03, DECwriter	DECLAH.PRD	180×72	B&W
LA75+, LA424	IBMGRL.PRD	60×72	B&W
LA75+, LA424	IBMGRM.PRD	120×72	B&W
LA75+, LA424	IBMGRH.PRD	120×216	B&W
LA75+, LA424	IBMGRVH.PRD	240×216	B&W
multiJET 2000	HPLSRL.PRD	75×75	B&W
multiJET 2000	HPLSRM.PRD	100×100	B&W
multiJET 2000	HPLSRH.PRD	150×150	B&W
multiJET 2000	HPLSRVH.PRD	300×300	B&W
DECLaser 1150/2150/2250/3250	PS.PRD	300×300	B&W
Desktop			
Laser Beam	HPLSRL.PRD	75×75	B&W
Laser Beam	HPLSRM.PRD	100×100	B&W
Laser Beam	HPLSRH.PRD	150×150	B&W
Laser Beam	HPLSRVH.PRD	300×300	B&W
Diablo			
S32	DIABLSL.PRD	70×70	B&W
C-150 Inkjet	DIABLCCM.PRD	120×120	COL
P Series, 34LQ	EPSON9L.PRD	60×72	B&W
P Series, 34LQ	EPSON9M.PRD	120×72	B&W
P Series, 34LQ	EPSON9H.PRD	120×216	B&W
P Series, 34LQ	EPSON9VH.PRD	240×216	B&W
Diconix			
150	EPSON9L.PRD	60×72	B&W
150	EPSON9M.PRD	120×72	B&W
150	EPSON9H.PRD	120×216	B&W
150	EPSON9VH.PRD	240×216	B&W
Dynax-Fortis			
DM20, DH45	BROTWNL.PRD	60×72	B&W
DM20, DH45	BROTWNM.PRD	120×72	B&W

Manufacturer/printer model	Definition	Resolution	B&W/CL
DM20, DH45	BROTWNH.PRD	120×216	B&W
DM20, DH45	BROTWNVH.PRD	240×216	B&W
Epson			
LQ, SQ, or Action Printer Models	EPSON2L.PRD	60×60	B&W
LQ, SQ, or Action Printer Models	EPSON2M.PRD	120×60	B&W
LQ, SQ, or Action Printer Models	EPSON2H.PRD	180×180	B&W
LQ, SQ, or Action Printer Models	EPSON2VH.PRD	360×360	B&W
LQ, SQ, or Action Printer Models	EPSON2CH.PRD	180×180	COL
LQ, SQ, or Action Printer Models	EPSON2CV.PRD	360×360	COL
EPL-6000/7000/7500	EPSON6L.PRD	75×75	B&W
EPL-6000/7000/7500	EPSON6M.PRD	100×100	B&W
EPL-6000/7000/7500	EPSON6H.PRD	150×150	B&W
EPL-6000/7000/7500	EPSON6VH.PRD	300×300	B&W
MX, FX, RX, JX, LX, and DFX	EPSON9L.PRD	60×72	B&W
MX, FX, RX, JX, LX, and DFX	EPSON9M.PRD	120×72	B&W
FX, RX, JX, LX, and DFX	EPSON9H.PRD	120×216	B&W
FX, RX, JX, LX, and DFX	EPSON9VH.PRD	240×216	B&W
MX, FX, RX, JX, LX, and DFX	EPSON9CL.PRD	60×72	COL
MX, FX, RX, JX, LX, and DFX	EPSON9CM.PRD	120×72	COL
FX, RX, JX, LX, and DFX	EPSON9CH.PRD	120×216	COL
FX, RX, JX, LX, and DFX	EPSON9CV.PRD	240×216	COL
GQ 3500 Native Mode	EPSONGQH.PRD	300×300	B&W
ActionLaser II/EPL-8000	HPLSRL.PRD	75×75	B&W
ActionLaser II/EPL-8000	HPLSRM.PRD	100×100	B&W
ActionLaser II/EPL-8000	HPLSRH.PRD	150×150	B&W
ActionLaser II/EPL-8000	HPLSRVH.PRD	300×300	B&W
Everex			
Laser Script LX	HPLSRL.PRD	75×75	B&W
Laser Script LX	HPLSRM.PRD	100×100	B&W
Laser Script LX	HPLSRH.PRD	150×150	B&W
Laser Script LX	HPLSRVH.PRD	300×300	B&W
Laser Script LX	PS.PRD	300×300	B&W
Facit			
4528	FAC4528L.PRD	60×60	B&W
4542, 4544	FAC4542L.PRD	70×70	B&W
B2400	EPSON2L.PRD	60×60	B&W
B2400	EPSON2M.PRD	120×60	B&W
B2400	EPSON2H.PRD	180×180	B&W
B3550C	EPSON9L.PRD	60×72	B&W
B3550C	EPSON9M.PRD	120×72	B&W
B3550C	EPSON9H.PRD	120×216	B&W
B3550C	EPSON9VH.PRD	240×216	B&W
Fortis			
DP600S	HPLSRL.PRD	75×75	B&W

414

Manufacturer/printer model	Definition	Resolution	B&W/CL
DP600S	HPLSRM.PRD	100×100	B&W
DP600S	HPLSRH.PRD	150×150	B&W
DP600S	HPLSRVH.PRD	300×300	B&W
DH45	BROTWNL.PRD	60×72	B&W
DH45	BROTWNM.PRD	120×72	B&W
DH45	BROTWNH.PRD	120×216	B&W
DH45	BROTWNVH.PRD	240×216	B&W
DM2210, DM2215	EPSON9L.PRD	60×72	B&W
DM2210, DM2215	EPSON9M.PRD	120×72	B&W
DM2210, DM2215	EPSON9H.PRD	120×216	B&W
DM2210, DM2215	EPSON9VH.PRD	240×216	B&W
DQ 4110, 4210, 4215	EPSON2L.PRD	60×60	B&W
DQ 4110, 4210, 4215	EPSON2M.PRD	120×60	B&W
DQ 4110, 4210, 4215	EPSON2H.PRD	180×180	B&W
DP600P	PS.PRD	300×300	B&W
Fujitsu			
24C	FUJI24CH.PRD	180×180	B&W
24C	FUJI24CV.PRD	360×180	B&W
24C	FUJ24CCH.PRD	180×180	COL
24C	FUJ24CCV.PRD	360×180	COL
24D	FUJI24DL.PRD	60×60	B&W
24D	FUJI24DM.PRD	90×90	B&W
24D	FUJI24DH.PRD	180×180	B&W
DL 1200/3600/4400/4800/5800	EPSON2L.PRD	60×60	B&W
DL 1200/3600/4400/4800/5800	EPSON2M.PRD	120×60	B&W
DL 1200/3600/4400/4800/5800	EPSON2H.PRD	180×180	B&W
RX 7200/7300E, PrintPartner 10	HPLSRL.PRD	75×75	B&W
RX 7200/7300E, PrintPartner 10	HPLSRM.PRD	100×100	B&W
RX 7200/7300E, PrintPartner 10	HPLSRH.PRD	150×150	B&W
RX 7200/7300E, PrintPartner 10	HPLSRVH.PRD	300×300	B&W
RX 7100PS	PS.PRD	300×300	B&W
GCC			
BLP II(S)	PS.PRD	300×300	B&W
GENICOM			
3180–3404 Series	GENICOML.PRD	72×72	B&W
3410, 3820, 3840	EPSON9L.PRD	60×72	B&W
3410, 3820, 3840	EPSON9M.PRD	120×72	B&W
3410, 3820, 3840	EPSON9H.PRD	120×216	B&W
3410, 3820, 3840	EPSON9VH.PRD	240×216	B&W
1040	EPSON2L.PRD	60×60	B&W
1040	EPSON2M.PRD	120×60	B&W
1040	EPSON2H.PRD	180×180	B&W
4440 XT	IBMGRL.PRD	60×72	B&W
4440 XT	IBMGRM.PRD	120×72	B&W

Manufacturer/printer model	Definition	Resolution	B&W/CL
4440 XT	IBMGRH.PRD	120×216	B&W
4440 XT	IBMGRVH.PRD	240×216	B&W
7170	HPLSRL.PRD	75×75	B&W
7170	HPLSRM.PRD	100×100	B&W
7170	HPLSRH.PRD	150×150	B&W
7170	HPLSRVH.PRD	300×300	B&W
Gorilla			
Banana	GORILLAM.PRD	60×63	B&W
Hermes			
Printer I	EPSON9L.PRD	60×72	B&W
Printer I	EPSON9M.PRD	120×72	B&W
Hewlett-Packard			
7600 Model 355, DesignJet	HP7600M.PRD	102×102	B&W
7600 Model 355, DesignJet	HP7600H.PRD	406×406	B&W
7600 Model 355	HP7600CM.PRD	102×102	COL
7600 Model 355	HP7600CH.PRD	406×406	COL
LaserJet/DeskJet—All Models	HPLSRL.PRD	75×75	B&W
LaserJet/DeskJet—All Models	HPLSRM.PRD	100×100	B&W
LaserJet/DeskJet—All Models	HPLSRH.PRD	150×150	B&W
LaserJet/DeskJet—All Models	HPLSRVH.PRD	300×300	B&W
LaserJet 4	HPLSRVVH.PRD	600×600	B&W
DeskJet 500C/550C, PaintJet XL300	HPDSKCL.PRD	75×75	COL
DeskJet 500C/550C, PaintJet XL300	HPDSKCM.PRD	100×100	COL
DeskJet 500C/550C, PaintJet XL300	HPDSKCH.PRD	150×150	COL
DeskJet 500C/550C, PaintJet XL300	HPDSKCVH.PRD	300×300	COL
PaintJet—All Models	HPPNTM.PRD	90×90	B&W
PaintJet—All Models	HPPNTH.PRD	180×180	B&W
PaintJet—All Models	HPPNTCM.PRD	90×90	COL
PaintJet—All Models	HPPNTCMT.PRD	90×90	COL
PaintJet—All Models	HPPNTCH.PRD	180×180	COL
QuietJet	HPQJTEL.PRD	96×96	B&W
QuietJet	HPQJTEM.PRD	192×96	B&W
QuietJet	HPQJTEH.PRD	192×192	B&W
QuietJet	HPQJTL.PRD	96×96	B&W
QuietJet	HPQJTM.PRD	192×96	B&W
QuietJet	HPQJTH.PRD	192×192	B&W
ThinkJet	HPTNKEM.PRD	192×96	B&W
ThinkJet	HPTNKM.PRD	192×96	B&W
LaserJet 4 (with Postscript)	PS.PRD	300×300	B&W
Hyundai			
HDP-910/920	EPSON9L.PRD	60×72	B&W
HDP-910/920	EPSON9M.PRD	120×72	B&W
IBM			
3852-1 Color Inkjet	IBM381CM.PRD	84×84	COL

416

Manufacturer/printer model	Definition	Resolution	B&W/CL
3852-2 Color Inkjet	IBM382CM.PRD	100×96	COL
3852 Color Inkjet	IBM38M.PRD	84×63	B&W
Color Printer	IBMCLRL.PRD	60×72	B&W
Color Printer	IBMCLRM.PRD	120×72	B&W
Graphics, Proprinter, 2380 series	IBMGRL.PRD	60×72	B&W
Graphics, Proprinter, 2380 series	IBMGRM.PRD	120×72	B&W
Graphics, Proprinter, 2380 series	IBMGRH.PRD	120×216	B&W
Graphics, Proprinter, 2380 series	IBMGRVH.PRD	120×216	B&W
Personal Printer 2390, ExecJet	EPSON2L.PRD	60×60	B&W
Personal Printer 2390, ExecJet	EPSON2M.PRD	120×60	B&W
Personal Printer 2390, ExecJet	EPSON2H.PRD	180×180	B&W
LaserPrinter 6p/10p	HPLSRL.PRD	75×75	B&W
LaserPrinter 6p/10p	HPLSRM.PRD	100×100	B&W
LaserPrinter 6p/10p	HPLSRH.PRD	150×150	B&W
LaserPrinter 6p/10p	HPLSRVH.PRD	300×300	B&W
IDS			
440	IDS440L.PRD	64×64	B&W
Prism, 560, 480, P132, P80	IDSM.PRD	84×84	B&W
Prism, 560, 480, P132, P80	IDSCM.PRD	84×84	COL
Integrex			
Colour Jet 132	INTE132L.PRD	60×60	B&W
JDL			
750	JDL750L.PRD	60×60	B&W
750	JDL750M.PRD	90×90	B&W
750	JDL750H.PRD	180×180	B&W
750	JDL750CL.PRD	60×60	COL
750	JDL750CM.PRD	90×90	COL
750	JDL750CH.PRD	180×180	COL
Kentek			
K30D	HPLSRL.PRD	75×75	B&W
K30D	HPLSRM.PRD	100×100	B&W
K30D	HPLSRH.PRD	150×150	B&W
K30D	HPLSRVH.PRD	300×300	B&W
Kodak Diconix			
150	EPSON9L.PRD	60×72	B&W
150	EPSON9M.PRD	120×72	B&W
150	EPSON9H.PRD	120×216	B&W
150	EPSON9VH.PRD	240×216	B&W
Ektaplus 7008	HPLSRL.PRD	75×75	B&W
Ektaplus 7008	HPLSRM.PRD	100×100	B&W
Ektaplus 7008	HPLSRH.PRD	150×150	B&W
Ektaplus 7008	HPLSRVH.PRD	300×300	B&W
Color 4	HPPNTM.PRD	90×90	B&W
Color 4	HPPNTH.PRD	180×180	B&W

Manufacturer/printer model	Definition	Resolution	B&W/CL
Color 4	HPPNTCM.PRD	90×90	COL
Color 4	HPPNTCMT.PRD	90×90	COL
Color 4	HPPNTCH.PRD	180×180	COL
Kyocera			
Ecosys a-Si FS-1500A	HPLSRL.PRD	75×75	B&W
Ecosys a-Si FS-1500A	HPLSRM.PRD	100×100	B&W
Ecosys a-Si FS-1500A	HPLSRH.PRD	150×150	B&W
Ecosys a-Si FS-1500A	HPLSRVH.PRD	300×300	B&W
Laser Computer			
190E, 240	EPSON9L.PRD	60×72	B&W
190E, 240	EPSON9M.PRD	120×72	B&W
190E, 240	EPSON9H.PRD	120×216	B&W
190E, 240	EPSON9VH.PRD	240×216	B&W
Laser Master			
Unity 1000/1200XL, WinPrinter 800	HPLSRL.PRD	75×75	B&W
Unity 1000/1200XL, WinPrinter 800	HPLSRM.PRD	100×100	B&W
Unity 1000/1200XL, WinPrinter 800	HPLSRH.PRD	150×150	B&W
Unity 1000/1200XL, WinPrinter 800	HPLSRVH.PRD	300×300	B&W
TrueTech 800/1000	PS.PRD	300×300	B&W
Malibu			
All Models	MALIBUL.PRD	60×60	B&W
Mannesmann Tally			
160	MAN160L.PRD	50×64	B&W
160	MAN160M.PRD	100×64	B&W
160	MAN160H.PRD	133×64	B&W
420, 440	MAN420L.PRD	60×60	B&W
Spirit 80, 81	MANSPRTL.PRD	80×72	B&W
Spirit 80, 81	MANSPRTM.PRD	160×72	B&W
Spirit 80, 81	MANSPRTH.PRD	160×216	B&W
905, 908, 910, 661, 735	HPLSRL.PRD	75×75	B&W
905, 908, 910, 661, 735	HPLSRM.PRD	100×100	B&W
905, 908, 910, 661, 735	HPLSRH.PRD	150×150	B&W
905, 908, 910, 661, 735	HPLSRVH.PRD	300×300	B&W
MT150/9,MT151/9	EPSON9L.PRD	60×72	B&W
MT150/9,MT151/9	EPSON9M.PRD	120×72	B&W
MT150/9,MT151/9	EPSON9H.PRD	120×216	B&W
MT150/9,MT151/9	EPSON9VH.PRD	240×216	B&W
MT150/24,151/24, 82	EPSON2L.PRD	60×60	B&W
MT150/24,151/24, 82	EPSON2M.PRD	120×60	B&W
MT150/24,151/24, 82	EPSON2H.PRD	180×180	B&W
All Postscript models	PS.PRD	300×300	B&W
Microtek			
TrueLaser	PS.PRD	300×300	B&W

418

Manufacturer/printer model	Definition	Resolution	B&W/CL
Mitsubishi			
DiamondColor Print 300PS	PSC.PRD	300×300	COL
CHC-S446i ColorStream/DS	PSC.PRD	300×300	COL
MPI			
All Models	MPIL.PRD	60×72	B&W
All Models	MPIM.PRD	120×72	B&W
All Models	MPIH.PRD	120×144	B&W
NEC			
P2200/3200/3300/5300/9300 models	NEC24L.PRD	60×60	B&W
P2200/3200/3300/5300/9300 models	NEC24M.PRD	120×60	B&W
P2200/3200/3300/5300/9300 models	NEC24H.PRD	180×180	B&W
P2200/3200/3300/5300/9300 models	NEC24VH.PRD	360×360	B&W
P2200, P5300, 24 pin models	NEC24CH.PRD	180×180	COL
P2200, P5300, 24 pin models	NEC24CVH.PRD	360×360	COL
8023	NEC8023L.PRD	72×72	B&W
8027A	NEC8027L.PRD	80×72	B&W
P2, P3, CP2, CP3, 9 pin models	NEC9L.PRD	60×60	B&W
P2, P3, CP2, CP3, 9 pin models	NEC9M.PRD	120×60	B&W
P2, P3, CP2, CP3, 9 pin models	NEC9H.PRD	120×120	B&W
P2, P3, CP2, CP3, 9 pin models	NEC9VH.PRD	240×240	B&W
P2, P3, CP2, CP3, 9 pin models	NEC9CL.PRD	60×60	COL
P2, P3, CP2, CP3, 9 pin models	NEC9CM.PRD	120×60	COL
P2, P3, CP2, CP3, 9 pin models	NEC9CH.PRD	120×120	COL
P2, P3, CP2, CP3, 9 pin models	NEC9CVH.PRD	240×240	COL
LC 890XL, SilentWriter 95	HPLSRL.PRD	75×75	B&W
LC 890XL, SilentWriter 95	HPLSRM.PRD	100×100	B&W
LC 890XL, SilentWriter 95	HPLSRH.PRD	150×150	B&W
LC 890XL, SilentWriter 95	HPLSRVH.PRD	300×300	B&W
All Postscript models	PS.PRD	300×300	B&W
NewGen			
TurboPS/400p/630En/660/840e/880	HPLSRL.PRD	75×75	B&W
TurboPS/400p/630En/660/840e/880	HPLSRM.PRD	100×100	B&W
TurboPS/400p/630En/660/840e/880	HPLSRH.PRD	150×150	B&W
TurboPS/400p/630En/660/840e/880	HPLSRVH.PRD	300×300	B&W
TurboPS/1200T	HPLSRL.PRD	75×75	B&W
TurboPS/1200T	HPLSRM.PRD	100×100	B&W
TurboPS/1200T	HPLSRH.PRD	150×150	B&W
TurboPS/1200T	HPLSRVH.PRD	300×300	B&W
North Atlantic Quantex			
All Models	NORTHL.PRD	72×72	B&W
All Models	NORTHM.PRD	120×72	B&W
All Models	NORTHH.PRD	144×72	B&W
Okidata			
Okimate 20	OKI20L.PRD	60×72	COL

419

Manufacturer/printer model	Definition	Resolution	B&W/CL
2410, 2350	OKI2410L.PRD	72×72	B&W
2410, 2350, 24 pin models	OKI24L.PRD	60×60	B&W
2410, 2350, 24 pin models	OKI24M.PRD	120×60	B&W
2410, 2350, 24 pin models	OKI24H.PRD	180×180	B&W
2410, 2350, 24 pin models	OKI24VH.PRD	363×363	B&W
2410, 2350, 24 pin models	OKI24CH.PRD	180×180	COL
ML-92,ML-93,ML-82,ML-83 (w/o P&P)	OKI9L.PRD	72×72	B&W
ML-92,ML-93,ML-82,ML-83 (w/o P&P)	OKI9M.PRD	144×72	B&W
ML-92,ML-93,ML-82,ML-83 (w/o P&P)	OKI9H.PRD	144×144	B&W
Above models (w/ Plug & Play)	EPSON9L.PRD	60×72	B&W
Above models (w/ Plug & Play)	EPSON9M.PRD	120×72	B&W
Above models (w/ Plug & Play)	EPSON9H.PRD	120×216	B&W
Above models (w/ Plug & Play)	EPSON9VH.PRD	240×216	B&W
Laserline (HP)	HPLSRL.PRD	75×75	B&W
Laserline (HP)	HPLSRM.PRD	100×100	B&W
Laserline (HP)	HPLSRH.PRD	150×150	B&W
Laserline (HP)	HPLSRVH.PRD	300×300	B&W
Pacemark 3410	EPSON9L.PRD	60×72	B&W
Pacemark 3410	EPSON9M.PRD	120×72	B&W
Pacemark 3410	EPSON9H.PRD	120×216	B&W
Pacemark 3410	EPSON9VH.PRD	240×216	B&W
Microline 184 Turbo	IBMGRL.PRD	60×72	B&W
Microline 184 Turbo	IBMGRM.PRD	120×72	B&W
Microline 184 Turbo	IBMGRH.PRD	120×216	B&W
Microline 184 Turbo	IBMGRVH.PRD	240×216	B&W
OL 810 LED	HPLSRL.PRD	75×75	B&W
OL 810 LED	HPLSRM.PRD	100×100	B&W
OL 810 LED	HPLSRH.PRD	150×150	B&W
OL 810 LED	HPLSRVH.PRD	300×300	B&W
OL 830	PS.PRD	300×300	B&W
Olympia			
NP	EPSON9L.PRD	60×72	B&W
NP	EPSON9M.PRD	120×72	B&W
NP	EPSON9H.PRD	120×216	B&W
NP	EPSON9VH.PRD	240×216	B&W
Output Technology			
All models	EPSON9L.PRD	60×72	B&W
All models	EPSON9M.PRD	120×72	B&W
All models	EPSON9H.PRD	120×216	B&W
All models	EPSON9VH.PRD	240×216	B&W
LaserMatrix 1000 Model 5	HPLSRL.PRD	75×75	B&W
LaserMatrix 1000 Model 5	HPLSRM.PRD	100×100	B&W
LaserMatrix 1000 Model 5	HPLSRH.PRD	150×150	B&W
LaserMatrix 1000 Model 5	HPLSRVH.PRD	300×300	B&W

Manufacturer/printer model	Definition	Resolution	B&W/CL
PMC			
DMP-85	NEC8027L.PRD	80 × 72	B&W
Panasonic			
All Models (9 pin printers)	PANASL.PRD	60 × 72	B&W
All Models (9 pin printers)	PANASM.PRD	120 × 72	B&W
All Models (9 pin printers)	PANASH.PRD	120 × 216	B&W
All Models (9 pin printers)	PANASVH.PRD	240 × 216	B&W
All Models (24 pin printers)	EPSON2L.PRD	60 × 60	B&W
All Models (24 pin printers)	EPSON2M.PRD	120 × 60	B&W
All Models (24 pin printers)	EPSON2H.PRD	180 × 180	B&W
KX-P4410/4430	HPLSRL.PRD	75 × 75	B&W
KX-P4410/4430	HPLSRM.PRD	100 × 100	B&W
KX-P4410/4430	HPLSRH.PRD	150 × 150	B&W
KX-P4410/4430	HPLSRVH.PRD	300 × 300	B&W
All Postscript models	PS.PRD	300 × 300	B&W
Postscript			
All models	PS.PRD	300 × 300	B&W
Printronix			
L2324	HPLSRL.PRD	75 × 75	B&W
L2324	HPLSRM.PRD	100 × 100	B&W
L2324	HPLSRH.PRD	150 × 150	B&W
L2324	HPLSRVH.PRD	300 × 300	B&W
QMS			
All Postscript models	PS.PRD	300 × 300	B&W
Quadram			
Quadjet	QUADRL.PRD	80 × 80	B&W
Quadjet	QUADRCL.PRD	70 × 72	COL
Qume			
All Postscript models	PS.PRD	300 × 300	B&W
Raster Devices			
All Postscript models	PS.PRD	300 × 300	B&W
Ricoh			
All Postscript models	PS.PRD	300 × 300	B&W
Riteman			
All models	EPSON9L.PRD	60 × 72	B&W
All models	EPSON9M.PRD	120 × 72	B&W
All models	EPSON9H.PRD	120 × 216	B&W
All models	EPSON9VH.PRD	240 × 216	B&W
Royal			
CJP 450	HPLSRL.PRD	75 × 75	B&W
CJP 450	HPLSRM.PRD	100 × 100	B&W
CJP 450	HPLSRH.PRD	150 × 150	B&W
CJP 450	HPLSRVH.PRD	300 × 300	B&W

421

Manufacturer/printer model	Definition	Resolution	B&W/CL
Samsung			
Finale' 8000	HPLSRL.PRD	75×75	B&W
Finale' 8000	HPLSRM.PRD	100×100	B&W
Finale' 8000	HPLSRH.PRD	150×150	B&W
Finale' 8000	HPLSRVH.PRD	300×300	B&W
Seikosha			
GP-100A	SEIKOL.PRD	60×63	B&W
SP-180AI/1600AI/2400/2415, BP-5460	EPSON9L.PRD	60×72	B&W
SP-180AI/1600AI/2400/2415, BP-5460	EPSON9M.PRD	120×72	B&W
SP-180AI/1600AI/2400/2415, BP-5460	EPSON9H.PRD	120×216	B&W
SP-180AI/1600AI/2400/2415, BP-5460	EPSON9VH.PRD	240×216	B&W
SL-230AI, LT-20	EPSON2L.PRD	60×60	B&W
SL-230AI, LT-20	EPSON2M.PRD	120×60	B&W
SL-230AI, LT-20	EPSON2H.PRD	180×180	B&W
Sharp			
JX 720	SHARPCM.PRD	60×63	B&W
JX-9500H	HPLSRL.PRD	75×75	B&W
JX-9500H	HPLSRM.PRD	100×100	B&W
JX-9500H	HPLSRH.PRD	150×150	B&W
JX-9500H	HPLSRVH.PRD	300×300	B&W
JX-9500PS	PS.PRD	300×300	B&W
Siemens			
PT90, PT88S	EPSON9L.PRD	60×72	B&W
PT90, PT88S	EPSON9M.PRD	120×72	B&W
PT90, PT88S	EPSON9H.PRD	120×216	B&W
PT90, PT88S	EPSON9VH.PRD	240×216	B&W
Smith-Corona			
D-200, D-300	EPSON9L.PRD	60×72	B&W
D-200, D-300	EPSON9M.PRD	120×72	B&W
D-200, D-300	EPSON9H.PRD	120×216	B&W
D-200, D-300	EPSON9VH.PRD	240×216	B&W
Star Micronics			
Delta,Radix,Gemini,SD,SR,NX,XR	STAR9L.PRD	60×72	B&W
Delta,Radix,Gemini,SD,SR,NX,XR	STAR9M.PRD	120×72	B&W
NX, and XR Series	STAR9H.PRD	120×144	B&W
NX, and XR Series	STAR9VH.PRD	240×144	B&W
SB-10	STAR24H.PRD	180×240	B&W
NB24-15,XB24-10/15,SJ-48,NX-2430	EPSON2L.PRD	60×60	B&W
NB24-15,XB24-10/15,SJ-48,NX-2430	EPSON2M.PRD	120×60	B&W
NB24-15,XB24-10/15,SJ-48,NX-2430	EPSON2H.PRD	180×180	B&W
LaserPrinter 4	HPLSRL.PRD	75×75	B&W
LaserPrinter 4	HPLSRM.PRD	100×100	B&W
LaserPrinter 4	HPLSRH.PRD	150×150	B&W
LaserPrinter 4	HPLSRVH.PRD	300×300	B&W

Manufacturer/printer model	Definition	Resolution	B&W/CL
LaserPrinter 4 Star Script	PS.PRD	300 × 300	B&W
Synergystex			
CF1000	HPLSRL.PRD	75 × 75	B&W
CF1000	HPLSRM.PRD	100 × 100	B&W
CF1000	HPLSRH.PRD	150 × 150	B&W
CF1000	HPLSRVH.PRD	300 × 300	B&W
Talaris			
1590-T Printstation	HPLSRL.PRD	75 × 75	B&W
1590-T Printstation	HPLSRM.PRD	100 × 100	B&W
1590-T Printstation	HPLSRH.PRD	150 × 150	B&W
1590-T Printstation	HPLSRVH.PRD	300 × 300	B&W
Tandy (Radio Shack)			
2100 Series	TAN2100L.PRD	60 × 60	B&W
2100 Series	TAN2100H.PRD	180 × 180	B&W
DMP-430/440	TAN430M.PRD	120 × 144	B&W
CGP-220	TANCGPCL.PRD	70 × 72	COL
CGP-220	TANCGPL.PRD	80 × 80	B&W
Most Tandy Printers	TANDYL.PRD	60 × 72	B&W
Most Tandy Printers	TANDYM.PRD	60 × 144	B&W
IBM Emulation	TANIBML.PRD	60 × 72	B&W
IBM Emulation	TANIBMM.PRD	120 × 72	B&W
IBM Emulation	TANIBMH.PRD	120 × 216	B&W
IBM Emulation	TANIBMVH.PRD	240 × 216	B&W
DMP-310	EPSON9L.PRD	60 × 72	B&W
DMP-310	EPSON9M.PRD	120 × 72	B&W
DMP-310	EPSON9H.PRD	120 × 216	B&W
DMP-310	EPSON9VH.PRD	240 × 216	B&W
LP 950	HPLSRL.PRD	75 × 75	B&W
LP 950	HPLSRM.PRD	100 × 100	B&W
LP 950	HPLSRH.PRD	150 × 150	B&W
LP 950	HPLSRVH.PRD	300 × 300	B&W
Tektronix			
PhaserII PXe/IIsd/III PXi	PSC.PRD	300 × 300	COL
Texas Instruments			
855/857/865	TI855CL.PRD	60 × 72	COL
855/857/865	TI855CM.PRD	120 × 72	COL
855/857/865	TI855CH.PRD	120 × 144	COL
855/857/865	TI855CVH.PRD	120 × 144	COL
855/857/865	TI855L.PRD	60 × 72	B&W
855/857/865	TI855M.PRD	120 × 72	B&W
855/857/865	TI855H.PRD	120 × 144	B&W
855/857/865	TI855VH.PRD	144 × 144	B&W
850	EPSON9L.PRD	60 × 72	B&W
850	EPSON9M.PRD	120 × 72	B&W

Manufacturer/printer model	Definition	Resolution	B&W/CL
TI MicroLaser Turbo/XL Turbo	HPLSRL.PRD	75×75	B&W
TI MicroLaser Turbo/XL Turbo	HPLSRM.PRD	100×100	B&W
TI MicroLaser Turbo/XL Turbo	HPLSRH.PRD	150×150	B&W
TI MicroLaser Turbo/XL Turbo	HPLSRVH.PRD	300×300	B&W
Toshiba			
1350	TOSH1350.PRD	180×180	B&W
24 Pin Models	TOSH24CE.PRD	360×360	COL
24 Pin Models	TOSH24CH.PRD	180×180	COL
24 Pin Models	TOSH24CV.PRD	360×180	COL
24 Pin Models	TOSH24H.PRD	180×180	B&W
24 Pin Models	TOSH24VH.PRD	360×180	B&W
24 Pin Models	TOSH24EH.PRD	360×360	B&W
Express Writer 301/311	EPSON2L.PRD	60×60	B&W
Express Writer 301/311	EPSON2M.PRD	120×60	B&W
Express Writer 301/311	EPSON2H.PRD	180×180	B&W
PageLaser GX200/GSX400	HPLSRL.PRD	75×75	B&W
PageLaser GX200/GSX400	HPLSRM.PRD	100×100	B&W
PageLaser GX200/GSX400	HPLSRH.PRD	150×150	B&W
PageLaser GX200/GSX400	HPLSRVH.PRD	300×300	B&W
Unisys			
AP 1327/9 Mod5, 1371, 115, 37	EPSON9L.PRD	60×72	B&W
AP 1327/9 Mod5, 1371, 115, 37	EPSON9M.PRD	120×72	B&W
AP 1327/9 Mod5, 1371, 115, 37	EPSON9H.PRD	120×216	B&W
AP 1327/9 Mod5, 1371, 115, 37	EPSON9VH.PRD	240×216	B&W
AP 1234	EPSON2L.PRD	60×60	B&W
AP 1234	EPSON2M.PRD	120×60	B&W
AP 1234	EPSON2H.PRD	180×180	B&W
AP 92/94 Mod 37 (HP)	HPLSRL.PRD	75×75	B&W
AP 92/94 Mod 37 (HP)	HPLSRM.PRD	100×100	B&W
AP 92/94 Mod 37 (HP)	HPLSRH.PRD	150×150	B&W
AP 92/94 Mod 37 (HP)	HPLSRVH.PRD	300×300	B&W
AP 94 (Postscript)	PS.PRD	300×300	B&W
Xante			
Accel-a-Writer 8000	HPLSRL.PRD	75×75	B&W
Accel-a-Writer 8000	HPLSRM.PRD	100×100	B&W
Accel-a-Writer 8000	HPLSRH.PRD	150×150	B&W
Accel-a-Writer 8000	HPLSRVH.PRD	300×300	B&W
Xerox			
2700/4045	XER2700L.PRD	77×77	B&W
2700/4045	XER2700H.PRD	154×154	B&W
4020 Inkjet	XER4020C.PRD	120×120	COL

424

Hidden key alert: If you accidentally pass the desired driver, you can scroll backward through the driver list by pressing "SHIFT" and "p" simultaneously (capitol "P").

Select port

PRINTERS is designed to operate a printer on LPT1 through LPT3, or COM1 or COM2. Left click on "Select Port" (or press "r") to scroll through available ports. Remember to connect your printer to the appropriate port before proceeding. By default, LPT1 is the selected port.

Manual codes

Most printers provide a suite of printer-specific functions and features (e.g., bold print, underlining, double-width print, double-height print, and so on). While it would be virtually impossible for any diagnostic to test each of these functions for every available printer, PRINTERS provides a means for you to test your printer's functions manually using printer codes (also referred to as *Escape Sequences* or *Escape Codes*). The Manual Codes feature provides a printer technician with almost unlimited versatility in checking and verifying the most subtle features of a printer's operation.

Information alert: In order to enter an Escape Code, you will need the User's Manual for your particular printer. The diskette's documentation lists Escape Codes for several popular printer models, but to test printer-specific functions, you will need printer-specific documentation.

To enter an Escape Code, left click on "Manual Codes" (or press "C") in the CONFIGURE menu. A text entry window will appear below the CONFIGURE menu, as in figure 13-3. You might enter up to 50 characters per line. Pressing the <TAB> key or right clicking anywhere on the display will abort the text entry routine. If you make a mistake in typing, simply backspace past the error and correct the mistake.

Example 1: Setting the Panasonic KX-P1124 to Underline mode

With the text entry routine running, reset the printer (by turning it off and on), then enter the following text, then press <ENTER>:

```
This is a test of default text
```

Your printer should print this text string in its default font and pitch. Now let's set the printer to its underline mode using the code ESC - 1 (from the PRINTERS documentation). Type the three keystrokes and press <ENTER>. Keep in mind that when

```
                PRINTERS v.1.00  The Printer Test and Repair Utility
Configure Impact    Ink Jet   Laser/LED                    About...   Quit

    CONFIGURE Menu

    Select Printer....  EPSON2L.PRD
    Select Port.......  LPT 1
    Manual Codes      ▶
    Print Intensity...  FULL
    Help Mode.........  ON
    Tutor Mode........  ON



    Enter Your ESC Code or Text String:
    _____

    [ENTER] Send String   [TAB] Cancel

                Enter customized ESC codes and command sequences
            Copyright (c) 1995 Dynamic Learning Systems.  All rights reserved
```

■ **13-3** *Entering manual escape codes.*

you press the <ESC> key, a backspace arrow will appear in that space:

```
<ESC>-1
```

The printer should now be in underline mode, so type the following text and press <ENTER>

```
This is the printer's underline mode
```

The printer should produce this text underlined. To turn the underline mode off, enter the code ESC – 0 (from PRINTERS documentation). Type the three keystrokes and press <ENTER>. Keep in mind that when you press the <ESC> key, a backspace arrow will appear in that space:

```
<ESC>-0
```

Now type the following text and press <ENTER>:

```
The underline mode is off
```

The type should no longer be underlined. To leave the text entry routine, press <TAB> or right click anywhere on the display.

Example 2: Setting the Panasonic KX-P1124 to Letter Quality (LQ) mode

With the text entry routine running, reset the printer (by turning it off and on), then enter the following text, then press <ENTER>:

```
This is the printer's default text
```

Your printer should print this text string in its default font and pitch. Now let's set the printer to its LQ mode using the code ESC x 1 (from PRINTERS documentation). Type the three keystrokes and press <ENTER>. Keep in mind that when you press the <ESC> key, a backspace arrow will appear in that space:

```
<ESC>x1
```

The printer should now be in letter quality mode, so type the following text and press <ENTER>:

```
This is the printer's LQ mode
```

The printer should produce this text in higher quality than the default. To turn the LQ mode off, enter the code ESC x 0 (from PRINTERS documentation). Type the three keystrokes and press <ENTER>. Keep in mind that when you press the <ESC> key, a backspace arrow will appear in that space:

```
<ESC>x0
```

Now type the following text and press <ENTER>:

```
The letter quality mode is off
```

The type should be back in its draft form. To leave the text entry routine, press <TAB> or right click anywhere on the display.

Print intensity

The print intensity setting allows you to set the overall darkness and lightness (i.e., contrast) of your test images. There are two choices: half and full. By default, all images are printed at full intensity (maximum contrast). At half intensity, the shading is lightened to reduce the image's contrast. You can toggle between full and half intensity by clicking on "Print Intensity" (or pressing "I"). For low-resolution devices (impact printers at 75 dpi or less), full intensity will typically yield superior results. For medium-to-high resolution devices (ink jet and almost all EP printers), half intensity is often better. Of course, you might experiment to find the best settings for your particular printer.

Help mode

PRINTERS is designed with two on-line documentation sources that are intended to provide instruction and guidance before and after a test is conducted. The help screens appear before the selected test starts, and will give insights into the purpose and objectives of the selected test. You might toggle the help mode on or off by left clicking on "Help Mode" (or pressing "H"). By default, the help mode is on. If you turn the help mode off, the selected test will start immediately.

Tutor mode

PRINTERS is designed with two on-line documentation sources that are intended to provide instruction and guidance before and after a test is conducted. The tutor screens appear after the selected test is complete, and will give advice on how to interpret the printed results. You can toggle the tutor mode on or off by left clicking on "Tutor Mode" (or pressing "T"). By default, the tutor mode is on. If you turn the tutor mode off, there will be no instruction provided when the test is finished.

Running the impact tests

The impact tests allow you to check the operations of almost any impact dot-matrix printer (and any 9-pin or 24-pin printer capable of Epson emulation). From the main menu bar, left click on "Impact" (or press "I"). The IMPACT DMP Test Menu will appear, as shown in figure 13-4. You can return to the main menu at any time by pressing the <ESC> key or right clicking anywhere in the display. There are seven functions available from the IMPACT DMP test menu:

428

☐ Preliminary Setup Information   Supplies initial information to help you set up and operate the printer safely.

☐ Carriage Transport Test   Allows you to test the Impact DMP carriage transport system.

☐ Paper Transport Test   Allows you to test the Impact DMP paper transport system.

☐ Paper Walk Test   Allows you to check friction feed paper transport systems in Impact DMPs.

☐ Print Head Test   Allows you to test the Impact DMP print head assembly.

☐ Clean Rollers   Allows you to check and clean the paper handling rollers.

☐ Print TEST PAGE   Provides a uniform test pattern for initial or final printer inspection.

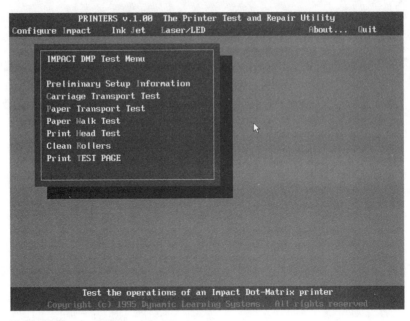

```
        PRINTERS v.1.00  The Printer Test and Repair Utility
Configure Impact    Ink Jet   Laser/LED          About...  Quit

    IMPACT DMP Test Menu

    Preliminary Setup Information
    Carriage Transport Test
    Paper Transport Test
    Paper Walk Test
    Print Head Test
    Clean Rollers
    Print TEST PAGE

        Test the operations of an Impact Dot-Matrix printer
        Copyright (c) 1995 Dynamic Learning Systems. All rights reserved
```

■ **13-4** *The Impact test menu.*

Preliminary Setup Information

You can access the preliminary setup information screen by left clicking on "Preliminary Setup Information" (or press "I"). This is not a test per se, but an information screen intended to provide helpful setup information. Novice troubleshooters will find it helpful to review this information before attempting any of the test sequences below. Experienced troubleshooters might find this to be a handy reminder. To leave the information screen, press <ESC> or right click anywhere in the display.

Carriage Transport Test

Impact printers are moving carriage devices; that is, the print head is carried left and right across the page surface. This movement is handled by the carriage transport mechanism. Proper printing of text and graphics demands that the print head be positioned precisely in both its left-to-right and right-to-left movement. The Carriage Transport Test is designed to test carriage alignment by generating a series of vertical lines, as shown in figure 13-5. The print head sweeps from left-to-right producing a series of vertical tic marks, then reverses direction and produces a right-to-left series of tic marks. Similarly, each line of tic marks is the result of two independent passes. Using this approach, we can check carriage alignment not just between lines, but within the same line.

The work screen

■ **13-5** *The Carriage test pattern.*

If the tic marks are not aligned precisely, there might be some mechanical slop in the carriage mechanics. Badly or erratically placed
tic marks might indicate a fault in the carriage motor driver circuitry, or in the carriage home sensor. If marks within the same
line are aligned precisely at the edges (but not elsewhere in the
line), there might be wiring problems in the print head or print
head cable. You might start this test by left clicking on "Carriage
Transport Test" (or press "C").

Paper Transport Test

There are two traditional means of moving paper through a printer: pull the paper (with a tractor feed), or push the paper (with a friction feed). Regardless of the means used, paper must be carried through a printer evenly and consistently; otherwise, print will overlap and cause distortion. The Paper Transport Test is designed to check the paper transport system's operation. The Paper Transport Test works with any transport type; however, it is intended primarily for tractor feed systems that pull the paper through. The test pattern counts off a number of marked passes. You must check each pass to see that they are spaced evenly apart. If not, there might be a problem with the transport mechanics, motor, or driving circuitry. To start this test, left click on "Paper Transport Test" (or press "P").

Paper Walk Test

Like the last test, the Paper Walk Test is designed to check the paper transport system's operation. While the Paper Transport Test is best utilized with tractor feed paper transports, the Paper Walk Test is intended primarily for friction feed systems, which push the paper through. The test pattern generates a series of evenly spaced horizontal lines. You must check each pass to see that they are spaced evenly apart. If not, there might be a problem with the transport mechanics, motor, or driving circuitry.

Another problem particular to friction feed paper transports is the tendency to "walk the page." Proper friction feed operation depends on roller pressure applied evenly across the entire page surface. Any damage, obstructions, or wear could cause excessive roller pressure that can allow the page to spin clockwise or counterclockwise. If you notice lines closer together on the left or right side of the image, the roller assembly might need adjustment or replacement, or there might be an obstruction in the paper path. To start this test, left click on "Paper Walk Test" (or press "W").

Print Head Test

Ideally, every pin on the print head should fire reliably. In actual practice, however, age, lack of routine maintenance, and heat buildup can affect firing reliability. This often results in horizontal white lines in the text where the corresponding print wires fail. The best way to stress-test an impact print head and detect problems is by printing a dense graphic. The impact print head test produces a large black rectangle, as shown in figure 13-6; this demands the proper operation of all print wires, and will often reveal

■ **13-6** *The Print Head test pattern.*

any age, damage, or heat and maintenance-related issues. You can adjust the speed and density of printing by adjusting print intensity and driver resolution under the CONFIGURE menu.

Examine the black rectangle for horizontal white lines. Consistent white lines indicate that a print wire is not firing. Check and clean the face of the print head to remove any accumulations of debris that might be jamming the print wire(s). If the problem persists, there might be a fault in the print head or print wire driver circuitry. If white lines appear only briefly or intermittently, there might be wiring problems with the print head cable, or within the print head itself. There might also be an intermittent problem in the corresponding print wire driver circuit. Finally, print intensity should be relatively consistent throughout each pass. Light (or *faded*) printing might indicate trouble with print head wear or spacing, ribbon quality, or the power supply. To start this test, left click on "Print Head Test" (or press "H").

Clean Rollers

To combat the dust and debris that naturally accumulate in the printer's mechanics, it is customary to periodically clean the main rollers that handle paper, usually the platen and other major rollers. This is especially important for friction feed paper transports where age and any foreign matter on the rollers can interfere with the paper path and "walk" the page. While it is possible (and sometimes more convenient) to rotate the rollers by hand (using the platen knob), the Clean Rollers function provides a paper advance that allows you to streamline the routine cleaning/rejuvenation of roller assemblies. Paper must be present in the printer to use this function.

The typical cleaning procedure involves rotating the rollers while wiping them gently with a clean cloth lightly dampened with water. You might use a bit of very mild household detergent to remove grime that resists water alone, but avoid using detergent regularly because chemicals applied to rubber and other synthetic roller materials can reduce their pliability. Never use harsh detergents or solvents! If you choose to try rejuvenating the rollers with a roller cleaning solvent, use extreme caution. First, work in a ventilated area (solvent fumes are dangerous). Second, it is impossible to predict how cleaning solvents will affect every possible roller material, so always try the solvent on a small edge patch of the roller in advance. To start this procedure, left click on "Clean Rollers" (or press "R"). A single cycle will advance the paper transport by two pages.

Print TEST PAGE

The TEST PAGE is typically the first and last test to be run on any printer. Initially, the test page will reveal problems with the print head, paper transport, or carriage transport. You can then proceed with more detailed tests to further isolate and correct the problem(s). When the repair is complete, the TEST PAGE is proof of the printer's operation, which you can keep for your records, or provide to your customer. The moving-head TEST PAGE pattern is illustrated in figure 13-7. To start this test, left click on "Print TEST PAGE" (or press "T").

Running the ink jet tests

The ink jet tests allow you to check the operations of almost any ink jet dot-matrix printer (and any printer capable of HP DeskJet emulation). From the main menu bar, left click on "Ink Jet" (or press "J"). The INK JET DMP Test Menu will appear, as shown in figure 13-8. You can return to the main menu at any time by press-

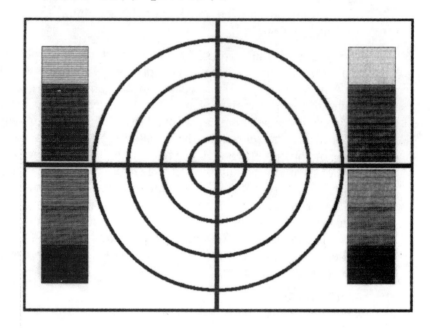

```
PRINTERS: The Printer Test and Alignment Utility  V.1.00
IMPACT TEST PAGE Pattern

A B C D E F G H I J K L M N O P Q R S T U V W X Y Z
A B C D E F G H I J K L M N O P Q R S T U V W X Y Z
a b c d e f g h i j k l m n o p q r s t u v w x y z
a b c d e f g h i j k l m n o p q r s t u v w x y z
1 2 3 4 5 6 7 8 9 0
1 2 3 4 5 6 7 8 9 0
` ~ ! @ # $ % ^ & * ( ) - _ = + ? . ) , <
` ~ ! @ # $ % ^ & * ( ) - _ = + ? . ) , <
```

■ **13-7** *The Test Page pattern.*

ing the <ESC> key or right clicking anywhere in the display. There
are seven functions available from the INK JET DMP test menu:

☐ Preliminary Setup Information   Supplies initial information to
help you set up and operate the ink jet printer safely.

☐ Carriage Transport Test   Allows you to test the Ink Jet DMP
carriage transport system.

☐ Paper Transport Test   Allows you to test the Ink Jet DMP
paper transport system.

☐ Paper Walk Test   Allows you to check friction feed paper
transport systems in Ink Jet DMPs.

☐ Print Head Test   Allows you to test the Ink Jet DMP print
head assembly.

```
                PRINTERS v.1.00  The Printer Test and Repair Utility
Configure Impact      Ink Jet    Laser/LED              About...   Quit

   ┌─────────────────────────────────────┐
   │ INK JET DMP Test Menu               │
   │                                     │
   │ Preliminary Setup Information       │
   │ Carriage Transport Test             │
   │ Paper Transport Test                │
   │ Paper Walk Test                     │
   │ Print Head Test                     │
   │ Clean Rollers                       │
   │ Print TEST PAGE                     │
   └─────────────────────────────────────┘




              Test the operations of an Ink Jet Dot-Matrix printer
            Copyright (c) 1995 Dynamic Learning Systems.  All rights reserved
```

■ **13-8** *The Ink Jet test menu.*

□ Clean Rollers   Allows you to check and clean the paper
  handling rollers.

□ Print TEST PAGE   Provides a uniform test pattern for initial
  or final printer inspection.

Preliminary Setup Information

You can access the preliminary setup information screen by left
clicking on "Preliminary Setup Information" (or press "I"). This is
not a test per se, but an information screen intended to provide
helpful setup information. Novice troubleshooters will find it helpful to review this information before attempting any of the ink jet
test sequences below. Experienced troubleshooters might find
this to be a handy reminder. To leave the information screen, press
<ESC> or right click anywhere in the display.

Carriage Transport Test

Like impact printers, ink jet printers are moving-carriage devices;
that is, the print head is carried left and right across the page surface. This movement is handled by the carriage transport mechanism. Proper printing of text and graphics demands that the print
head be positioned precisely in both its left-to-right and right-to-
left movement. The Carriage Transport Test is designed to test
carriage alignment by generating a series of vertical lines (similar

to figure 13-5). The print head sweeps from left to right producing a series of vertical tic marks, then it reverses direction and produces a right-to-left series of tic marks. Similarly, each line of tic marks is the result of two independent passes. Using this approach, we can check carriage alignment not just between lines, but within the same line.

If the tic marks are not aligned precisely, there might be some mechanical slop in the carriage mechanics. Badly or erratically placed tic marks might indicate a fault in the carriage motor driver circuitry, the mechanical home sensor, or the optical position encoder. If marks within the same line are aligned precisely at the edges (but not elsewhere in the line), there might be wiring problems in the ink jet print cartridge or print head cable. To start this test, left click on "Carriage Transport Test" (or press "C").

Paper Transport Test

There are two traditional means of moving paper through a printer: pull the paper (with a tractor feed), or push the paper (with a friction feed). Regardless of the means used, paper must be carried through a printer evenly and consistently; otherwise, print will overlap and cause distortion. The Paper Transport Test is designed to check the paper transport system's operation. While the Paper Transport Test will work with any transport type, it is intended primarily for tractor feed systems, which pull the paper through. The test pattern counts off a number of marked passes. You must check each pass to see that they are spaced evenly apart. If not, there might be a problem with the transport mechanics, motor, or driving circuitry. To start this test, left click on "Paper Transport Test" (or press "P").

Paper Walk Test

Like the last test, the Paper Walk Test is designed to check the paper transport system's operation. While the Paper Transport Test is best utilized with tractor feed paper transports, the Paper Walk Test is intended primarily for friction feed systems, which push the paper through. Today's ink jet printers utilize friction feed systems almost entirely. The test pattern generates a series of evenly spaced horizontal lines. You must check each pass to see that they are spaced evenly apart. If not, there might be a problem with the transport mechanics, motor, or driving circuitry.

Another problem particular to friction feed paper transports is the tendency to "walk the page." Proper friction feed operation depends on roller pressure applied evenly across the entire page surface. Any damage, obstructions, or wear might cause excessive roller pressure that can allow the page to spin clockwise or counterclockwise. This is especially evident in inexpensive ink jet systems. If you notice lines closer together on the left or right side of the image, the roller assembly might need adjustment or replacement, or there might be an obstruction in the paper path. To start this test, left click on "Paper Walk Test" (or press "W").

Print Head Test

Ideally, every nozzle on the ink cartridge should fire reliably. In actual practice, however, cartridge age, lack of routine maintenance, low ink levels, and circuit defects can affect firing reliability. This often results in horizontal white lines in the text where the corresponding print nozzles fail. The best way to stress-test an ink jet print head and detect problems is by printing a dense graphic. The ink jet print head test produces a large black rectangle (similar to figure 13-6); this demands the proper operation of all print nozzles, and will often reveal any age, damage, or maintenance-related issues. You can adjust the speed and density of printing by adjusting print intensity and driver resolution under the CONFIGURE menu. Remember that this is a very dark image that will require a substantial amount of ink to form. If the current ink cartridge is marginal, you might need to install a new ink cartridge before proceeding.

Examine the black rectangle for horizontal white lines. Consistent white lines indicate that a print nozzle is not firing. Check and clean the face of the print cartridge to remove any accumulations of dried ink or debris that might be jamming the print nozzle(s). If the problem persists, there might be a fault in the print cartridge or print nozzle driver circuitry. If white lines appear only briefly or intermittently, there might be wiring problems with the print head cable, or within the print cartridge itself. There might also be an intermittent problem in the corresponding print nozzle driver circuit. Finally, print intensity should be relatively consistent throughout each pass. Light (or *faded*) printing might indicate poor paper selection, low ink levels, or trouble with the print nozzle power supply. To start this test, left click on "Print Head Test" (or press "H").

Clean Rollers

To combat the dust and debris that naturally accumulate in the printer's mechanics, it is customary to periodically clean the main rollers that handle paper, usually the platen and other major rollers. This is especially important for delicate ink jet friction feed paper transports where age and any foreign matter on the rollers can interfere with the paper path and "walk" the page. Few contemporary ink jet designs allow you to rotate the rollers by hand (using a platen knob), so the Clean Rollers function provides a paper advance that allows you to streamline the routine cleaning/rejuvenation of roller assemblies. Paper must be present in the printer to use this function.

The typical cleaning procedure involves rotating the rollers while wiping them gently with a clean cloth lightly dampened with water. You might use a bit of very mild household detergent to remove grime that resists water alone, but avoid using detergent regularly because chemicals applied to rubber and other synthetic roller materials can reduce their pliability. Never use harsh detergents or solvents! If you choose to try rejuvenating the rollers with a roller cleaning solvent, use extreme caution. First, work in a ventilated area (solvent fumes are dangerous). Second, it is impossible to predict how cleaning solvents will affect every possible roller material, so always try the solvent on a small edge patch of the roller in advance. To start this procedure, left click on "Clean Rollers" (or press "R"). A single cycle will advance the paper transport by two pages.

Print TEST PAGE

The TEST PAGE (such as the one shown in figure 13-7) is typically the first and last test to be run on any printer. Initially, the test page will reveal problems with the print cartridge, paper transport, or carriage transport. You can then proceed with more detailed tests to further isolate and correct the problem(s). When the repair is complete, the TEST PAGE is proof of the printer's operation, which you can keep for your records, or provide to your customer. To start this test, left click on "Print TEST PAGE" (or press "T").

Running the Laser/LED tests

The Laser/LED (EP) tests allow you to check the operations of almost any electrophotographic (EP) printer (any printer capable of HP LaserJet emulation). From the main menu bar, left click on "Laser/LED" (or press "L"). The LASER/LED Test Menu will appear, as shown in figure 13-9. You can return to the main menu at any time by pressing the <ESC> key or right clicking anywhere in

■ 13-9 *The EP test menu.*

the display. There are seven functions available from the INK JET DMP test menu:

☐ Preliminary Setup Information   Supplies initial information to help you set up and operate the EP printer safely.

☐ Toner Test   Allows you to check the effects of low/expired toner or poor toner distribution.

☐ Corona Test   Allows you to quickly identify the location of fouling on the Primary or Transfer corona wires.

☐ Drum and Roller Test   Allows you to identify the source of repetitive defects in the printer.

☐ Fuser Test   Allows you to check for low or inconsistent fusing.

☐ Paper Transport Test   Allows you to quickly and conveniently check the paper transport without wasting time or toner.

☐ Print TEST PAGE   Provides a uniform test pattern for initial or final printer inspection.

Preliminary Setup Information

You can access the preliminary setup information screen by left clicking on "Preliminary Setup Information" (or press "I"). This is not a test per se, but an information screen intended to provide helpful setup information for EP testing. Novice troubleshooters will find it helpful to review this information before attempting any of the test sequences below. Experienced troubleshooters might find this to be a handy reminder. To leave the information screen, press <ESC> or right click anywhere in the display.

Toner Test

Toner is the raw material (the media) that is used to form printed images (the "ink" of the EP printer). As a consequence, low, expired, poor-quality, or badly distributed toner will have an impact on print quality. The Toner Test is designed to test toner condition by printing a full-page black graphic, as shown in figure 13-10.

PRINTERS: The Printer Test and Alignment Utility V.1.00
TONER Cartridge Test

Copyright (c) 1995 by Dynamic Learning Systems   All rights reserved worldwide
Dynamic Learning Systems P.O. Box 805, Marlboro, MA 01752 USA
Tel. 508-366-3487   Fax 508-898-9995 BBS 508-366-7683   CIS: 73652,3205

■ **13-10** *The Toner test pattern.*

Light streaks appearing vertically along the page (usually on either side of the image) are typical of a low-toner cartridge. An overall light image might be compensated for by increasing the print intensity wheel (on the printer itself), but it might also suggest a poor-quality toner cartridge or expired toner. If a fresh toner cartridge fails to correct the problem and print intensity is already set high, the fault might be in the printer's high-voltage power supply. Remember that EP printing technology is heavily dependent on paper type and quality. Light splotches might be the result of damp or coated paper. Try a supply of fresh, dry 20-lb. xerography-grade paper. To start this test, left click on "Toner Test" (or press "n").

Corona Test

EP technology relies on high voltage to produce the electrostatic fields that charge the EP drum, and attract toner off the drum to the page. These fields are established by a Primary corona and a Transfer corona respectively. A *corona* is really nothing more than a length of thin wire, but it has a critical impact on the evenness of the electric field it produces. Later EP engines replace corona wires with charge rollers, but the effect is the same as corona wires. The Corona Test pattern shown in figure 13-11 is designed to highlight corona problems.

The dust and paper particles in the air tend to be attracted to high-voltage sources, such as corona wires. Over time, an accumulation of foreign matter will weaken the field distribution and affect the resulting image. Fouling at one or more points on the primary corona will cause toner to always be attracted to those corresponding points on the drum; this results in black vertical streaks that can be seen against a white background. Conversely, fouling at one or more points on the transfer corona will prevent toner at those corresponding points from being attracted off the drum and onto the page; this results in white vertical streaks that can be seen against a black background. To start this test, left click on "Corona Test" (or press "C").

Drum and Roller Test

In spite of the high level of refinement in today's EP printers, image formation is still a delicate physical process. Paper must traverse a torturous course through a series of roller assemblies in order to acquire a final, permanent image. With age, wear, and accidental damage, the various rotating elements of an EP printer can succumb to marks or other slight damage. While such damage is incidental, the results can be seen in the printed output.

PRINTERS: The Printer Test and Alignment Utility V.1.00
PRIMARY and TRANSFER CORONA Test Pattern

Transfer Corona:

Primary Corona:

■ **13-11** *The Corona test pattern.*

Consider the EP drum itself. It has a circumference of about 3.75".
This means that any one point on the drum will approach every
page surface at least twice. If a nick were to occur on the drum
surface, it would appear in the final page at least twice, separated
by about 3.75". Because each of the major roller assemblies have a
slightly different circumference, it is possible to quickly identify
the source of a repetitive defect simply by measuring the distance
between instances. The Drum and Roller Test pattern shown in
figure 13-12 is designed to help you correlate measured distances
to problem areas. For example, a defect that occurs every 3.75"
can be traced to the EP drum, an error that occurs every 3" or so
can be traced to the fusing rollers, and a defect that occurs every
2" can often be related to the development roller. To start this test,
left click on "Drum and Roller Test" (or press "D").

442

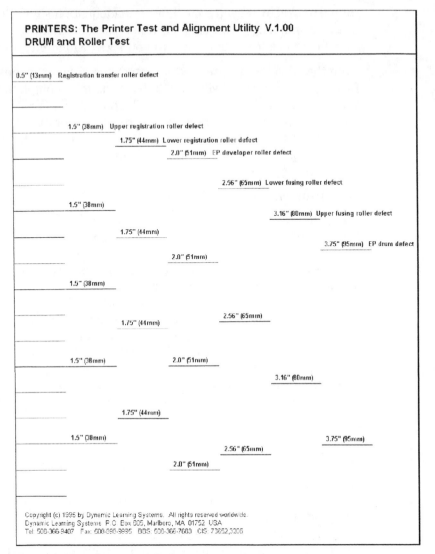

■ 13-12 *The Drum and Roller test pattern.*

Fuser Test

Fusing is a vital part of the image formation process; it uses heat and pressure to literally melt toner into the paper fibers. To ensure proper fusing, the upper fusing roller must reach and maintain a fairly narrow range of temperatures. When temperature is marginally low, fusing might not be complete. The Fuser Test is intended to check the stability of a fusing system's temperature control by running a series of full-page test graphics. You can then run your thumb briskly over each page. Low or inadequate fusing will result in toner smudging (you can see toner on your thumb). You can then take the appropriate steps to optimize fusing temperature.

Note that this test is intended to detect marginal or inconsistent fusing performance; it will not catch a serious or complete fusing system failure because the printer will generate an error message and halt if the fusing assembly fails to reach proper fusing temperature in 90 seconds or so, or falls below a minimum temperature for a prolonged amount of time. To start this test, left click on "Fuser Test" (or press "F").

Paper Transport Test

The paper transport system of an EP printer is a highly modified friction feed system designed to carry paper through the entire image formation system. There are also a series of time-sensitive sensors intended to track each page as it passes (and detect jam conditions). When you work on any aspect of the printer's mechanics, you are affecting paper transport. The Paper Transport Test is intended to eject five sheets of blank paper (saving toner and printing time). You can use this test to help you troubleshoot problems in the paper path, or verify that any repairs you might have made do not obstruct the paper path. Once a repair is complete, it is recommended that you verify the paper path before running test patterns. To start this test, left click on "Paper Transport Test" (or press "P").

Print TEST PAGE

The TEST PAGE is typically the first and last test to be run on any Laser/LED printer. Initially, the Laser/LED test page (shown in figure 13-13) will reveal problems with the toner cartridge, paper transport, fusing system, coronas, or writing mechanism. You can then proceed with more detailed tests to further isolate and correct the problem(s). When the repair is complete, the TEST PAGE is proof of the printer's operation, which you can keep for your records, or provide to your customer. To start this test, left click on "Print TEST PAGE" (or press "T").

About PRINTERS

To learn about the PRINTERS program, click on "About" in the main menu bar (or press "A"). An information box will appear in the middle of the display. To clear the information box, press the <ESC> key or click the right mouse button anywhere in the display.

444

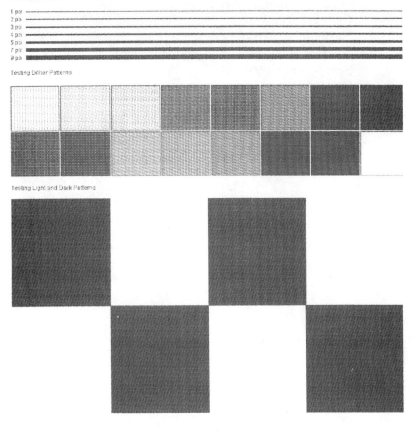

PRINTERS: The Printer Test and Alignment Utility V.1.00
LASER/LED (EP) Test Page Pattern

Testing Line Widths

1 pix
2 pix
3 pix
4 pix
5 pix
7 pix
9 pix

Testing Dither Patterns

Testing Light and Dark Patterns

■ **13-13** *The EP Test Page pattern.*

Quitting PRINTERS

To quit the PRINTERS program and return to DOS, click on "Quit" in the main menu bar (or press "Q"). After a moment, the DOS prompt will appear.

Making PRINTERS better

Dynamic Learning Systems is dedicated to providing high-quality, low-cost diagnostic utilities. As a result, we are always interested in ways to improve the quality and performance of our products. We welcome your comments, questions, or criticisms:

Dynamic Learning Systems
P.O. Box 805
Marlboro, MA 01752
Tel: 508-366-9487
Fax: 508-898-9995
BBS: 508-366-7683
CompuServe: 73652,3205
Internet: sbigelow@cerfnet.com
WWW: http://www.dlspubs.com/home.htm

Parts & supplies vendors

Printer Manufacturers

Dot-Matrix Printer Manufacturers
Advanced Matrix Technology, Inc.
747 Calle Plano
Camarillo, CA 93012-8598
Tel: 800-992-2264
Fax: 805-494-3087

Citizen America, Inc.
2450 Broadway #600
P.O. 4003
Santa Monica, CA 90411
Tel: 310-453-0614
Fax: 310-453-2814

CIE America, Inc. (C.Itoh)
2701 Dow Ave.
Tustin, CA 92680
Tel: 800-877-1421
Fax: 714-757-4488

Digital Equipment Corp.
111 Powdermill Rd.
Maynard, MA 01754-2571
Tel: 800-777-4343
Fax: 800-234-2298

Epson America, Inc.
20770 Madrona Ave.
Torrance, CA 90509
Tel: 800-289-3776

Mannesmann Tally Corp.
8301 S 180th St.
P.O. Box 97018
Kent, WA 98064-9718
Tel: 800-843-1347
Fax: 206-251-5520

Panasonic Communications &
Systems Co.
2 Panasonic Way
Secaucus, NJ 07094
Tel: 800-742-8086

Printer Systems International
85 Wells Ave. #200
Newton, MA 02159
Tel: 617-928-3580
Fax: 617-928-3581

Samsung Electronics America
105 Challenger Rd.
Ridgefield Park, NJ 07660
Tel: 800-446-0262
Fax: 201-229-4110

Star Micronics America, Inc.
70 Ethel Rd. W
Piscataway, NJ 08854
Tel: 800-506-7727
Fax: 908-572-3300

Ink Jet Printer Manufacturers

CalComp
2411 W LaPalma Ave.
Anaheim, CA 92801
Tel: 714-821-2000
Fax: 800-451-7568

Cannon Computer Systems, Inc.
2995 Redhill Ave.
Costa Mesa, CA 92626
Tel: 800-423-2366
Fax: 800-922-9068

447

Citizen America Corp.
2450 Broadway #600
P.O. Box 4003
Santa Monica, CA 90411
Tel: 310-453-0614
Fax: 310-453-2814

CIE America, Inc. (C.Itoh)
2701 Dow Ave.
Tustin, CA 92680
Tel: 800-877-1421
Fax: 714-757-4488

Digital Equipment Corp.
111 Powdermill Rd.
Maynard, MA 01754
Tel: 800-777-4843
Fax: 800-234-2298

Epson America, Inc.
20770 Madrona Ave.
Torrance, CA 90503
Tel: 310-782-0770
Fax: 310-782-4248

Hewlett-Packard Co.
5301 Stevens Creek Blvd. #51L
Santa Clara, CA 95052-8059
Tel: 800-527-3753
Fax: 800-333-1917

Lexmark International, Inc.
740 New Circle Rd. NW
Lexington, KY 40511
Tel: 800-358-5835
Fax: 800-232-9534

Olivetti Office USA
765 US Hwy 202
Bridgewater, NJ 08807
Tel: 800-243-4002

Tektronix, Inc.
26600 SW Parkway
Mailstop 63-630
Wilsonville, OR 97070
Tel: 800-835-6100
Fax: 503-682-2780

Texas Instruments
P.O. Box 149149
Austin, TX 78714
Tel: 800-848-3927
Fax: 512-250-7329

Laser/LED Printer Manufacturers

Brother International Corp.
200 Cottontail Lane
Somerset, NJ 08875
Tel: 800-276-7746
Fax: 908-764-4481

Citizen America Corp.
2450 Broadway
P.O. Box 4003
Santa Monica, CA 90411
Tel: 310-453-0614

Dataproducts Corp.
6219 De Soto Ave.
Woodland Hills, CA 91365
Tel: 800-980-0374
Fax: 703-860-9084

Digital Equipment Corp.
111 Powdermill Rd.
Maynard, MA 01754
Tel: 800-777-4843
Fax: 800-234-2298

Epson America, Inc.
20770 Madrona Ave.
Torrance, CA 90503
Tel: 800-289-3776
Fax: 310-782-0770

GCC Technologies, Inc.
209 Burlington Rd.
Bedford, MA 01730
Tel: 800-422-7777
Fax: 617-275-1115

Hewlett-Packard Co.
5301 Stevens Creek Blvd. #51L
Santa Clara, CA 95052-8059
Tel: 800-527-3753
Fax: 800-333-1917

Kyocera Electronics, Inc.
100 Randolph Rd.
Somerset, NJ 08875
Tel: 800-232-6797
Fax: 908-560-8380

LaserMaster Corp.
6900 Shady Oak Rd.
Eden Prairie, MN 55344
Tel: 800-950-6868

Lexmark International, Inc.
55 Railroad Ave.
Greenwich, CT 06836
Tel: 800-891-0331
Fax: 606-232-2380

NEC Technologies, Inc.
1414 Massachusetts Ave.
Boxboro, MA 01719
Tel: 800-632-4636
Fax: 800-366-0476

NewGen Systems Corp.
17550 Newhope St.
Fountain Valley, CA 92708
Tel: 800-756-0556

Okidata
532 Fellowship Rd.
Mt. Laurel, NJ 08054
Tel: 800-654-3282
Fax: 609-778-4184

Panasonic Communications &
  Systems Co.
Two Panasonic Way
Secaucus, NJ 07094
Tel: 800-222-0584

QMS, Inc.
One Magnum Pass
Mobile, AL 36689
Tel: 800-523-2696
Fax: 334-633-4866

Samsung Electronics America, Inc.
Information Systems Division
105 Challenger Rd.
Ridgefield Park, NJ 07660
Tel: 800-446-0262

Sharp Electronics, Corp.
Sharp Plaza
Mahwah, NJ 07430
Tel: 800-526-0522

Star Micronics America, Inc.
420 Lexington Ave. #2702
New York, NY 10170
Tel: 800-447-4700

Tandy Corp.
1500 One Tandy Center
Fort Worth, TX 76102
Tel: 817-390-3011

Texas Instruments, Inc.
5701 Airport Rd.
Temple, TX 76502
Tel: 800-848-3927
Fax: 800-443-2984

Xante Corp.
2559 Emogene St.
Mobile, AL 36606
Tel: 800-926-8839
Fax: 334-476-9421

Printer Parts Vendors

Comdisco Parts
Tel: 800-635-2211
Fax: 708-980-1435

Computer Parts Unlimited
5069 Maureen Lane
Moorpark, CA 93021
Tel: 805-532-2500
Fax: 805-532-2599

Impact
10435 Burnet Rd. #114
Austin, TX 78758
Tel: 800-777-4323
Fax: 512-832-9321

449

National Parts Depot, Inc.
31 Elkay Drive
Chester, NY 10918
Tel: 914-469-4800
Fax: 914-469-4855

NIE International
3000 East Chambers
Phoenix, AZ 85040
Tel: 602-470-1500
Fax: 602-470-1540

PC Service Source
2350 Valley View Lane
Dallas, TX 75234
Tel: 214-406-8583
Fax: 214-406-9081

Peak
9200 Berger Rd.
Columbia, MD 21046
Tel: 800-950-6372
Fax: 410-312-6067

ProAmerica
650 International Pkwy, Suite 180
Richardson, TX 75081
Tel: 800-888-9600
Fax: 214-690-8648

Unicomp, Inc.
2501 W Fifth St.
Santa Ana, CA 92703
Tel: 800-359-5092
Fax: 714-571-1909

General Parts Vendors

Active Electronics, Inc.
11 Cummings Park
Woburn, MA 01801
Tel: 800 677-8899

Consolidated Electronics, Inc.
705 Watervliet Ave.
Dayton, OH 45420
Tel: 800-543-3568

Dalbani Corp. of America
2733 Carrier Ave.
Los Angeles, CA 90040
Tel: 800-325-2264

Eiger Electronics
91 Toledo St.
Farmingdale, NY 11735
Tel: 800-835-8316

ESP (Electronic Service Parts)
2901 E. Washington St.
Indianapolis, IN 46201
Tel: 800-382-9976

Howard W. Sams & Co.
2647 Waterfront Parkway East Dr.
Indianapolis, Indiana 46214
Tel: 800-428-7267

Lee Products Co.
800 East 80th St.
Minneapolis, MN 55420

Mill Electronics, Inc.
2026 McDonald Ave.
Brooklyn, NY 11223
Tel: 800-346-8994

NTE Electronics, Inc.
44 Farrand St.
Bloomfield, NJ 07003
Tel: 800-631-1250

Union Electronic Distributors
16012 S. Cottage Grove
So. Holland, IL 60473
Tel: 800-648-6657

450

Glossary

ACK (Acknowledge) A handshaking signal sent from printer to computer indicating that the printer has successfully received a character.

active component A semiconductor-based electronic device (i.e., a diode, transistor, or integrated circuit) designed to provide a specific function in a circuit or system.

anode The positive electrode of a two-terminal electronic device.

ASCII (American Standard Code for Information Interchange) A standard set of binary codes that define basic letters, numbers, and symbols used by a printer.

auto feed A rarely used signal from the computer that allows the printer to automatically advance paper upon receiving a carriage return <CR> character.

bail roller A set of small light rollers that rest on the page as it leaves the platen. Bail rollers help to keep the paper flat and even against the platen.

base One of three electrodes on a bipolar transistor.

baud rate The rate of serial data transmissions, which is measured in bits-per-second (bps).

belt A synthetic rubber or plastic band (usually notched on one side) used to transfer mechanical force between a primary and secondary pulley. In a moving-carriage printer, the moving belt carries the carriage back and forth.

bidirectional A printer mode where printing is supported in both the left-to-right and right-to-left passes of the print head.

binary A number system consisting of only two digits.

buffer *See* data buffer.

451

BUSY A parallel handshaking signal sent from printer to computer indicating that the printer cannot accept any more characters.

capacitance The measure of a device's ability to store an electric charge, measured in farads, microfarads, or picofarads.

capacitor A device used to store an electrical charge.

carriage A stabilized platform that the print head is mounted on. The carriage is moved back and forth across the page by the carriage transport system.

carriage transport The motor, mechanics, and driving circuitry that moves a carriage back and forth across a page.

cathode The negative electrode of a two-terminal electronic device.

Centronics The de-facto industry standard that defines the signals and timing needed to transfer parallel data bits from computer to printer. Recent modifications allow high-speed bidirectional communication.

CGL (carrier generation layer) A lower coated layer of an EP drum, which is an important attribute to the drum's photosensitive coating.

cleaning An important step in the electrophotographic process that removes residual toner particles from the drum prior to electrical erasure and conditioning.

collector One of three electrodes on a bipolar transistor.

conditioning The process of physically cleaning, electrically erasing, and uniformly charging the photosensitive EP drum.

continuity The integrity of a connection measured as a very low resistance by an ohmmeter.

corona A field of concentrated electrical charge produced by a large voltage potential. Corona wires form one electrode of this voltage potential.

CPI (characters per inch) The number of characters that will fit onto one inch of horizontal line space, also called *character pitch*.

CPL (characters per line) The number of characters that will fit on a single horizontal line.

CPS (characters per second) The rate at which characters are delivered to the page surface by a printer.

CPU (central processing unit) *See* microprocessor.

CTL (carrier transport layer) The upper (outer) coated layer of an EP drum's photosensitive material.

CTS (Clear To Send) A serial handshaking line at the computer usually connected to the RTS line of a printer.

daisy wheel An obsolete "typewriter-style" printer that formed characters as complete letters molded onto a single replaceable type wheel. Each character was held on its own "petal," so the type wheel resembled a daisy, thus the term *daisy wheel.*

data Any of eight parallel data lines that carry binary information from computer to printer.

data buffer Temporary memory where characters from the computer are stored by the printer prior to printing. The term *buffer* denotes a relatively small amount of memory (8KB or 16KB) indicative of dot-matrix or older ink-jet printers.

DCD (Data Carrier Detect) A serial handshaking line usually found in serial modem interfaces.

453

development The step in an EP process where toner fills in the latent image written to the photosensitive drum.

development roller The roller in the toner reservoir that passes toner to the photosensitive drum.

device driver A small program (loaded when a PC first starts) that allows the printing application to use all of the printer's special features.

dielectric A material that offers insulation against electrical charge. All materials have some dielectric strength. Dielectrics also form a main internal part of a capacitor.

diode A two-terminal electronic device used to conduct current in one direction only.

dot pitch Also known as dots per inch (dpi). The number of printed dots that can be placed along one linear inch. This is the foundation of printer resolution.

driver An amplifier used to convert low-power signals into high-power signals.

drum An aluminum cylinder coated with light-sensitive organic material that is used to hold latent electrophotographic images.

DSR (Data Set Ready) The primary computer signal line for hardware handshaking over a serial interface. It is connected to the DTR line at the printer.

DTR (Data Terminal Ready) The primary serial printer signal for hardware handshaking over a serial interface. It is connected to the DSR pin at the computer.

ECP (Enhanced Capabilities Port) A parallel port convention conforming to IEEE standard 1284 that allows fast parallel port data transfer (800KB/s to 2MB/s) between PCs and printers. For full performance, the PC's LPT port, the parallel cable, and printer must all be IEEE 1284-compliant.

ECU (Electronic Control Unit) The general electronic assembly used to control a printer consisting of main logic, memory, drivers, a power supply, and a control panel.

EIA (Electronic Industries Association) An organization dedicated to developing and maintaining standards for the electronics industry.

emitter One of three electrodes on a bipolar transistor.

emulation An operating "sleight-of-hand" that allows a printer to use the command set of another printer, thus behaving like that other printer. For example, most current laser printers emulate one of the Hewlett-Packard LaserJet products.

enclosure The plastic housings that provide a printer with its cosmetic appearance.

encoder An electro-optical device used to relay the speed and direction of the print head back to main logic.

EP (electrophotographic) A process of creating images using the forces of high voltage to attract or repel media (toner) as needed to form the desired image.

EPP (Enhanced Parallel Port) An advanced parallel port convention conforming to IEEE standard 1284 that allows fast parallel port data transfer (800KB/s to 2MB/s) between PCs and printers, as well as multiple devices attached to the same parallel port. For full performance, the PC's LPT port, the parallel cable, and printer must all be IEEE 1284-compliant.

EPROM (erasable programmable read only memory) A flexible type of permanent memory that can be erased and rewritten to many times (with the proper programming equipment).

flyback voltage An undesired inductive effect that occurs frequently with motors and solenoids where high-voltage spikes are produced when inductive devices are turned on and off. Such spikes can damage improperly designed circuitry. Diodes are typically used to provide "flyback protection."

font A character set of particular size, style, and spacing.

friction The unwanted force produced when two materials rub together. When properly planned and balanced, friction can be useful (such as to push paper through the printer), but generally, friction results in mechanical wear that will eventually demand repair or replacement of the worn part(s).

fuser The assembly that applies heat and pressure to a page in order to melt the toner image and permanently fix it to the page.

gates Integrated circuits used to perform simple logical operations on binary data in digital systems.

gear ratio The relationship of gear sizes that allows designers to achieve desired rotating speeds and torque in printer rollers.

GND (ground) A common electrical reference point for electronic data signals.

GPIB (General Purpose Interface Bus) A parallel communication interface intended primarily for networked instrumentation, also known as IEEE 488. GPIB made a relatively brief appearance in some Hewlett-Packard printers, but has largely given way to parallel or serial communication.

IEEE (Institute of Electrical and Electronics Engineers) A professional organization dedicated to the engineering profession, as well as developing and supporting design standards.

inductance The measure of a device's ability to store a magnetic charge, measured in henries, millihenries, or microhenries.

inductor A device used to store a magnetic charge.

initialization Restoring default or start-up conditions to the printer due to fault or power-up.

interlock A cut-off switch used to detect potentially hazardous situations, such as open covers and protective devices, and cuts

455

off main lower, high-voltage, or laser output to reduce the possibility of personal injury.

ITU (International Telecommunication Union) An international consortium charged with developing communications standards, particularly involved with data communication through modems and other serial means.

laser A device producing a narrow intense beam of coherent, single-wavelength light waves.

LED (light emitting diode) A semiconductor device designed such that photons of light are liberated when its p-n junction is forward biased.

LPI (lines per inch) The number of horizontal lines that fit into one inch of vertical page space, also known as *line pitch*.

lubricant Oils or grease applied between moving parts in order to reduce friction.

magnetic coupling The transfer of magnetic signals between two conductors through air or a core material.

microprocessor A complex programmable logic device that will perform various logical operations and calculations based on predetermined program instructions. Also called a *CPU*.

mirror Also called the *hexagonal mirror* or *scanner mirror*. A rotating mirror assembly driven by the scanner motor used to sweep the modulated laser beam across a drum surface.

motor An electromechanical device used to convert electrical energy into mechanical motion.

MPBF (mean pages between failures) A measure of a printer's reliability expressed as a number or printing cycles or pages printed.

MTBF (mean time between failures) A measure of a device's reliability expressed as time or an amount of use.

multimeter A versatile test instrument used to test such circuit parameters as voltage, current, and resistance.

NLQ (near letter quality) High-quality dot-matrix characters formed with high-density print heads, or by making multiple printing passes to fill in spaces between dots.

nozzle A fine aperture on an ink jet print head that allows a dot of ink to be ejected, but holds in the liquid ink supply through capillary action.

OPC (organic photoconductive chemicals) This is the term often used to describe the various layers of light-sensitive chemicals applied to an EP drum.

optoisolator A light-driven switch that can be operated by physically interrupting a light source (i.e., a paper-out sensor), or by switching a light source on and off (such as electrical isolation between signals).

ozone An irritating gas released when air is ionized by high voltage (such as the high voltage from a primary or transfer corona).

Paper Error (PE) A parallel handshaking signal sent from the printer to tell the computer that paper is exhausted.

paper transport The printer mechanism that is responsible for carrying paper through the printer. There are two classical methods for carrying paper: friction feed and tractor feed.

parallel The communication method that sends data to the printer a byte at a time. Also known as *Centronics*.

parity An extra bit added to a serial data word used to check for errors in communication.

passive component An energy storage or dissipation component, such as a resistor, capacitor, or inductor.

PCL (printer control language) The set of commands used by a printer to execute its essential functions. Most printers use a standard language or one of its variations created by Hewlett-Packard.

phase A motor winding; one of the coils in a motor that is selectively energized to develop rotating force. Also called a *phase winding*.

photosensitive A material or device that reacts electrically when exposed to light.

piezoelectric The property of certain materials to vibrate when voltage is applied to them.

platen The main rubber (or synthetic) roller that not only forms the basis of paper handling, but also provides a pliable foundation for impact printers.

potentiometer An adjustable resistor. In EP printers, the potentiometer is often used to adjust printing contrast levels (darkness).

PPM (pages per minute) The rate at which a printer generates complete pages. This rate will vary depending on the amount of graphics or control codes there are in the pages being printed.

pressure roller In a friction feed printer, the pressure roller is used to clamp the page to the platen so that rotating the platen will push the page through.

printer A device that transcribes data from a PC into written words and images.

print head An electromechanical device (used with moving-carriage printers) that forms images as a series of vertical dot patterns. The two popular print head technologies use moving wires to form dots by "firing" them against a page, or ink nozzles to "spray-paint" dots on the page.

pulley A wheel connected to one or more other wheels through a belt that transfers force from one pulley to another. To improve contact, the belt and pulleys can be notched or grooved.

RAM (random access memory) A temporary memory device used to store digital information. Printers typically use RAM to receive characters and commands from the PC, and as a temporary scratch pad for internal CPU work (*see* data buffer).

rectification The action of allowing current to flow in only one direction. Electronic circuits use this principle to turn ac into pulsating dc.

registration EP printers hold a sheet of paper until the leading edge of a drum image aligns with the leading edge of the page. Registration ensures that the top of the form and the beginning of an image are always aligned properly.

regulator An electronic device or circuit arrangement used to control the output of voltage and current from a power supply.

resistance The measure of a device's ability to limit electrical current, measured in ohms, kilohms, or megohms.

resistor A device used to limit the flow of electrical current.

resolution The number of individual dots that a printer can place along the horizontal or vertical axis, often expressed as dots per inch (*see* dot pitch). For example, a typical laser printer can achieve resolutions of 300×300 dpi (300 dpi vertical and 300 dpi horizontal). Higher resolution allows you to achieve higher levels of detail in the image.

rheostat *See* potentiometer.

ribbon A length of inked fabric wound onto a spool or packed into a cassette that serves as the media for an impact printer. Thermal transfer printers can also use a plastic ribbon with solid ink wound onto a spool.

roller A thin cylinder or circle (usually coated with rubber or a synthetic material) used for handling paper transport in a printer.

ROM (read only memory) A permanent memory device used to store digital information.

rotor The rotating part of a motor assembly.

RS-232 The formal name for a serial port.

RTS (Request To Send) A printer serial handshaking line usually connected to the CTS line of a computer.

Rx (Receive Data) This is the serial input line. The printer's Rx line is connected to the computer's Tx line.

Select A parallel control signal from the computer that prepares the printer to receive data.

separation pawls An assembly of plastic "claws" in an EP printer that help to keep the charged page from wrapping around the drum after receiving the toner image.

solenoid An electromechanical device consisting of a coil of wire wrapped around a core that is free to move.

stator The stationary (case) portion of a motor assembly.

stepping motor A variation on the induction motor that allows the rotor to turn in repeatable fixed steps through the use of properly sequenced square wave phase signals.

Strobe A handshaking line from the computer that tells the printer to accept valid parallel data on its data lines.

thick-film TDM A type of thermal dot-matrix print technology which is best employed as a moving-carriage head assembly.

thin-film TDM A type of thermal dot-matrix print technology that is best employed as a line print head assembly.

tinning The technique of removing accumulated grime from a hot soldering tip and replacing it with a layer of fresh solder to achieve a good heat transfer to the solder joint.

toner A fine powder of plastic, iron, and pigments used to form images in electrostatic printing systems.

torque The rotational force generated by a motor. Torque can be adjusted through careful gear ratios.

transformer A device used to step the voltage and current levels of ac signals.

transistor A three-terminal electronic device whose output signal is proportional to its input signal. A transistor can act as an amplifier or a switch.

turns ratio In a transformer, the ratio of primary windings to secondary windings.

Tx (Transmit Data) This is the data output line for serial devices. The computer's Tx line is connected to the printer's Rx line.

unidirectional A printer mode where printing is supported in only one direction (usually left-to-right passes) of the print head.

VLSI (very large-scale integration) IC fabrication technology that allows the construction of ICs with over 100 individual logic gates. These are often seen as ASICs and controllers in a printer ECU.

wettable In soldering terminology, a surface that solder will adhere to properly (such as wire and IC leads).

winding *See* phase.

wire stroke The distance that an impact dot-matrix print wire will travel to the page surface (usually about 0.5 mm).

Bibliography

Olivetti JP 360 Service & Parts Manual. Italy: Ing. C. Olivetti & C., S.p.A., 1994.

Service Manual: DMP 203 Dot-Matrix Printer. Ft. Worth, TX: Tandy Corporation, 1992.

Service Manual: LP800 Laser Printer. Ft. Worth, TX: Tandy Corporation, 1993.

HP DeskJet Family Printers Service Manual. USA: Hewlett-Packard Company, 1990.

HP 2227A/2228A QuietJet Series Printer Service Manual. USA: Hewlett-Packard Company, 1988.

LaserJet Series II Printer and LaserJet Series III Printer Combined Service Manual. USA: Hewlett-Packard Company, 1990.

Davenport, Keith B. "Taming the Hot Heads." *Byte* (1987):221-22.

Guardado, Julio. "Color Thermal Transfer Printing." *Byte* (1987):209-212.

SDP 340 Ink Jet Printer Interface Considerations. Illinois: Singer Data Products, Inc., 1988.

The Basics of Better Product Marking & Coding. Illinois: Videojet Systems International, Inc., 1990.

Matisoff, Bernard S. *Wiring and Cable Designer's Handbook*. Pennsylvania: McGraw-Hill, 1987.

Douglas-Young, John. *Illustrated Encyclopedic Dictionary of Electronics*. New York: Parker Publishing Co., 1981.

House, Kim G. & Marble, Jeff. *Printer Connections Bible*. Indiana: Howard W. Sams & Co., 1985.

Sclater, Neil. *Electrostatic Discharge Protection for Electronics*. Pennsylvania: McGraw-Hill, 1990.

Index

Illustrations are indicated in **boldface**.

466

memory, printer memory, *continued*
  RAM error message, 246, 337-338
  ROM, 242
  ROM checksum (or similar error), 246, 248-249, **248**, 336-337
  SRAM, 243
  temporary memory, 243, 245
  testing memory circuits, 245-246
  troubleshooting, 245-249
metal oxide semiconductors (MOSFETs), 47
microprocessor operations, 267
missing print, 390-391
missing right-hand text, 349-350, **350**
motors, 37-39, **38**
  bin feeder motor, 316
  carriage motor drivers, 257, **258**, **259**
  line feed motor drivers, 259-260, **260**, **261**
  magnetic motor (development unit), 313-314, **314**
  main motor (paper transport), 314-316, **315**, **316**
  polygon motor (laser/scanner system), 311, 313, **313**
  stepping motors, 38
  torque, 38
moving head or serial print heads, 61, 64-66, **64**, **65**
moving-carriage printer, block diagram, **13**
multimeters, 108-117, **109**
  capacitor testing, 113-115, **114**
  continuity testing, 113, **113**
  current measurement, 111, **112**
  diode testing, 115-116, **115**
  resistance measurement, 111, **112**, 113
  semiconductor testing, 115-116, **115**
  transistor testing, 116, **117**
  voltage measurement, 110, **110**

N

near letter quality (NLQ) printing, 60
NEC PinWriter P3200 dot matrix printer, **53**
NEC SuperScript Color 3000 printer, **385**
nibble mode, 229
No EP Cartridge message, 380-381
no printer, 399-400
no printing or incorrect printing, 390
number systems and conversion, 222-223, **223**

O

Offending Command error, 389-390
Okidata 24-pin impact printer, **145**
Okidata Microline 182 printer, **283**
on-line information services, 132-134
optical sensors, 275-277, **276**, **277**
optoisolators, 50-51, **50**, 276-277, **277**
organic photoconductive chemicals (OPC), 78, 95
oscilloscopes, 118-125, **119**
  calibration, 121-122, **122**
  controls, 119-120
  specifications for oscilloscopes, 120-121
  start-up procedures, 121-122

  time and frequency measurements, 124-125, **125**
  voltage measurement, 123-124, **124**
overlapping print, 355-356, **357**
ozone hazards, 141

P

page "walk" test sheet, **288**
pages per minute (PPM) rating, 7
paper feed area, 322-324
Paper Out message, 376-379, **377**, **378**
paper transport systems, 7-8, 12-16, **13**, **14**, **15**, **16**, 284-299
  bunching or gathering of paper at tractor-feed, 296-297
  carriage won't return to home position, 279-280
  cleaning and lubrication, 331
  drive system circuitry, **292**
  drive systems, 290-293, **290**, **292**, **293**
  erratic carriage motion, 280-281
  exit area, 325-327
  feed paths, **298**
  friction-feed paper systems, 284-287, **287**
  line feed motor signals, **293**
  line feed system logic timing, **293**
  main motor, EP printers, 314-316, **315**, **316**
  page "walk" test sheet, **288**
  paper feed area, 322-324
  Paper Out message, 376-379, **377**, **378**
  paper-advance inoperative, 294, **295**, 296, **296**
  paper-out alarm error, 277, **278**, 279
  registration/transfer area, 324-325
  slipping or "walking" of paper at tractor-feed, 297-298
  tears or wrinkles in paper, 298-299
  tractor-feed paper systems, 287-290, **289**
  troubleshooting, 294-299
paper-out alarm error, 277, **278**, 279
papers
  ink jet printing, 75-76
  thermal printing papers, 68-69
parallel interface, 26, 224-228, **225**
  troubleshooting, 235, **237**, 238-239
parity bits, 231, 234
partial page printed only, 391-392, 401
parts and supplies vendors, 447-450
PC Toolbox, 472-473
peak inverse voltage (PIV) in diodes, 46
permanent memory, 242
piezoelectric pumps, ink jet printing, 71, **72**
pitch or character pitch, 4
plastic leaded chip carriers (PLCC), **51**, 52
pliers, 98
polygon motor (laser/scanner system), 311, 313, **313**
poor print quality, 149-152, **150**
PostScript graphics print too slow, 389
potentiometers, 40
power requirements, 3

471

About the author

Stephen J. Bigelow is the founder and president of Dynamic Learning Systems, which is a technical writing, research, and publishing company that specializes in electronic and PC service topics. Bigelow is the author of nine feature-length books for McGraw-Hill, and almost 100 major articles for mainstream electronics magazines, such as *Popular Electronics*, *Electronics NOW*, *Circuit Cellar INK*, and *Electronic Service & Technology*. Bigelow is also the editor and publisher of *The PC Toolbox*, a premier PC service newsletter for computer enthusiasts and technicians. He is an electrical engineer with a BSEE from Central New England College in Worcester, MA.

The PC Toolbox™/PRINTERS

Use this form when ordering *The PC Toolbox™* or the registered version of **PRINTERS**.
You may tear out or photocopy this order form.

YES! Please accept my order as shown below: (check any one)

_____ Send me the registered version of **PRINTERS** for **$20** (US)
Massachusetts residents please add $1 sales tax.

Keep the software, but start my 1 year subscription (6 issues) to *The PC Toolbox™* for
_____ **$39** (US). I understand that I have an unconditional 90 day money-back guarantee
with the newsletter.

_____**A special offer for readers of "Troubleshooting and Repairing Computer
Printers"!** I'll take the registered version of **PRINTERS** *and* the 1 year subscription to *The PC
Toolbox™* (6 issues) for *only* **$49** (US). I understand that the newsletter has an unconditional
90 day money-back guarantee.

SPECIFY YOUR DISK SIZE: (check any one)

_____ **3.5"** High-Density (1.44MB)   _____ **5.25"** High-Density (1.2MB)

PRINT YOUR MAILING INFORMATION HERE:

Name: Company:

Address:

City, State, Zip:

Country:

Telephone: (   )   Fax: (   )

PLACING YOUR ORDER:

By FAX: Fax this completed order form (24 hrs/day, 7 days/week) to 508-898-9995

By Phone: Phone in your order (Mon-Fri; 9am-4pm EST) to 508-366-9487

___ MasterCard   Card: ___ ___ ___ ___ ___ ___ ___ ___ ___ ___ ___ ___ ___ ___ ___ ___

___ VISA   Exp: ___/___   Sig: _____

Or by Mail: Mail this completed form, along with your check, money order, PO, or credit card info to: